Springer Series in Materials Science

Volume 342

The Springer Series in Materials Science covers the complete spectrum of materials research and technology, including fundamental principles, physical properties, materials theory and design. Recognizing the increasing importance of materials science in future device technologies, the book titles in this series reflect the state-of-the-art in understanding and controlling the structure and properties of all important classes of materials.

José Joaquim Costa Cruz Pinto ·
José Reinas dos Santos André

Analytical Molecular Dynamics of Amorphous Condensed Matter

Thermal and Non-equilibrium Response Behavior

 Springer

José Joaquim Costa Cruz Pinto
Department of Chemistry
University of Aveiro
Aveiro, Baixo Vouga, Portugal

José Reinas dos Santos André
UDI, UTC—Engineering and Technology
Polytechnic Institute of Guarda
Guarda, Portugal

ISSN 0933-033X ISSN 2196-2812 (electronic)
Springer Series in Materials Science
ISBN 978-3-031-56519-9 ISBN 978-3-031-56517-5 (eBook)
https://doi.org/10.1007/978-3-031-56517-5

This Springer imprint is published by the registered company Springer Nature Switzerland AG
The registered company address is: Gewerbestrasse 11, 6330 Cham, Switzerland

Paper in this product is recyclable.

*To the most grateful memories of my father
João, mother Suzana, and sister Maria João,
and of all those who taught me at school,
high school, and at the Universities of Lisbon
(IST), Nancy (ENSIC), and Manchester
(UMIST). I most warmly thank my dear wife
Ana Nery for all her support during my deep
immersion into this project.*

*"A vida é fruto arrancado
de planta ainda mal florida ...
... e vento, que nos sopra de dentro
e nos puxa para o imprevisto."*

José Joaquim Costa Cruz Pinto

*To my parents José and Maria Augusta,
brothers Francisco, Maria de Lurdes, Maria
José and António, and to my wife Raquel and
daughters Cláudia and Filipa, I thank all
their love and support, always and during the
development of the work.*

José Reinas dos Santos André

Preface

This textbook is concerned with the description and analytical modeling of the temperature-dependent equilibrium and (linear and nonlinear) non-equilibrium viscoelastic response dynamics of amorphous condensed matter in general, far from rupture conditions. Though the focus and most of the examples are on the dynamics of the mechanical response (stress relaxation, creep, and cyclic stressing/straining) of amorphous polymers—whose treatment also includes in Part I the classical and most recent phenomenological models of linear and nonlinear viscoelasticity—the actual physical scope of the more advanced analysis of Part II of the book covers both more general and more specific aspects, like the thermal equilibrium and non-equilibrium dynamics at the "microscale," the still challenging (and even "mysterious"[1]) problem of the glass transition and glass transition temperature, how partial crosslinking or crystallization limits the response, and the relative universality of the behavior among different classes of amorphous materials.

Many valuable and widely available textbooks on the mechanical (viscoelastic) response of materials, namely polymers, discuss wide domains of the behavior— often from small stresses/strains to fracture. The benefits of a more specific and detailed physical and mechanistic discussion of the thermal and stress/strain linear and nonlinear viscoelastic response are thus often limited, where the greatest present challenges persist in the materials' knowledge and engineering fields, where realistic physical representation and predictability remain of paramount relevance and difficulty. By contrast, other very significant publications feature advanced presentations or reviews, and specialized discussions accessible to advanced researchers, without however succeeding in providing comprehensive physical (mechanistic) insight and,

---

[1] Consider P. W. Anderson's statement in *Science,* **267**, 1615 (1995), that "*the deepest and most interesting unresolved problem in solid state theory is probably the theory of the nature of glass and the glass transition*" and the one by E.-J. Donth in his 2001 book on "*The Glass Transition— Relaxation Dynamics in Liquids and Disordered Solids,*" that "*comprehensive and microscopic answers*" (to the problems of the crossover region, and the universality and super-Arrhenius nature of the behavior) "*still remain to be invented and require a high level of abstraction.*"

above all, significant predictive power with respect to all known characteristics of the dynamics of amorphous condensed matter. Further, despite all recent advances, there appears to be no common opinion on what theoretical developments might be necessary to solve the general problem of amorphous materials behavior.

So, this textbook intends (1) to bridge a gap between the above two sets of texts, in a way accessible to a readership of both undergraduate (Part I and the first two chapters of Part II) and graduate students and junior researchers (the remainder of Part II), where detailed physical discussion is emphasized (as throughout the whole book) and (2) in Part II, outline and give substantial development to the possibility of a new, mainly analytical, approach to this field (a *cooperative theory of materials dynamics*—CTMD), involving a limited number of meaningful and quantifiable physical parameters (actual properties), with reference to, and comparison with, experimental polymer stress relaxation data. Nevertheless, it will be made clear that such new approach and development still falls short of providing final answers. Many open questions remain and are identified, and researchers are encouraged to tackle them. Part III, which may be expected to be omitted in the first reading, discusses a few domains where new applications or extensions of CTMD might be explored.

Having mostly students and junior researchers in mind, the emphasis will be both on detailed physical discussions and reasonably expanded derivations throughout, with reference to an adequate but not overwhelming number of recent contributions to specialized literature. Where relevant or advantageous, readers will also find support by a limited number of theoretical or practical highlights and proposed study questions, and three Appendixes on more specific or difficult subjects.

Ílhavo, Aveiro, Portugal
Guarda, Portugal
José Joaquim Costa Cruz Pinto
José Reinas dos Santos André

The original version of the book has been revised: The author corrections have been updated. A correction to the book can be found at https://doi.org/10.1007/978-3-031-56517-5_16

Acknowledgements

José Joaquim C. Cruz Pinto expresses his gratitude to the late Professors Sebastião Formosinho and Hugh Burrows of the Department of Chemistry of the University of Coimbra, and colleagues at the Universities of Minho, Aveiro and Lisbon (IST) for many enlightening and fruitful discussions. My former students at Aveiro and Minho are also thanked for their patience and hard work, especially my co-author José André, José Martins at the Department of Polymer Engineering of the University of Minho, and all my colleagues and co-workers at the Department of Chemistry/CICECO (Materials Research Centre) and its Laboratory of Thermal Analyses at the University of Aveiro. The financial support over the years by Gulbenkian Foundation, the French Embassy in Lisbon, INVOTAN, JNICT, and Foundation for Science and Technology (FCT) is also gratefully acknowledged.

José Reinas dos Santos André expresses his gratitude to his research supervisor and co-author for their support, enthusiasm, and friendship. The financial support from PRODEP III during his doctoral degree at the University of Aveiro and the ongoing support from the Interior Development Unit—UDI is also gratefully acknowledged.

Aveiro, Portugal	José Joaquim Costa Cruz Pinto
Guarda, Portugal	José Reinas dos Santos André

Contents

About the Authors

José Joaquim Cruz Pinto was born in Lisbon, Portugal, in 1948, graduated in Chemical Engineering in IST, University of Lisbon (1971) and in ENSIC ("Section Spéciale"), University of Nancy, France (1972), obtained his Ph.D. in Chemical Engineering at UMIST, Manchester (1975–79), and his Full Professorship in 1994. Taught and conducted research at the Universities of Mozambique (1972–75), Minho (1978–97), and Aveiro (1997–2011) in Chemical Engineering (since 1972) and Polymer Materials Science (since 1979). Since his retirement in 2011, among his hobbies and projects, he keeps his interest in the most challenging but entertaining subject of the cooperative dynamics of polymers and other materials. He lives at Ílhavo, Portugal, married to Ana Nery.

José Reinas dos Santos André was born in Guarda, Portugal, in 1962, graduated in Industrial Chemistry at the University of Coimbra in 1985, obtained his M.Sc. in Materials Science and Engineering at the University of Minho in 1992, and his Ph.D. in the same field at the University of Aveiro in 2004. He teaches and does research at the Polytechnic Institute of Guarda, Inland Development Research Unit, since 1987, where he is Coordinating Professor. He lives in V. N. Foz Côa, Portugal, married to Raquel André with two daughters, Cláudia and Filipa André.

Abbreviations

CM	Coupling Model
CTMD	Cooperative Theory of Materials Dynamics
ETG	Equilibrium Theory of Glasses
FDT/FDR	Fluctuation-Dissipation Theorem/Relationship
KWW	Kohlrausch–Williams–Watts
MCT	Mode Coupling Theory
NETG	Non-equilibrium Theory of Glasses
NMR	Nuclear Magnetic Resonance
PC	Polycarbonate
PMMA	Poly(methylmethacrylate)
PP	Polypropylene
RFOT	Random First-Order Theory
RRKM	Rice–Ramsperger–Kassel–Marcus Theory
SLS	Standard Linear Solid
TSE/TSS	Time-Stress Equivalence/Superposition
TST	Transition State Theory
TTE/TTS	Time-Temperature Equivalence/Superposition
UHMWPE	Ultra-High Molecular Weight Polyethylene
VTF	Vogel–Tammann–Fulcher
WLF	Williams–Landel–Ferry

List of Figures

Chapter 1
Introduction and Scope

As in any textbook dealing with the response (namely mechanical) behavior of materials—here assumed far from rupture conditions—one starts in Chap. 2 by recalling the classical concepts of *instantaneous elastic*, *delayed elastic* and *viscous* responses, as resulting from the absence or presence, to varying degrees, of *time dependency*, whose practical realizations may be found in very rigid solids (like common glass, ceramics and many metals at sufficiently low temperatures and stresses, well below yield or rupture), most polymers (both solid and as melts), and many liquids at room temperature, respectively. Classical phenomenological representations and formulations of such variable behavior commonly use familiar ideal *linear springs* and *Newtonian dampers* (or *dashpots*), in varying combinations, which are mentioned and dealt with in most textbooks—such as those of [1–6] and many older ones or, in greater formal detail, in Tschoegl's book [7], for example.

The second and third chapters of the book, at the start of Part I, are quickly directed to the physical discussion and mathematical formulation of the behavior of what may be classified and modeled as a *standard linear solid* or a *standard linear liquid*, in the latter case to account for the effects of irreversible viscous flow, as the simplest representations of typical behavior of greatly idealized materials, when subject to stress relaxation, creep and cycling stress/strain, all in the so-defined *linear viscoelastic domain*, *i.e.* where the specific response amplitudes (stress per unit applied strain—the *stress relaxation modulus*—or strain per unit applied stress—the *creep compliance*) do not depend on the excitation amplitudes, but only on *time*. To the benefit of the students' physical reasoning abilities, equal relevance is given to the detailed qualitative (descriptive, physical) and quantitative (mathematical, model) descriptions of the above types of response by such idealized models.

Part I, in Chap. 3, will then introduce the concept and utility of the *response* (relaxation and retardation) *spectra*, treated in most classical bibliography on linear viscoelasticity as the obvious but most often difficult generalization of those (standard linear solid or liquid) models, to close in Chap. 4 by (1) reviewing in detail the earliest and most recent attempts at physically model the *non-linear behavior* when

© The Author(s), under exclusive license to Springer Nature Switzerland AG 2024
J. J. Cruz Pinto and J. R. dos Santos André, *Analytical Molecular Dynamics of Amorphous Condensed Matter*, Springer Series in Materials Science 342,
https://doi.org/10.1007/978-3-031-56517-5_1

the response amplitudes depend on the excitation ones in addition to the elapsed time—to show that the extremely simplistic nature of most of the available models may easily be substantially improved, though not yet satisfactorily—and, in Chap. 5, (2) by specifying how to treat the experimental data in stress relaxation or creep situations, to account for the initially very fast raising strain or stress steps. On the first issue, the question of which kind or nature of parameters should be sought to characterize the behaviour—desirably a limited number of quantifiable (or experimentally measurable) parameters, with actual specific physical meaning—will be given the appropriate emphasis, as an important requirement and goal of any contribution to the physical understanding and quantitative description of the behavior of amorphous materials, which is the main target of Part II of the book.

Part II will start (in Chap. 6) with the detailed physical discussion and illustration of the eleven most important characteristics of the dynamics of amorphous condensed matter in general. Four of them, widely studied in most undergraduate courses and textbooks, are (1) the *wide range of the response times*, (2) the *non-linearity to the excitation intensity* (strain or stress), leading, as already mentioned, to the need of considering earlier and more advanced *non-linear viscoelastic models* and the corresponding relaxation and retardation spectra (dealt with in Chap. 4), (3) the great *sensitivity to temperature*, particularly in polymers, and (4) the non-negligible *influence of molecular packing*. The two largely competing (or, perhaps better, complementary) views on the actual physical basis and relative importance of the last two characteristics—*temperature-* versus *packing-dominated dynamics*—will warrant an initial discussion centered on an analysis of the classical and so far widely shared views on WLF (Williams-Landel-Ferry) [8] and VTF (Vogel-Tammann-Fulcher) [9] temperature dependences, well described in most classical textbooks, in addition to more specific *free volume theories* [10]. This highly important subject is revisited later in the light of the new molecular formulations developed and tested in Chaps. 8, 9, 11 and 12.

Of the above, Chap. 8 concentrates on a proposed new look at the dynamics of materials' response at the *micro (or molecular) scale*,[1] using and expanding on the modern concepts of *primitive relaxors* [11], their *cooperativity* [3, 12] and its consequences, and the meaning and values of properties such as the *crossover temperature and frequency* [12], but such as yielding fresh physical and quantitative insight into the older concepts of the *ergodicity-making* and *ergodicity-breaking temperatures* [13] and, within Chap. 9, the *fragility index* [14].

Chapter 9 gives a detailed outline on how to possibly tackle the problem of the distribution of response times within this *cooperative view of the dynamics*, and formulate the actual development of non-linear responses, such as to yield greater physical quantitative insight into the still outstanding problem of the relative roles of temperature and molecular packing on the behavior, and even into the nature and variability of the *glass transition* and *glass transition temperatures* with the

---

[1] [leading to a *cooperative theory of materials dynamics* (CTMD)].

timescale of the observation or testing, while requiring only a limited number of specific, physically meaningful, known or experimentally quantifiable, parameter values, as also shown in Chaps. 8, 10 and 11.

Chapter 10 considers the relationship and comparison of this approach with two of the other most recent ones, namely Ngai's coupling model (CM) or correlation [11], and Götze's mode-coupling theory (MCT) [15], with respect to both physical insight and predictive power. Additional comparisons with a significantly wider range of alternative theories would do justice to many relevant contributions but would exceed the intended reasonable bounds for the present book. Nevertheless, various characteristics that are currently sought in alternative or future theoretical approaches will be briefly mentioned in a final section of the chapter.

Chapter 11 then concentrates on what theory and experiments say, supporting (and to what extent), or not, the approach suggested, to close with useful conclusions, details on the computability of this analytical (*vs.* modern numerical dynamic simulations) approach, and Chap. 12 closes Part II with the identification of outstanding questions, and of possible ways forward in the continuing search to understand, model and predict the behavior of amorphous condensed matter.

The overarching goal of most of Part II of this book is to guide students and prospective researchers into a cautious but courageous frame of mind for the critical study of these very old but extremely important and still challenging subjects, recognizing that final answers to many open or even "mysterious" questions will not be easily forthcoming. Finally, in Part III, three prospective (and conjectural) future applications and extensions of CTMD are suggested.

Wherever deemed necessary or useful, the reader is directed to a limited number of specific *Highlights* and *Study Questions*. A few *Appendixes* are included at the end of the book to provide additional clarification, insight, or more comprehensive solutions of the problems discussed in Part II.

References

1. I.M. Ward, J. Sweeney, *An Introduction to the Mechanical Properties of Solid Polymers*, 2nd edn. (Wiley, 2004)
2. M.T. Shaw, W.J. MacKnight, *Introduction to Polymer Viscoelasticity* (Wiley-Interscience, New Jersey, 2005)
3. J.L. Halary, F. Lauprêtre, L. Monnerie, *Mécanique des Matériaux Polymères* (Éditions Belin, Paris, 2008)
4. R. Lakes, *Viscoelastic Materials* (Cambridge University Press, New York, 2009)
5. C.M. Roland, *Viscoelastic Behavior of Rubbery Materials* (Oxford University Press, Oxford, 2011)
6. C. Lexcellent, *Linear and Non-Linear Mechanical Behavior of Solid Materials* (Springer, Cham, Switzerland, 2018)
7. N.W. Tschoegl, *The Phenomenological Theory of Linear Viscoelastic Behavior—An Introduction* (Springer, Berlin, Heidelberg, New York, London, Paris, Tokyo, 1989)
8. M.L. Williams, R.F. Landel, J.D. Ferry, J. Am. Chem. Soc. **77**, 3701 (1955); J.D. Ferry, *Viscoelastic Properties of Polymers*, 3rd edn. Chap. 11 (Wiley, New York, 1980)

9. H. Vogel, Phys. Z. **222**, 645 (1921); G.S. Fulcher, J. Am. Ceram. Soc. **8**, 339 (1923); V.G. Tammann, W.Z. Hesse, Anorg. Allg. Chem. 156, 245 (1926)

10. T.G. Fox, P.J. Flory, J. Apply. Phys. **21**, 581 (1950); J. Phys. Chem. **55**, 221 (1951); J. Polym. Sci. **14**, 315 (1954); A.K. Doolittle, J. Appl. Phys., **22**, 1031, 1471 (1951); M.H. Cohen, D. Turnbull, J. Chem. Phys. **31**, 1164 (1959); D. Turnbull, M. H. Cohen, J. Chem. Phys. **34**, 120 (1961); **52**, 3038 (1970); M. Goldstein, J. Phys. Chem. **77**, 667 (1973); J.P. Johari, E. Whalley, Faraday Symp. Chem. Soc. **6**, 23 (1973); M.H. Cohen, G.S. Grest, Phys. Rev. B **20**, 1077 (1979) + Erratum Phys. Rev. B, **26**, 6313 (1982); **24**, 4091 (1981); G.S. Grest, M.H. Cohen, Adv. Chem. Phys. **48**, 455 (1981); M.H. Cohen, G.S. Grest, Ann N.Y. Acad. Sci. **371**, 199 (1981); M.H. Cohen, G.S. Grest, J. Non-Cryst. Solids. **612/62**, 749 (1984)

11. K.L. Ngai, J. Non-Cryst. Solids, **353**, 709 (2007); *Relaxation and Diffusion in Complex Systems* (Springer, New York, 2011)

12. E.-J. Donth, *The Glass Transition—Relaxation Dynamics in Liquids and Disordered Materials* Springer Series in Materials Science 48. (Springer, Berlin, Heidelberg, 2001)

13. C. Angell, Cuur. Opin. Solid State & Mat., **1**, 578 (1996)

14. W.T. Laughlin, D.R. Uhlmann, J. Phys. Chem. **76**, 2317 (1972); C.A. Angell in *Relaxation in Complex Systems*, eds. by K.L. Ngai, G.B. Wright (U. S. Dept. Commerce, Springfield, 1985); J. Non-Cryst. Solids, **131–133**, 13 (1991); Science, **267**, 1924 (1995)

15. W. Götze, *Complex Dynamics of Glass Forming Liquids—A Mode-Coupling Model* International Series of Monographs on Physics—143. (Oxford Science Publications, Oxford, 2009)

Part I
Classical Formulations

Chapter 2
Mechanical Behavior of Time-Dependent, Viscoelastic, Materials

2.1 Basic Definitions

In *tensile stressing*, one applies two equal opposing forces along a single axis of a material test specimen, such as to extend it along the same direction. The corresponding *longitudinal or tensile stress*, σ, is the applied force divided by the cross-sectional area of the test specimen (cf. Fig. 2.1).

The *longitudinal strain*, ε, would be accurately defined by the integral of all relative differential extensions of the specimen along the direction of the applied tensile force, taking as reference any longitudinal segment within the material with initial length l_0, $\varepsilon = \int_{l_0}^{l} \frac{dl}{l} = \ln(l/l_0)$. For small elongations, however, $l \approx l_0$, and this definition expands as

$$\varepsilon = \ln[1 + (l - l_0)/l_0] = \frac{(l - l_0)}{l_0} - \frac{1}{2}\left[\frac{(l - l_0)}{l_0}\right]^2 + \cdots ,$$

and so it simplifies as $\varepsilon \approx \frac{(l - l_0)}{l_0}$, which is the *nominal tensile strain* generally adopted in engineering and also here.[1]

As Fig. 2.2 illustrates, in *shear stressing* one applies two equal and opposite forces along parallel directions, such as to result in the relative sliding of planes within the material along distances proportional to their distance. The corresponding *shear stress*, τ, is the applied force per unit area parallel to the opposing forces and the *shear strain*, γ, is the ratio between the relative sliding of parallel planes and their distance, d, $\gamma = \frac{x}{d} = \tan\theta$. For small shear strains, $\gamma \approx \theta$ (in radians).

---

[1] In the introductory chapters of several general references [1, 2], the reader will find accurate and detailed presentations of the concepts of stress and strain. [When dealing with highly deformable materials, such as elastomers, the extension ratio, $\frac{l}{l_0} = 1 + \varepsilon$ (with ε as defined above), is the definition used to formulate their behavior [3]].

© The Author(s), under exclusive license to Springer Nature Switzerland AG 2024

J. J. Cruz Pinto and J. R. dos Santos André, *Analytical Molecular Dynamics of Amorphous Condensed Matter*, Springer Series in Materials Science 342, https://doi.org/10.1007/978-3-031-56517-5_2

Fig. 2.1 Qualitative sketch
illustrating a tensile stress
and elongation[2]

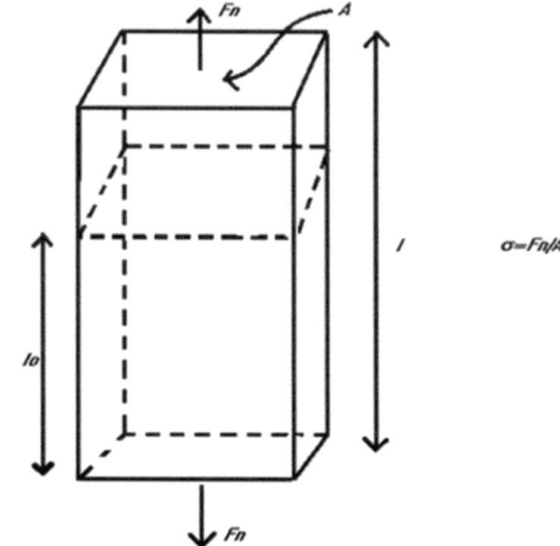

Fig. 2.2 Shear stress and
strain

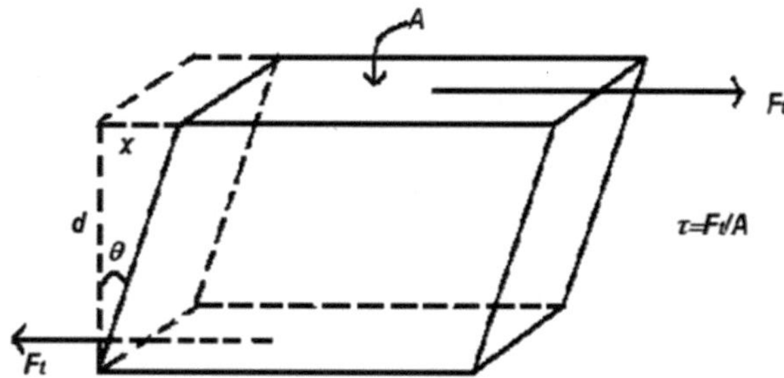

The *tensile elastic* or *Young's modulus*, E, is the initial slope of the stress–strain curve characterizing the material, $E = (d\sigma/d\varepsilon)_{\varepsilon=0}$. A *secant modulus* is sometimes used, $(E_s)_\varepsilon = \sigma(\varepsilon)/\varepsilon$, which only coincides with E when $\sigma(\varepsilon)$ is linear between 0 and ε, defining a region of perfect *linear elastic* behavior (cf. Fig. 2.3). It may also be defined a *tangent modulus* by $(E_t)_\varepsilon = d\sigma/d\varepsilon$. When the stress–strain relationship is exactly linear, the material is classified as *Hookean*.

2.2 Examples from Polymers and Other Materials Properties

The various classes of materials (metals, ceramics, polymers, …) have widely different properties, namely mechanical, depending on chemical type, manufacture and processing technology, temperature, and timescale (or testing speed). Examples at about room temperature (with their typical Young's moduli in

---

[2] An associated transversal contraction is not represented.

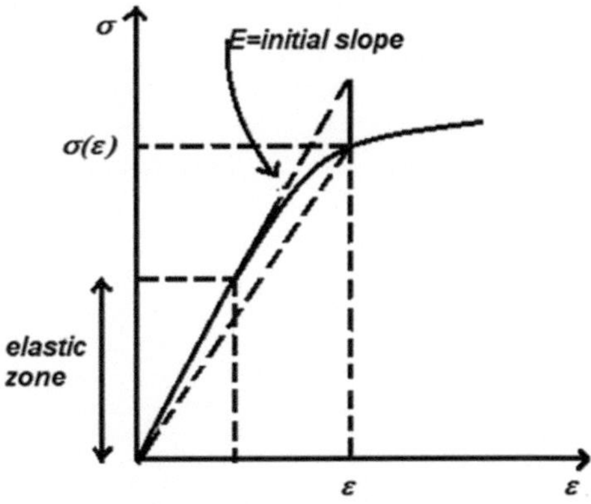

Fig. 2.3 The elastic or Young's modulus

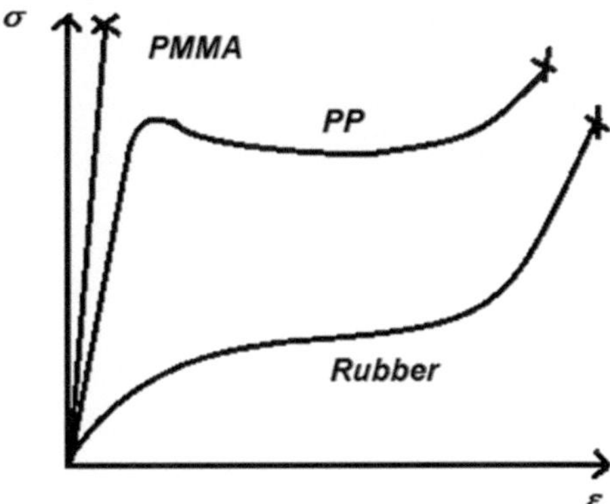

Fig. 2.4 Typical uniaxial tensile stress–strain curves for three types of polymer materials

Pa and maximum % strains) are mild steel $(2 \times 10^{11}, 1)$, common glass $(6 \times 10^{10}, 1)$, PMMA[3] $(3 \times 10^9, 2\,\text{to}\,7)$, PP[4] $(1 \times 10^9, 500\,\text{to}\,900)$, ordinary rubber $(2 \times 10^6, 600\,\text{to}\,800)$.

Their stress–strain curves (cf. Fig. 2.4) have different shapes. In addition, in the case of polymers, their mechanical properties are strongly dependent on the average and distribution of molar masses and, in crystallizable or crosslinked materials, on their degrees of crystallinity or crosslinking, respectively. Further, for a wide range of materials, in particular polymers, the mechanical and other physical behavior vary very strongly with *temperature* and the so-called *timescale* of the observation or testing. These two factors are the main variables treated at length in Part II of this book (for amorphous materials of whatever nature), and Figs. 2.5 and 2.6 give a first qualitative representation of their effect on the stress–strain curves.

The most important observation to make from the latter two Figures is that, in qualitative terms, nearly the same wide range of changes in mechanical behavior may

---

[3] Poly(methylmethacrylate).

[4] Poly(propylene).

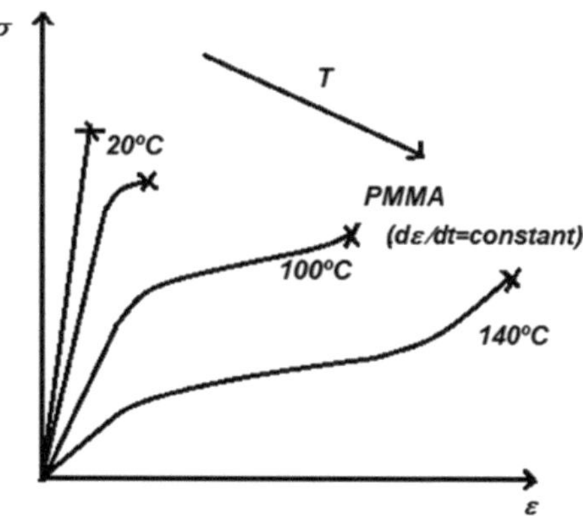

Fig. 2.5 Effect of temperature on the uniaxial tensile stress–strain curves of a typical PMMA

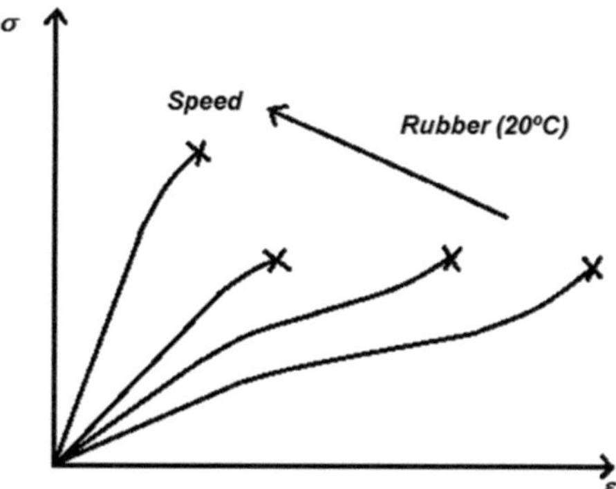

Fig. 2.6 Effect of timescale or testing speed—straining rate ($d\varepsilon/dt$)—on the uniaxial tensile stress–strain curves of a typical rubber at room temperature

be obtained for many (or perhaps most) materials upon decreased/increased temperatures at constant timescale and shortened/lengthened timescales at constant temperature, respectively. In addition to a further related comment in the next section, this subject will be given extended and detailed discussion, and quantitative formulation in Part II of the book.

2.3 Time-Dependency or Viscoelasticity

2.3.1 General

We may distinguish two extreme types of *ideal materials*: perfect *linear elastic* and *Newtonian viscous liquids*.

The linear elastic (or *Hookean*) materials have definite shape, and suffer nearly *instantaneous* small deformations under stress, with also nearly instantaneous recovery upon removal of the stresses, where the elastic energy stored upon stressing is the same released upon recovery. Quantitatively, as referred in Sect. 2.1, there is a perfect proportionality between the applied stress and strain—$\sigma = E\varepsilon$ in tension and $\tau = G\gamma$ in shear (G being the *shear modulus*), which is the result of the fact that the material's constitutive "particles" suffer only very small displacements relative to their equilibrium positions within a field of very localized forces. Knowing the corresponding potential energy of interaction (more easily accessible in the case of very simple structures), it would be possible to estimate their elastic (Young's) modulus [4].

Newtonian viscous liquids, by contrast, have no definite shape and suffer permanent (irreversible) deformations that grow with *time* when subject to stresses, meaning that the energy received is entirely dissipated (in the form of heat). Their behavior is quantitatively characterized by a proportionality of the applied stress to the strain rate, $\tau = \eta \times d\gamma/dt$, where η is the *viscosity* of the liquid[5,6]. The mentioned strain rate determines the so-called *timescale* of the testing or observation.

Real materials generally have an intermediate behavior, only part of the energy absorbed during deformation being recoverable and show significant *time-dependency*. They are thus considered *viscoelastic*, prominent examples being most polymer materials, where the relative importance of their individual elastic and viscous features shows very significant variations with *temperature* and rate of stressing/straining (as of heating/cooling)—which determine the mentioned *timescale*. The basic physical reason for this intermediate behavior is that, in addition to the universal elasticity of most materials, their deformation may require significant, long-range, rearrangements of their constitutive "particles", which will require (not only stress but) *time*.[7]

2.3.2   Uniaxial Tensile Stress Relaxation

Stress relaxation is the gradual decrease of the stress upon a material to keep constant a given "instantaneously" applied strain.[8] In a tensile situation, we will thus have, for $t \geq 0$, $\varepsilon = \varepsilon_0 = constant$ and $\sigma(t, \varepsilon_0) = \varepsilon_0 \times E_r(t, \varepsilon_0)$, where $E_r(t, \varepsilon_0)$ defines the corresponding (tensile) *stress relaxation modulus* (in Pa).[9] The test may

---

[5] For some liquids (extreme examples being polymer melts), the viscosity is not constant, depending on the timescale through the shear rate, and they thus are non-Newtonian.

[6] In tension, the corresponding (σ, ε) behavior enable the analogous definition of an *extensional viscosity*, η_e, such that $\sigma = \eta_e \times d\varepsilon/dt$.

[7] [The longer the lower the temperature, as we will see].

[8] With time-dependent materials (as most of them are), such instantaneous deformation is purely ideal and thus inaccurate, regardless of the instrumentation used. Chap. 5 of the book addresses this issue.

[9] In a shear situation, we would have $\gamma = \gamma_0 = constant$ and $\tau(t, \gamma_0) = \gamma_0 \times G_r(t, \gamma_0)$, where $G_r(t, \gamma_0)$ is the corresponding shear *relaxation modulus*.

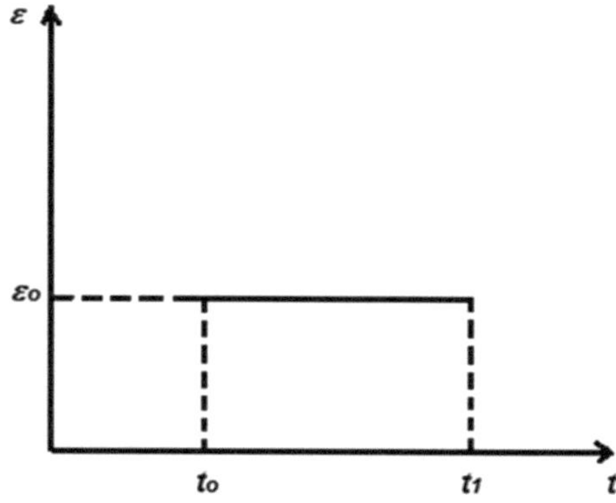

Fig. 2.7 The strain in a uniaxial tensile stress relaxation test between times t_0 and t_1, followed by sudden (ideal) strain reduction to zero

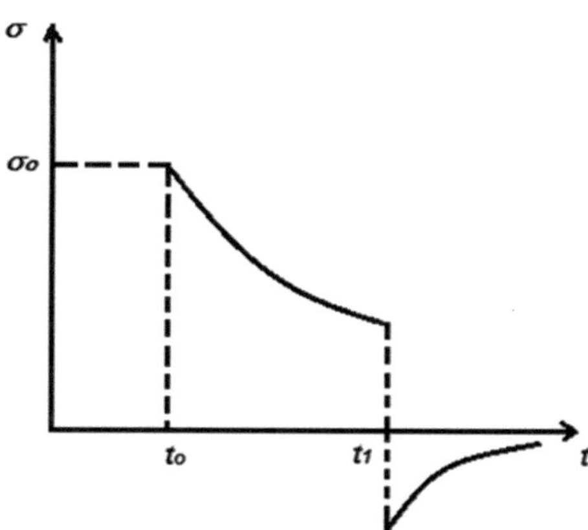

Fig. 2.8 The stress in a uniaxial tensile stress relaxation test between times t_0 and t_1, followed by recovery after sudden (ideal) strain reduction to zero at t_1

be illustrated and its typical results shown for a polymeric viscoelastic material by Figs. 2.7, 2.8 and 2.9.

The negative branch of the stress curve in Fig. 2.8 for $t > t_1$ stems from the fact that the fast reduction to zero of the applied strain requires that the material be subjected to a compressive stress, which will in turn progressively relax. If the applied strain ε_0 remained for a very long time, the resulting measured relaxation modulus would follow the semi-quantitative plots of Fig. 2.9, for the indicated polymer types. The plateau for very short times[10] corresponds to the initial elastic modulus, $E_0 = \sigma(t \to 0^+)/\varepsilon_0$, characterizing the instantaneous elasticity of what we may call the material's "backbone".[11] As to the long-time plateau (where reasonably defined, in the absence of significant viscous flow), it corresponds in amorphous polymers to the typical behavior of elastomers, with $E \sim 1$ MPa, where most of the macromolecular

---

[10] Note the log-scale of the elapsed time.

[11] [Which, in the case of polymers, might be due to a mix of the van der Walls or other intermolecular interactions and the actual macromolecular chains' backbone rigidity, depending on the degree of molecular orientation, crystallinity or crosslinking].

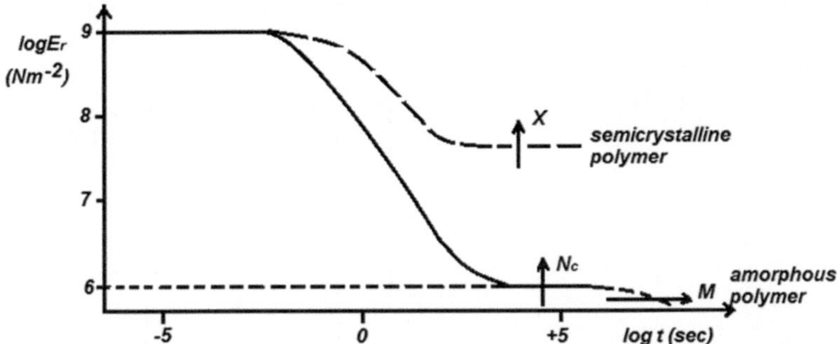

Fig. 2.9 Typical uniaxial tensile stress relaxion moduli of an amorphous and a semicrystalline or crosslinked polymer at a constant temperature just above its *glass transition temperature, T_g* [12]

chains are significantly or fully extended. The Figure also shows the plots for typical semi-crystalline in growing degrees (X), as well as, for amorphous polymers, the effects of growing molar masses (M) on the delay of significant viscous flow in the log-time scale used and of the degree of crosslinking (N_c). [13]

This type of behavior is easy to qualitatively justify at a microscopic scale. For very short times (immediately after imposing the chosen strain), despite the reasonably high temperature $(T > T_g)$, the "particles" or polymer chain segments did not have enough time to change their positions or orientations, and so what we obtain is the response of the stiff "backbone" of the material. With time, various delayed elastic motions and "particle" or polymer molecular segment reorientations (and some irreversible flow, when possible) will then occur under the set constant overall strain, thereby relaxing the actual measured stress. We leave a more detailed physical (mechanistic) interpretation of this behavior to Chaps. 3, 4 and 9, where kinetic formulations of growing accuracy and generality will be described or newly proposed.

2.3.3   Uniaxial Tensile Creep

Creep is the gradual deformation of a material under constant stress. In tension, we will have $\sigma = \sigma_0 = constant$ and $\varepsilon(t, \sigma_0) = \sigma_0 \times D(t, \sigma_0)$, where $D(t, \sigma_0)$ defines the corresponding (tensile) *creep compliance* (in Pa^{-1}). [14] The test may be illustrated, and its typical results shown for the indicated viscoelastic polymer materials by Figs. 2.10, 2.11 and 2.12.

---

[12] [Here loosely "defined" as the temperature at which the material begins to show significant wide scale deformations when tested (stressed or heated) at some chosen or conventional speed]. Part II of the book will define and treat at length the physics of this thermal property.

[13] [In the latter case, through the average number of molecular chains between crosslinks, N_c].

[14] In a shear situation, we would have $\tau = \tau_0 = constant$ and $\gamma(t, \tau_0) = \tau_0 \times J(t, \tau_0)$, where $J(t, \gamma_0)$ is the *shear compliance*.

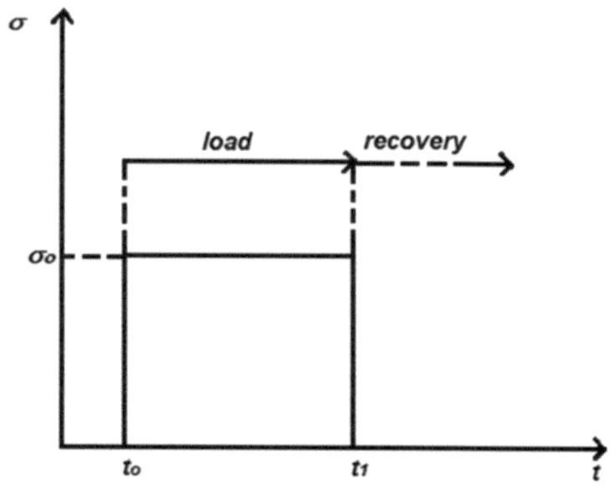

Fig. 2.10 The stress in a uniaxial tensile creep test between t_0 and t_1, followed by sudden (ideal) stress reduction to zero at t_1

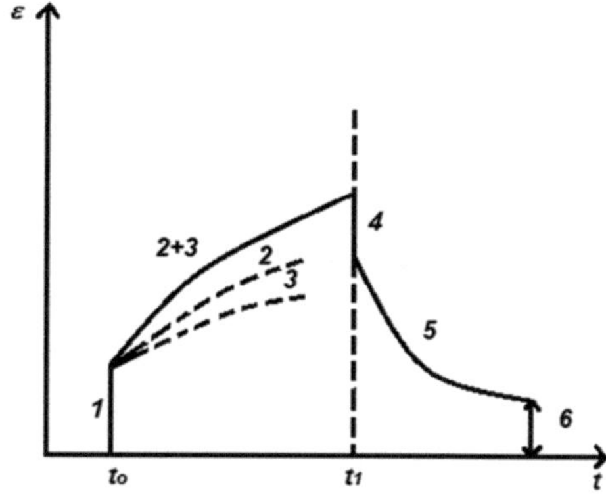

Fig. 2.11 The strain in a uniaxial tensile creep test between times t_0 and t_1, followed by recovery after sudden (ideal) stress reduction to zero at t_1

Fig. 2.12 Typical uniaxial tensile creep compliances of an amorphous and a semicrystalline or crosslinked polymer at a constant temperature just above its *glass transition temperature*

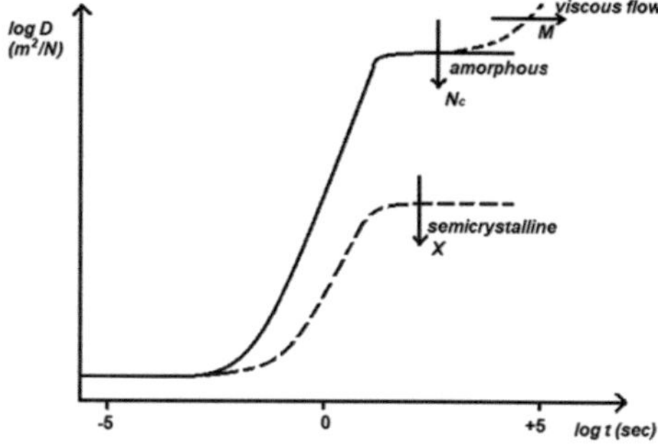

The creep and recovery curves of Figs. 2.10 and 2.11 have the following components:

1. *Instantaneous elastic strain*, corresponding to the initial fast extension of inter-atomic or intermolecular distances (and possibly also additional opening of bond angles in polymer materials, depending on the degree of macromolecular orientation).

2. *Delayed elastic strain* or *primary creep*, corresponding to further time-dependent but reversible motions and rearrangements of the constitutive "particles", for example the uncoiling of chains in polymer materials.
3. *Viscous flow* or *secondary creep*, corresponding to irreversible relative translation of "particles" (or molecular chains in polymers).
4. *Instantaneous elastic recovery* of the strain imposed in step 1.
5. *Delayed elastic recovery* of the strain component 2.[15]
6. *Permanent irreversible strain* equivalent to the amount of viscous flow of component 3.

If the applied stress σ_0 remained for a very long time, the resulting measured creep compliance would follow the semi-quantitative plots of Fig. 2.12, for the indicated polymer types. The plateau for very short times corresponds to the initial elastic compliance, $D_0 = \varepsilon(t \to 0^+)/\sigma_0$ (characterizing the instantaneous elasticity of what we called the material's "backbone"). As to the long-time plateau (where reasonably defined, in the absence of significant viscous flow), it corresponds in amorphous polymers to the typical behavior of elastomers, with $D = 1/E \sim 10^{-6}$ Pa, where most of the macromolecular chains are significantly or fully extended. The Figure also shows the plots for typical semi-crystalline or amorphous crosslinked polymers in growing degrees (X or N_c, respectively), as well as the effect of growing molar masses (M) in amorphous polymers on the delay of significant viscous flow in the log-time scale used.

As for stress relaxation, Chaps. 3, 4 and 9 will describe and propose kinetic formulations of creep of growing accuracy and generality.

Study Question 2.1 Physically explain the general mechanism of creep of a polymer material at a microscopic scale, by analogy and adequate adaptation of the short discussion of stress relaxation given in the last paragraph of Sect. 2.3.2.

2.3.4   *Phenomenological Analogues of Pure Elasticity and Time-Dependency*

Figure 2.13 shows the obvious best and simplest symbols or paradigms of the *elastic* and *viscous* behavior—the *Hookean elastic spring* and the *viscous damper* or "*dashpot*", respectively, the latter (generally in its Newtonian, constant viscosity variety) enabling to phenomenologically portray the *time-dependency* shown by most materials. So, it is not surprising that classical developments in the theoretical formulation of viscoelasticity[16] concentrated on combinations of both springs and "dashpots" of varying complexity. We will not dwell in those developments beyond the

---

[15] The timescale (or kinetics) of this recovery step is generally not the simple reversal of that of step 2. The initial recovery is generally faster than the final creep.

[16] [In its *linear* domain, where relaxation moduli are strain-independent and creep compliances are stress-independent (Cf. Sect. 2.3.5)].

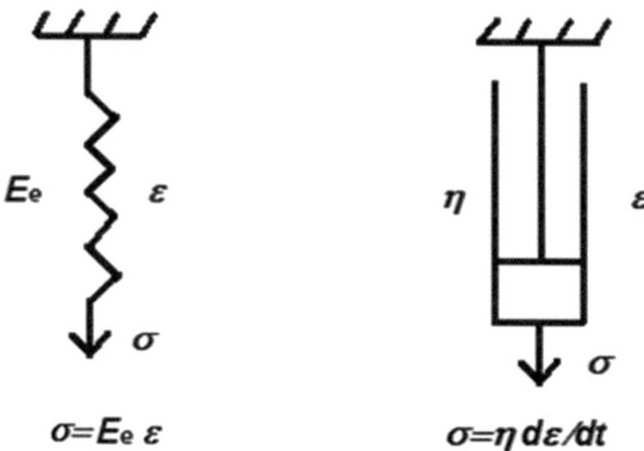

Fig. 2.13 Phenomenological analogues of pure elastic and viscous behavior

minimum justified as background to the specific formulations of Part II of the book, which will extend well into the non-linear domain. References [1, 2, 5] provide very comprehensive coverage of the above type of phenomenological representations.

2.3.5 General Features of Linear and Non-linear Viscoelasticity

In formal terms, the behavior of a viscoelastic material is said to be *linear* if it can be interpreted within varying levels of accuracy by means of the mentioned combinations of linear elements with constant elastic, E, and viscous, η, parameters, irrespective of the stress or strain values. In that linear domain, the relationship $\sigma(\varepsilon, t)$ may be expressed as $\sigma = \varepsilon \times g(t)$ for all ε and σ. Figure 2.14 illustrates the three main formal types of mechanical behavior, where the various points of the various theoretical time-labelled curves would be obtained by testing the material such as to reach each pair of (ε, σ) values after the time specified by the label.[17]

An equally relevant but more practical[18] characterization of the behavior is the one implied by the plots of Figs. 2.15 and 2.16 for uniaxial tensile stress relaxation and

---

[17] So, these curves provide mere formal definitions of the concepts of elastic, linear and non-linear viscoelastic behavior. The constant t (*isochronous*) curves would require one test specimen for each measured pair of (ε, σ) values or repeated testing of each specimen after lengthy ($t_{recovery} > 4t_{creep}$) and questionable material recovery—completely impractical due to materials' variability among the different or imperfectly recovered test specimens necessary to draw each of the curves, leading to data scatter.

[18] A different test specimen will now be necessary to obtain each of the curves. Individual curve scatter will thus be much reduced but, of course, test specimens' variability will still limit curve distinguishability at low strains and stresses, making difficult the estimates of the ε_l and σ_l linear limits, indicated in the Figures.

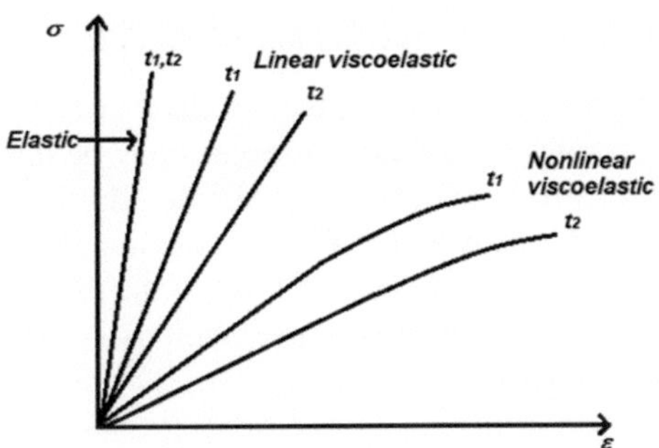

Fig. 2.14 Generic types of mechanical behavior

creep, respectively. The material is linear below a given strain ε_l in stress relaxation, and stress σ_l in creep, and non-linear above those values. These are not clear-cut limits, and there will be a narrow *linear viscoelastic domain*, generally bordered by a diffuse transition region. The *non-linear viscoelastic domain* will be one of the emphases of Chap. 9.

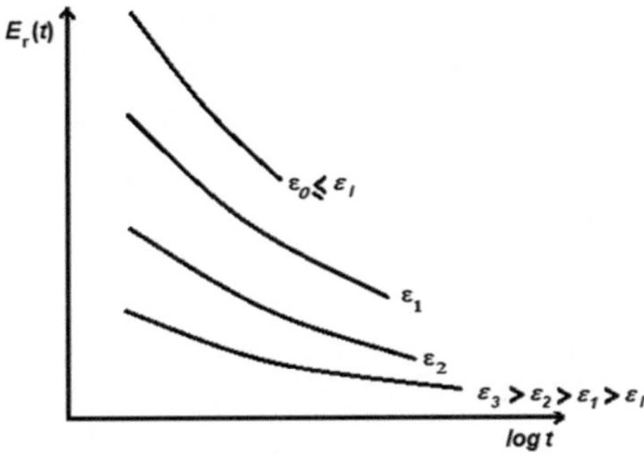

Fig. 2.15 Uniaxial tensile relaxation moduli in linear ($\varepsilon_0 \leq \varepsilon_l$) and non-linear ($\varepsilon_0 > \varepsilon_l$) viscoelasticity

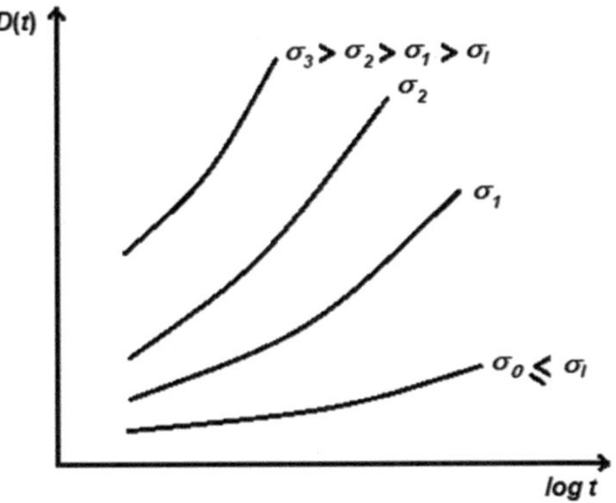

Fig. 2.16 Uniaxial tensile creep compliances in linear ($\sigma_0 \leq \sigma_l$) and non-linear ($\sigma_0 > \sigma_l$) viscoelasticity

Boltzmann's Superposition Principle

Within the narrow domain of *linear viscoelasticity*, the mentioned principle is valid, whereby a material's response may be directly calculated from all excitations that it has been subject to by simple addition of their respective individual responses. This is because, within that domain, all ensuing responses become independent of the actual instantaneous level of the excitation (strain or stress).

So, for tensile strain excitations $\Delta\varepsilon_i\left(t_i'\right)$ applied at times t_i', the stress at time t may be calculated by

$$\sigma(t) = \sum_i E_r\left(t - t_i'\right)\Delta\varepsilon_i\left(t_i'\right)$$

$$= \sum_i E_r\left(t - t_i'\right)\left[\Delta\varepsilon_i\left(t_i'\right)/\Delta t_i'\right]\Delta t_i', \text{ or}$$

$$\sigma(t) = \int_{-\infty}^{t} E_r\left(t - t'\right)\frac{d\varepsilon\left(t'\right)}{dt'}dt' \tag{2.1}$$

for a continuous variation of the applied strain, where the $-\infty$ lower limit means that the result should include the response to all past excitations imparted on the material. In the example of stress relaxation of Figs. 2.7 and 2.8,

$$t \leq t_1, \quad \sigma_{r1}(t) = \varepsilon_0 E_r(t - t_0)$$

$$t \geq t_1, \quad \sigma_{r2}(t) = \varepsilon_0 E_r(t - t_0) - \varepsilon_0 E_r(t - t_1)$$

and it may be seen that, for $t \to t_1^+$, $\sigma_{r2}(t_1) = \varepsilon_0 E_r(t - t_0) - \varepsilon_0 E_r(t = 0)$, which will be negative, because $E_r(t)$ strongly decreases with t.

Likewise, for tensile stress excitations $\Delta\sigma_i\left(t_i'\right)$ applied at times t_i', the strain at time t may be calculated by

$$\varepsilon(t) = \sum_i D\left(t - t_i'\right)\Delta\sigma_i\left(t_i'\right)$$

$$= \sum_i D\left(t - t_i'\right)\left[\Delta\sigma_i\left(t_i'\right)/\Delta t_i'\right]\Delta t_i', \text{ or}$$

$$\varepsilon(t) = \int_{-\infty}^{t} D\left(t - t'\right)\frac{d\sigma\left(t'\right)}{dt'}dt' \tag{2.2}$$

for a continuous variation of the applied stress. In the example of creep and recovery of Figs. 2.10 and 2.11,

$$t \leq t_1 \quad (\text{creep}), \quad \varepsilon_c(t) = \sigma_0 D(t - t_0)$$

$$t \geq t_1 \text{ (creep + recovery)}, \varepsilon_{cr}(t) = \sigma_0 D(t - t_0) - \sigma_0 D(t - t_1).$$

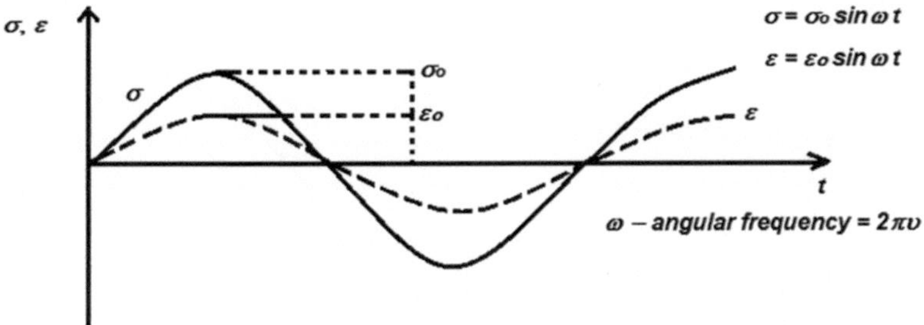

Fig. 2.17 Dynamic stress and strain in an ideal linear elastic material

After any time t, the material will thus have recovered the same deformation that it would have suffered if it had been under creep during time $t - t_1$ under the same stress. Note however that, because $D(t)$ grows with t at decreasing rate, one should not view the recovery as the exact reversal of creep, the initial recovery being faster than the final creep; this faster recovery will generally apply to both the linear and non-linear domains.

Study Question 2.2 Apply Boltzmann's superposition principle to obtain the stress response of a linear viscoelastic material to an increasing strain at strictly constant rate and its stress relaxation modulus for times within the duration of the test. [Assume that there are no other non-linear effects, such as yield and rupture].

To obtain the response to any general time-dependent excitation in the linear viscoelastic domain, the alternative to Boltzmann's superposition principle is by integrating the system's constitutive differential equation.[19] The method will also be used in Chap. 5 when evaluating the effect of non-instantaneous initial strain steps in stress relaxation situations.

Given the narrow domain of linear viscoelasticity for most materials, Boltzmann's superposition principle has limited, only approximate, applicability. In the reviews of Chap. 4 and new developments of Chap. 9, most of the discussion will concentrate on non-linear behavior.

2.3.6 Dynamic Straining or Stressing

At least as important as the response to constant strain or stress is the response of viscoelastic materials to cyclic (for example sinusoidal) excitations. A linear linear elastic material will respond to a periodic tensile stress by showing a periodic strain of the same frequency, keeping a constant stress/strain ratio equal to its Young's modulus, as illustrated in Fig. 2.17.

---

[19] Cf. the examples treated in Chaps. 3 and 5.

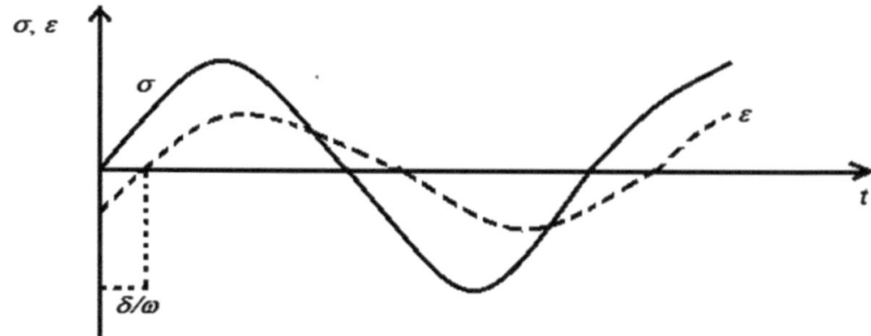

Fig. 2.18 Dynamic stress and strain in a linear viscoelastic material

By contrast, a viscoelastic material will respond by showing a strain out of phase with the stress but varying with the same frequency, as illustrated in Fig. 2.18, the phase lag increasing with the viscous nature of the material.[20] As what drives the response at the molecular scale is the local stress, it is always the *strain* that is *delayed relative to the stress.*

Study Question 2.3 Consider a sinusoidal strain excitation, $\varepsilon = \varepsilon_0\sin(\omega t)$, and show that a stress response of the form $\sum_i \alpha_i \sin(\omega t + \delta_i)$ is indeed equivalent to $A\sin(\omega t + \delta)$ whatever the number of terms in the series, and specify how to obtain A and δ as functions of the α_i and δ_i. Consider the first two terms of the series, then a third, and generalize.

This behavior is due to internal friction opposing the motions at the molecular scale, as in any viscous medium, and originates a continuous dissipation of energy as heat at a rate that we will see below to depend on the material, and the amplitude and frequency of the oscillations. Excessive amplitudes and/or frequencies with a high number of cycles, and failing adequate temperature control, may thus lead to material failure by ordinary *fatigue* and *thermal fatigue*, situations that are not considered in this book.[21] Therefore, meaningful dynamic data on materials' behavior requires the use of amplitudes well within the linear viscoelastic domain with adequate temperature control.

Storage and Loss Modulus and Compliance, and Loss Factor

Any viscoelastic simple system or real material will thus respond to a sinusoidal (for example, tensile) stress, $\sigma(t) = \sigma_0\sin(\omega t + \delta)$ by a sinusoidal strain $\varepsilon(t) =$

---

[20] Even though real materials (unlike some of the simple systems analyzed in Chap. 3) are characterized by a wide spectrum of response times that lead to responses equivalent to extended sums of sinusoidal elementary ones (one per single response time), those sums are easily understood as equivalent to one single sinusoidal response with a specific phase lag—$\sum_i \alpha_i \sin(\omega t - \delta_i) = A\sin(\omega t - \delta)$ (cf. *Study Question 2.3*).

[21] [And are deliberately avoided in ordinary dynamic mechanical testing].

$\varepsilon_0\sin(\omega t)$, and it is easily seen that the stress shows two components, one in phase and the other $\pi/2$ out of phase with the strain, as

$$\sigma(t) = \varepsilon_0 E'\sin(\omega t) + \varepsilon_0 E''\cos(\omega t) \tag{2.3}$$

with $E' = (\sigma_0/\varepsilon_0)\cos\delta$ and $E'' = (\sigma_0/\varepsilon_0)\sin\delta$. E' is called the *tensile storage modulus* and E'' the corresponding *loss modulus*, their ratio $\frac{E''}{E'} = \tan\delta_E$ being a *loss factor*.

This decomposition of the stress in two components $\pi/2$ out of phase (which have different properties) suggests their combined representation in the complex plane, as currently done in other domains of physics, such that a complex stress and strain might be formulated as $\sigma^* = \sigma_0 e^{i(\omega t+\delta)}$ and $\varepsilon^* = \varepsilon_0 e^{i(\omega t)}$, enabling to define a *tensile complex modulus* as

$$E^* = \frac{\sigma^*}{\varepsilon^*} = (\sigma_0/\varepsilon_0)e^{i\delta} = E' + iE'' \tag{2.4}$$

E' and E'' are therefore also named the real and imaginary parts of the complex tensile modulus, respectively, which determine the loss factor, $\tan\delta_E$.[22] Typical values of these properties for a polymer may be $E' \sim 1\,\mathrm{GPa}$, $E'' \sim 10\,\mathrm{MPa}$, yielding $\tan\delta_E \sim 0.01$ and $|E^*| \sim E'$.

Alternatively, representing the stress and strain by $\sigma(t) = \sigma_0\sin(\omega t)$ and $\varepsilon(t) = \varepsilon_0\sin(\omega t - \delta)$, corresponding to a translation of the origin of time, we obtain

$$\varepsilon(t) = \sigma_0 D'\sin(\omega t) - \sigma_0 D''\cos(\omega t) \tag{2.5}$$

with $D' = (\varepsilon_0/\sigma_0)\cos\delta$ and $D'' = (\varepsilon_0/\sigma_0)\sin\delta$, where D' is the *tensile storage compliance* and D'' the corresponding *loss compliance*, their ratio $\frac{D''}{D'} = \tan\delta_D$ being another measure of the *loss factor*.[23] Again, adopting the complex notation, one has $\sigma^* = \sigma_0 e^{i(\omega t)}$ and $\varepsilon^* = \varepsilon_0 e^{i(\omega t-\delta)}$, and a *tensile complex compliance* may be defined as

$$D^* = \frac{\varepsilon^*}{\sigma^*} = (\varepsilon_0/\sigma_0)e^{-i\delta} = D' - iD'' \tag{2.6}$$

---

[22] Analogous behavior and definitions apply to shear—τ, γ, G', G'' and $\tan\delta_G$.

[23] While $\tan\delta_E = \tan\delta_D$ for all systems characterized by a single response time (for example, a standard linear solid—cf. Chaps. 3 and 11), in real systems (characterized by a response time spectrum) such equality is not expected and generally not verified. Most literature rightly point out that the response of any real material to a sinusoidal excitation is always a sinusoidal of the same frequency but out of phase (strain always lagging behind the stress in dynamic mechanical excitations) but, with exceptions (cf. [6]), they generally do not mention that the lags obtained under stress and strain excitation do not have and are not expected to be identical. Chap. 11's dynamic mechanical calculations quantitatively illustrate the issue. In our opinion, this fact and the lower sensitivity of $\tan\delta$ plots to fine structural features of the materials than those of the loss properties E'' and D'' (cf. Chap. 11) do not lend strong support to the widespread use of $\tan\delta$ as a characterizing parameter instead of E'' and D''.

Fig. 2.19 Sketch of a stress–strain hysteresis cycle

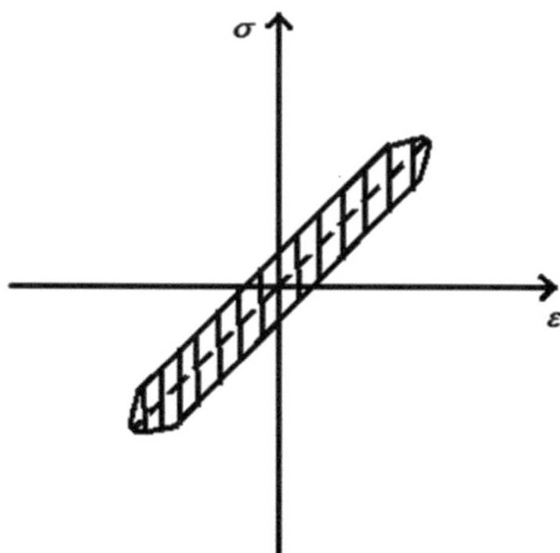

with $D^* = 1/E^*$,[24] but D' will only be the reciprocal of E' for a linear elastic material ($\delta = 0$) and D'' will only be the reciprocal of E'' for a Newtonian liquid ($\delta = \pi/2$). A related observation is that the relaxation moduli of a viscoelastic material are not the reciprocal of its creep compliances at the same values of time; in fact, the kinetics of the stress relaxation and creep processes widely differ, but the relaxation modulus and creep compliance may be related [1, 2, 5], as shown in Sect. 2.3.7.

Energy Dissipation

As implied in the third paragraph of Sect. 2.3.6, while in a linear elastic material (because there is no phase lag between stress and strain) the mechanical energy absorbed by the material during each quarter-cycle is entirely released in the following quarter-cycle[25] with no internal dissipation, within a viscoelastic material energy is dissipated, its stress and strain values at controlled temperature following a cyclic path (or *hysteresis cycle*), resembling that of Fig. 2.19. This results in non-zero values of the integral $\oint \sigma d\varepsilon$, which gives the energy dissipated per cycle and unit initial volume, E_d.

A straightforward calculation of E_d gives

$$E_d = \int_0^{t=\frac{2\pi}{\omega}} \sigma\left(\frac{d\varepsilon}{dt}\right)dt = \omega\varepsilon_0^2 \int_0^{\frac{2\pi}{\omega}} \left[E'\sin(\omega t)\cos(\omega t) + E''\cos^2(\omega t)\right]dt$$

$$\text{or } E_d = \pi\varepsilon_0^2 E'' = \pi\sigma_0\varepsilon_0 \sin\delta = \pi\sigma_0^2 D'' \tag{2.7}$$

---

[24] Analogous behavior and definitions also apply to shear—τ, γ, J', J'' and $\tan\delta_J$.

[25] [The material's stress and strain following up and down the broken straight-line in the diagram of Fig. 2.19, being zero the value of $\oint \sigma d\varepsilon$].

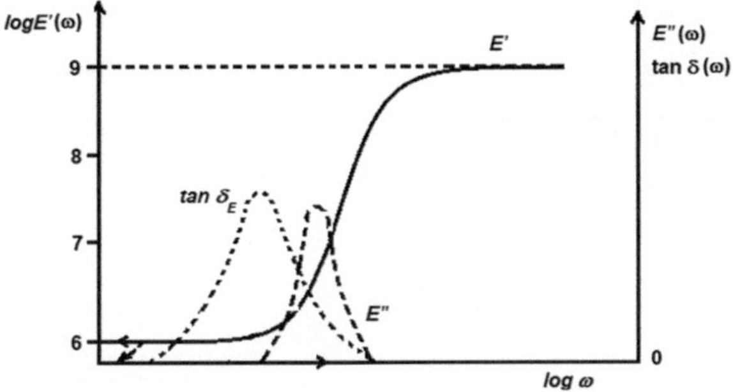

Fig. 2.20 Typical frequency-dependent uniaxial tensile storage and loss moduli, and loss factor of an amorphous (polymer) material

and the direct proportionality of E_d to E'' and D'' justifies their classification as *loss* properties.[26]

Typical Dynamic Behavior of Viscoelastic Materials

From the experimental study of the dynamic mechanical behavior of a range of real materials,[27] namely polymers, their frequency-dependent behavior at a constant temperature slightly above the corresponding glass transition temperature may arguably[28] be extrapolated to qualitatively follow the diagrams of Figs. 2.20 and 2.21. As a matter of fact, even using different techniques and instrumentation, it is very difficult to experimentally cover a wide range of frequencies with any individual material, contrary to what the plots of Figs. 2.20 and 2.21 suggest (assuming an amorphous un-crosslinked polymer), and so, in the literature, we only find abundant temperature-dependent data at limited ranges of frequencies, rather than frequency-dependent data at constant temperatures. Both types of *predicted* data representations will nevertheless be considered in Chap. 11 (though neglecting viscous flow) in the application of the theoretical formulations developed in Chap. 9. Viscous flow effects may be found in the storage modulus (but not, as we will see, in the storage compliance) as drops in the low modulus/low frequency region. Viscous effects on all loss properties are also expected but are not illustrated in the qualitative Figures shown here.

---

[26] Note that the proportionality of E_d to the square of the amplitudes explains the possibility of thermal fatigue and the requirement of low amplitudes (well within the linear viscoelastic domain), combined with good temperature control, to obtain meaningful dynamic mechanical data.

[27] [By various techniques and instrumentation of *dynamic mechanical analysis*] At a different range of frequencies, polar materials may also be studied by techniques and instrumentation of *dielectric spectroscopy*.

[28] Cf. Chap. 3 for the expected influence of viscous flow on the values of various properties, including the storage modulus.

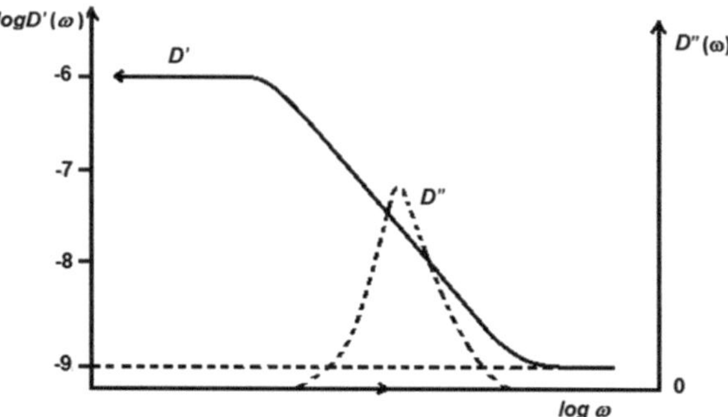

Fig. 2.21 Typical frequency-dependent uniaxial tensile storage and loss compliances of an amorphous (polymer) material

2.3.7 Formal Interrelations of the Viscoelastic Functions in the Linear Viscoelastic Domain

Stress relaxation, creep and periodic dynamic responses are obvious manifestations of one and the same material's viscoelastic behavior. Interrelations of the various viscoelastic functions are thus indeed expected in any of the (linear and non-linear) domains but, while in the non-linear domain they are (in the case of $E_r(t)$ and $D(t)$) still themes of current and future research,[29] simple interrelations may be easily established in the linear case.

From 2.1, integrating by parts,

$$\sigma(t) = \underbrace{\left[\varepsilon(t')E_r(t-t')\right]_{-\infty}^t}_{E_0\varepsilon(t)} - \underbrace{\int_{-\infty}^t \varepsilon(t')\frac{dE_r(t-t')}{dt'}dt'}_{\int_0^\infty \varepsilon(t-x)\frac{dE_r(x)}{dx}dx} \tag{2.8}$$

and, applying the Laplace transform[30] [7] to both sides,

$$\begin{aligned}
\mathcal{L}[\sigma(t)] &= E_0\mathcal{L}[\varepsilon(t)] \\
&\quad + \mathcal{L}[\varepsilon(t)]\{s\mathcal{L}[E_r(t)] - E_0\} \\
&= s\mathcal{L}[\varepsilon(t)]\mathcal{L}[E_r(t)]
\end{aligned} \tag{2.9}$$

---

[29] Chapter 9 will try to "scratch" the surface of the problem.

[30] $\mathcal{L}[f(t)] = \int_0^\infty e^{-st}f(t)dt.$

where one took $\varepsilon(t' < 0) = 0$, and the convolution theorem[31] and the transform of a derivative[32] were applied to the integral of the right-hand side. A similar treatment to 2.2 yields

$$\varepsilon(t) = \underbrace{\left[\sigma(t')D(t-t')\right]^{t}_{-\infty}}_{D_0\sigma(t)} - \underbrace{\int_{-\infty}^{t} \sigma(t')\frac{dD(t-t')}{dt'}dt'}_{\int_{0}^{\infty} \sigma(t-x)\frac{dD(x)}{dx}dx} \qquad (2.10)$$

and

$$\begin{aligned} \mathcal{L}[\varepsilon(t)] &= D_0\mathcal{L}[\sigma(t)] \\ &\quad + \mathcal{L}[\sigma(t)]\{s\mathcal{L}[D(t)] - D_0\} \\ &= s\mathcal{L}[\sigma(t)]\mathcal{L}[D(t)] \end{aligned} \qquad (2.11)$$

and, combining 2.9 and 2.11, $\mathcal{L}[E_r(t)]\mathcal{L}[D(t)] = 1/s^2$ or, by Laplace inversion and the convolution theorem,

$$\int_{0}^{t} E_r(t')D(t-t')dt' = \int_{0}^{t} D(t')E_r(t-t')dt' = t \qquad (2.12)$$

which interrelates the two basic viscoelastic time-dependent functions.

Study Question 2.4 From the initial and final values theorems (cf. Footnote 32), show that $D_0 = 1/E_0$ and $D_\infty = 1/E_\infty$, results that are used throughout the book.

Study Question 2.5 (to be attempted later, only after studying Chap. 3) Prove that a *standard linear solid* (defined in Sect. 3.1.3) obeys 2.12. [Consider the interrelationships of its elastic and compliance constants and its relaxation and retardation times].

Other interrelationships of these functions may be obtained. From 2.8, dividing both sides by the initial stress, σ_0, and assuming the strain $\varepsilon(t')$ is adjusted as to maintain that stress constant, the result is

$$E_0D(t) + \int_{0}^{t} D(t-t')\frac{dE_r(t')}{dt'}dt' = 1 \qquad (2.13)$$

---

[31] $\mathcal{L}\left[\int_{0}^{t} f(t')g(t-t')dt'\right] = \mathcal{L}\left[\int_{0}^{t} g(t')f(t-t')dt'\right] = \mathcal{L}[f(t)]\mathcal{L}[g(t)]$, if $f(t < 0) = g(t < 0) = 0$, the convolution theorem.

[32] $\mathcal{L}\left[\frac{df(t)}{dt}\right] = s\mathcal{L}[f(t)] - f(0^+)$, from which one may conclude that $\lim_{s\to\infty} sF(s) = f(0^+)$ and $\lim_{s\to 0} sF(s) = f(\infty)$, the initial and final values theorems, respectively, where $F(s) = \mathcal{L}[f(t)]$.

and from (2.10), dividing both sides by the initial strain, ε_0, while assuming the stress is adjusted as to maintain that strain constant, the results is

$$D_0 E_r(t) + \int_0^t E_r(t - t')\frac{dD(t')}{dt'}dt' = 1 \tag{2.14}$$

Study Question 2.6 (to be attempted later, only after studying Chap. 3) Prove that a *standard linear solid* (defined in Sect. 3.1.3) obeys (2.13) and (2.14). [Consider the interrelationships of its elastic and compliance constants and its relaxation and retardation times].

Applying now Boltzmann's superposition principle to the complex representations of sinusoidal excitations, one may write [1, 2, 5]

$$\sigma^*(t) = \int_{-\infty}^t E_r(t - t')\left[d\varepsilon^*(t')/dt'\right]dt'$$

$$= i\omega\varepsilon_0 \int_{-\infty}^t E_r(t - t')e^{i\omega t'}dt'$$

$$= i\omega\varepsilon^*(t) \int_0^\infty E_r(\Delta t)e^{-i\omega\Delta t}d(\Delta t) \tag{2.15}$$

$$\varepsilon^*(t) = \int_{-\infty}^t D(t - t')\left[d\sigma^*(t')/dt'\right]dt'$$

$$= i\omega\sigma_0 \int_{-\infty}^t D(t - t')e^{i\omega t'}dt' = i\omega\sigma^*(t) \int_0^\infty D(\Delta t)e^{-i\omega\Delta t}d(\Delta t) \tag{2.16}$$

where $\Delta t = t - t'$ and, from the definitions of $E^*(\omega)$ and $D^*(\omega)$, the storage and loss properties may be related to the time-dependent moduli and compliances as

$$E'(\omega) = \omega \int_0^\infty E_r(t)\sin(\omega t)dt \tag{2.17}$$

$$E''(\omega) = \omega \int_0^\infty E_r(t)\cos(\omega t)d \tag{2.18}$$

$$D'(\omega) = \omega \int_0^\infty D(t)\sin(\omega t)dt \qquad (2.19)$$

$$D''(\omega) = \omega \int_0^\infty D(t)\cos(\omega t)dt \qquad (2.20)$$

all of them Fourier transforms [8], that may be inverted to obtain $E_r(t)$ and $D(t)$ as

$$E_r(t) = \frac{2}{\pi} \int_0^\infty \left[\frac{E'(\omega)}{\omega}\right] \sin(\omega t)d\omega = \frac{2}{\pi} \int_0^\infty \left[\frac{E''(\omega)}{\omega}\right] \cos(\omega t)d\omega \qquad (2.21)$$

$$D(t) = \frac{2}{\pi} \int_0^\infty \left[\frac{D'(\omega)}{\omega}\right] \sin(\omega t)d\omega = \frac{2}{\pi} \int_0^\infty \left[\frac{D''(\omega)}{\omega}\right] \cos(\omega t)d\omega \qquad (2.22)$$

These interrelationships, valid only within the linear viscoelastic domain, are mathematically very challenging and prone to numerical errors when interconverting stress relaxation and creep experimental data. They thus have mainly theoretical interest, although calculation packages have been developed for their application. Practical solutions to the problem are more often obtained via approximations or the relaxation and retardation spectra.[33] All such interrelationships may be found in classical comprehensive references [5, 9, 10].

References

1. I.M. Ward, *Mechanical Properties of Solid Polymers*, 2nd edn. (Wiley, 1985)
2. I.M. Ward, J. Sweeney, *An Introduction to the Mechanical Properties of Solid Polymers*, 2nd edn. (Wiley, 2004)
3. L.R.G. Treloar, *The Physics of Rubber Elasticity*, 3rd edn. (Clarendon Press, Oxford, 1975)
4. R.A.L. Jones, *Soft Condensed Matter, Oxford Master Series in Condensed Matter Physics* (Oxford University Press, 2002)
5. N.W. Tschoegel, *The Phenomenological Theory of Linear Viscoelastic Behavior—An Introduction* (Springer, Berlin, Heidelberg, New York, London, Paris, Tokyo, 1989)
6. J.C. Duncan, *Principles and applications of mechanical thermal analysis*, 119–163, in *Principles and Applications of Thermal Analysis,* ed. by P. Gabbott (Blackwell Pubs., Oxford, 2008)

---

[33] These spectra will be introduced in Chaps. 3 and 4, and in Part II of the book they will play a major role in the discussion. However, as for the viscoelastic functions, the interrelationship of the relaxation and retardation spectra is still an open question in the non-linear viscoelastic domain.

7. D. Fleisch, *A Student's Guide to Laplace Transforms* (Cambridge University Press, Cambridge, UK, 2022)
8. J.F. James, *A Student's Guide to Fourier Transforms with Applications in Physics and Engineering*, 3rd edn. (Cambridge University Press, Cambridge, 2011)
9. B. Gross, *Mathematical Structure of the Theories of Viscoelasticity* (Hermann et Cie, Paris, 1953)
10. J.D. Ferry, *Viscoelastic Properties of Polymers* (Wiley, New York, 1980)

Chapter 3
Review of Classical Phenomenological Models of Linear Viscoelasticity

3.1 Simple Classical Models

The linear viscoelastic behavior of materials has since long been modeled by various phenomenological models [1–3] taken to approximately represent how materials respond to mechanical excitations of low intensity. Only the most representative and useful of those simple models in relation to the subjects of Part II of this book will be dealt with here.

3.1.1 Maxwell and Modified Maxwell Models

One of the simplest models is the so-called *Maxwell model* or *unit*, made of an elastic Hookean spring in series with a Newtonian damper or "dashpot", represented by just the right part of Fig. 3.1.

Study Question 3.1 Using the basic relationships obeyed by Hookean springs and Newtonian dashpots (cf. Chap. 2), write the mechanical *constitutive* 1st order ordinary differential *equation* of a Maxwell unit as just defined, and show that it yields (1) a single exponential decay of its *stress relaxation modulus* and (2) a linear increase of its *creep compliance*, in stress relaxation and creep tests, respectively. Define the characteristic *relaxation time* of the unit and (with the discussion of Chap. 2 in mind) specify *three* main *reasons*[1] why the results obtained are not representative of the behavior of real materials.

---

[1] In Section "Dynamic Testing" (Study Question 3.4), the reader will be able to identify a *fourth* reason.

© The Author(s), under exclusive license to Springer Nature Switzerland AG 2024
J. J. Cruz Pinto and J. R. dos Santos André, *Analytical Molecular Dynamics of Amorphous Condensed Matter*, Springer Series in Materials Science 342,
https://doi.org/10.1007/978-3-031-56517-5_3

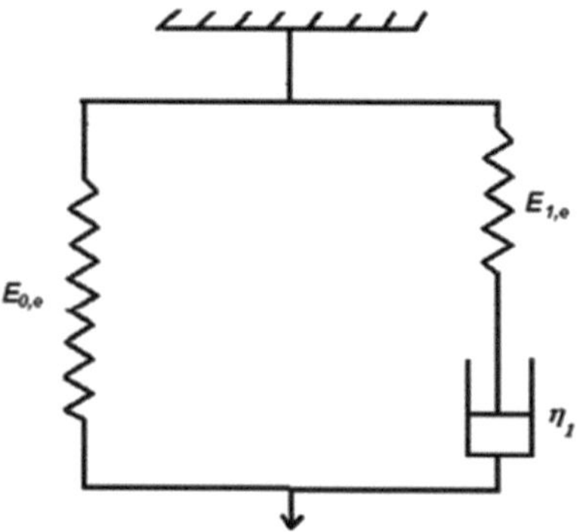

Fig. 3.1 The modified Maxwell model

The next simplest model is the modified Maxwell one of Fig. 3.1 itself, which resolves two of the above three main deficiencies.[2]

Basic Constitutive Equation

The total stress is now the sum of the stress in the left spring with that on the right one, where the latter is the same acting on the dashpot, and therefore equal to the product of the dashpot's viscosity by its strain rate. Bearing in mind that the dashpot's strain is the total strain subtracted of the right spring strain, easily related to its own stress and elastic constant, the model's *constitutive equation* may be written as

$$\sigma = E_{0,e}\varepsilon + \eta_1 \frac{d}{d\varepsilon}\left(\varepsilon - \frac{\sigma - E_{0,e}\varepsilon}{E_{1,e}}\right), \quad \text{or}$$

$$\frac{\eta_1}{E_{1,e}}\frac{d\sigma}{dt} + \sigma = \eta_1 \frac{E_{0,e} + E_{1,e}}{E_{1,e}}\frac{d\varepsilon}{dt} + E_{0,e}\varepsilon, \tag{3.1}$$

where we will soon see that the coefficients of $\frac{d\sigma}{dt}$ and $\frac{d\varepsilon}{dt}$ are related to the characteristic response times in stress relaxation (relaxation time) and creep (retardation time), respectively.

Uniaxial Tensile Stress Relaxation

After imposing and kept constant a strain ε_0 at $t = 0$,[3] (3.1) yields for the stress relaxation modulus $E_r(t) = \frac{\sigma(t)}{\varepsilon_0} = E_{0,e} + E_{1,e}e^{-t/(\eta_1/E_{1,e})}$,[4] or

---

[2] [And the above *fourth* reason as well].

[3] Cf. Footnote 8 of Chap. 2.

[4] The initial condition is $t = 0$, $\sigma(t = 0) = \sigma_0 = (E_{0,e} + E_{1,e})\varepsilon_0$, because only the two springs will be able to deform instantaneously (by an equal amount ε_0), the dashpot then behaving as a rigid body.

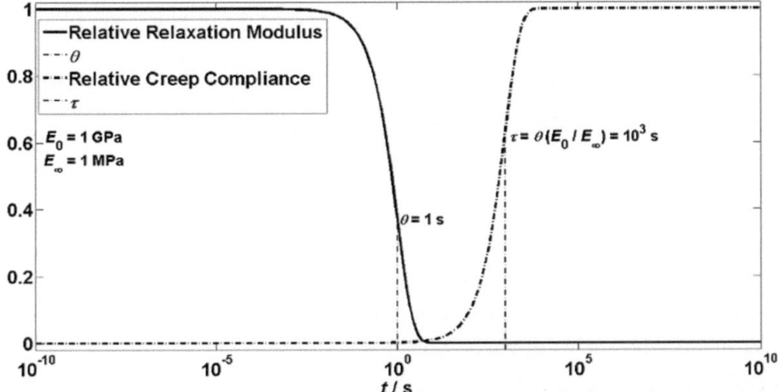

Fig. 3.2 Time-dependent relative uniaxial tensile relaxation modulus and relative creep compliance of the modified Maxwell model—later shown to be identical to those of the modified Voigt–Kelvin model (cf. Sections "Uniaxial Tensile Stress Relaxation" and "Uniaxial Tensile Creep"), jointly representative of a standard linear solid with the specified elastic and response properties (cf. Sect. 3.1.3)

$$E_r(t) = E_\infty + (E_0 - E_\infty)e^{-t/\theta}, \tag{3.2}$$

where the instantaneous and long-time values of the relaxation modulus are $E_0 = E_{0,e} + E_{1,e}$ and $E_\infty = E_{1,e}$ and the characteristic *relaxation time* is the same obtained in the previous study question, $\theta = \eta_1/E_{1,e}$. Figure 3.2 shows a plot of the relative stress relaxation modulus[5] for an identical model with $\theta = 1\,$s (solid curve) as a function of time, and it is now found that the stress and the relaxation modulus do not completely relax to zero.

Uniaxial Tensile Creep

After imposing and kept constant a stress σ_0 at $t = 0$,[6] (3.1) may be worked out to yield for the creep compliance $D(t) = \left(E_{0,e} + E_{1,e}\right)^{-1}\left[1 + \frac{E_{1,e}}{E_{0,e}}\left(1 - e^{-t/\tau}\right)\right]$,[7] or

$$D(t) = D_0 + (D_\infty - D_0)\left(1 - e^{-t/\tau}\right), \tag{3.3}$$

---

[5] {Defined as $[E_r(t) - E_\infty]/(E_0 - E_\infty)$}.

[6] Cf. Footnote 8 of Chap. 2.

[7] The initial condition is now $t = 0$, $\varepsilon(t = 0) = \varepsilon_0 = \left(E_{0,e} + E_{1,e}\right)^{-1}\sigma_0$, again because only the two springs will be able to deform instantaneously (by the same ε_0), the dashpot then behaving as a rigid body.

where now the characteristic *retardation time* is $\tau = \left(\frac{\eta_1}{E_{1,e}}\right)(E_{0,e} + E_{1,e})/E_{0,e} = \eta_1\left(\frac{1}{E_{0,e}} + \frac{1}{E_{1,e}}\right) = \theta(E_0/E_\infty)$, and the instantaneous and long-time compliances are $D_0 = \left(E_{0,e} + E_{1,e}\right)^{-1}$ and $D_\infty = 1/E_{0,e}$.

An important observation is that *retardation times* are always *longer* (often much longer) *than* the corresponding[8] *relaxation times*, which entails that *stress relaxation is faster than creep*, a result of very significant practical and theoretical relevance (valid for a wide range of different physical responses [4]) that we will further discuss later in the book. The relative creep compliance[9] of the same modified Maxwell model is also plotted in Fig. 3.2 as a function of time (broken curve) and it is seen that the creep rate eventually drops down to zero at long times.

Dynamic Testing

In treating the response to a sinusoidal excitation, we have seen in Chap. 2 that it is practical to combine both its (in-phase and out-of-phase) components in a complex response. So, taking a strain excitation $\varepsilon^* = \varepsilon_0 e^{i\omega t}$, whose response will be $\sigma^* = \sigma_0 e^{i(\omega t + \delta)}$, and substituting them in (3.1), we may simplify and write $\sigma_0 e^{i\delta} + \sigma_0 \omega i e^{i\delta} = E_{0,e}\varepsilon_0 + i E_{0,e}\varepsilon_0 \omega\tau$, or $E' - E''\omega\theta + i\left(E'' + E'\omega\theta\right) = E_{0,e} + i E_{0,e}\omega\tau$. Two equations then result from the equality of the real and imaginary parts of the left and right-hand sides, from which (considering the relationship between the relaxation and retardation times of the preceding section) the storage and loss modulus, and the loss factor, may be obtained as

$$E'(\omega\theta) = E_{0,e} + \frac{E_{1,e}(\omega\theta)^2}{1 + (\omega\theta)^2} = \frac{E_\infty + E_0(\omega\theta)^2}{1 + (\omega\theta)^2}, \tag{3.4}$$

$$E''(\omega\theta) = \frac{E_{1,e}\omega\theta}{1 + (\omega\theta)^2} = \frac{(E_0 - E_\infty)\omega\theta}{1 + (\omega\theta)^2}, \tag{3.5}$$

$$\tan\delta_E(\omega\theta) = \frac{E''}{E'} = \frac{(E_0 - E_\infty)\omega\theta}{E_\infty + E_0(\omega\theta)^2}. \tag{3.6}$$

Their plots as functions of the frequency, $\nu = \omega(2\pi)^{-1}$, in Hz are shown in Figs. 3.3 and 3.4 by the solid curves.

Study Question 3.2 Using an analogous complex representation of the strain response to an oscillating stress, $\sigma^* = \sigma_0 e^{i\omega t}$ (and recalling Sect. 2.3.5), carefully work out the dynamic components of the compliance of the modified Maxwell model as

---

[8] In real materials, the same basic structures at the molecular scale are surely involved in stress relaxation, as well as in creep or any other physical responses.

[9] {Defined as $[D(t) - D_0]/(D_\infty - D_0)$}.

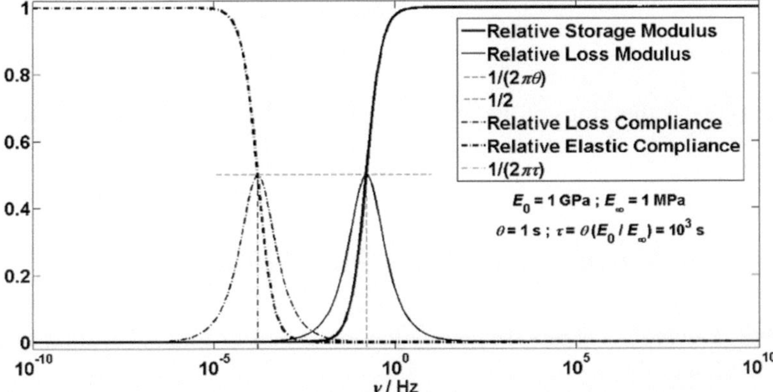

Fig. 3.3 Frequency-dependent $\left[\nu = \omega(2\pi)^{-1}\right]$ relative uniaxial tensile storage and loss modulus, and relative uniaxial tensile storage and loss compliance of the modified Maxwell model—later shown to be identical to those of the modified Voigt–Kelvin model (cf. Section "Dynamic Testing"), jointly representative of a standard linear solid with the specified elastic and response properties (cf. its Sect. 3.1.3)

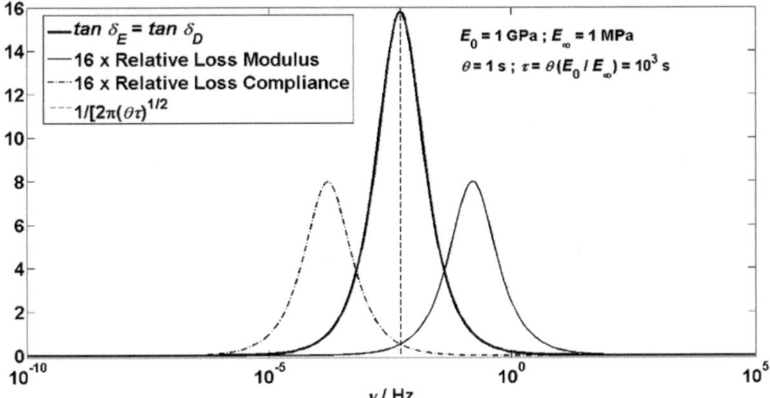

Fig. 3.4 Frequency-dependent $\left[\nu = \omega(2\pi)^{-1}\right]$ relative uniaxial tensile loss modulus and loss compliance, and loss factor of the modified Maxwell model—later shown to be identical to those of the modified Voigt–Kelvin model (cf. Section "Dynamic Testing"), jointly representative of a standard linear solid with the specified elastic and response properties (cf. its Sect. 3.1.3)

$$D'(\omega\theta) = \frac{1}{E_{0,e}} - \frac{(\omega\tau)^2}{1+(\omega\tau)^2}\frac{E_{0,e}^{-2}}{E_{0,e}^{-1}+E_{1,e}^{-1}} = \frac{D_\infty + D_0(\omega\tau)^2}{1+(\omega\tau)^2}, \qquad (3.7)$$

$$D''(\omega\theta) = \frac{E_{0,e}^{-2}}{E_{0,e}^{-1}+E_{1,e}^{-1}}\frac{\omega\tau}{1+(\omega\tau)^2} = \frac{(D_\infty - D_0)\omega\tau}{1+(\omega\tau)^2}, \qquad (3.8)$$

$$\tan \delta_D(\omega\theta) = \frac{(D_\infty - D_0)\omega\tau}{D_\infty + D_0(\omega\tau)^2}. \tag{3.9}$$

Their plots as functions of the frequency, $\nu = \omega(2\pi)^{-1}$, in Hz are also shown in Figs. 3.3 and 3.4 by the broken curves.

3.1.2   Voigt–Kelvin and Modified Voigt–Kelvin Models

The *Voigt–Kelvin unit* is represented by the lower part of Fig. 3.5, with a spring in parallel with a dashpot.

Study Question 3.3 Using the basic relationships obeyed by Hookean springs and Newtonian dashpots (cf. Chap. 2), write the mechanical *constitutive* 1st order ordinary differential *equation* of a Voigt–Kelvin unit (by formulating the total stress on the system), and show that (1) it does not allow any stress relaxation and (2) yields a creep compliance growing from zero at a rate gradually decreasing to zero at very long times. As for Study Question 3.1, discuss these results.

The next simplest model is the modified Maxwell one of Fig. 3.5 itself.

Basic Constitutive Equation

The total strain is now the sum of the strain of the top spring with that of the Voigt–Kelvin unit, and so

$$\varepsilon = \frac{\sigma}{E'_{0,e}} + \frac{\sigma - \eta'_1 \frac{d}{dt}\left(\varepsilon - \frac{\sigma}{E'_{0,e}}\right)}{E'_{1,e}}, \quad \text{or}$$

$$\sigma + \frac{\eta'_1}{E'_{0,e} + E'_{1,e}}\frac{d\sigma}{dt} = \frac{E'_{0,e}E'_{1,e}}{E'_{0,e} + E'_{1,e}}\left(\varepsilon + \frac{\eta'_1}{E'_{1,e}}\frac{d\varepsilon}{dt}\right), \tag{3.10}$$

Fig. 3.5 The modified Voigt–Kelvin model

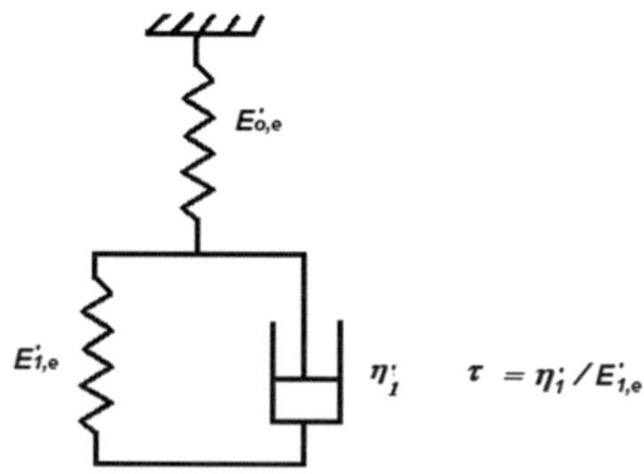

where we will soon see that $\theta = \frac{\eta'_1}{E'_{0,e}+E'_{1,e}}$ and $\tau = \frac{\eta'_1}{E'_{1,e}}$ are the relaxation and retardation times of this model.

Uniaxial Tensile Stress Relaxation

After imposing and kept constant a strain ε_0 at $t = 0$,[10] (3.10) yields for the stress relaxation modulus $E_r(t) = \frac{\sigma(t)}{\varepsilon_0} = \frac{E'_{0,e}E'_{1,e}}{E'_{0,e}+E'_{1,e}} + \frac{E'^2_{0,e}}{E'_{0,e}+E'_{1,e}} e^{-t/[\eta'_1/(E'_{0,e}+E'_{1,e})]}$,[11] or

$$E_r(t) = E_\infty + (E_0 - E_\infty)e^{-t/\theta}, \tag{3.11}$$

with $\theta = \eta'_1/(E'_{0,e} + E'_{1,e})$, and where the instantaneous and long-time values of the relaxation modulus are $E_0 = E'_{0,e}$ and $E_\infty = \frac{E'_{0,e}E'_{1,e}}{E'_{0,e}+E'_{1,e}}$.

The plot of the relative stress relaxation modulus for an identical model with $\theta = 1$ s as a function of time, is the same shown in Fig. 3.2 by the solid curve.

Uniaxial Tensile Creep

The result for the creep compliance is just the sum of the compliance of the top spring, $1/E'_{0,e}$ with that of the Voigt–Kelvin unit obtained in Study Question 3.3 or by integration of (3.10) with the initial condition $t = 0$, $\varepsilon(t = 0) = \varepsilon_0 = \sigma_0/E'_{0,e}$, to yield

$$D(t) = D_0 + (D_\infty - D_0)\left(1 - e^{-t/\tau}\right), \tag{3.12}$$

with $D_0 = 1/E'_{0,e} = 1/E_0$ and $D_\infty = \frac{1}{E'_{0,e}} + \frac{1}{E'_{1,e}} = 1/E_\infty$. The relative creep compliance of the modified Voigt–Kelvin model is the same plotted in Fig. 3.2 as a function of time by the broken curve.

Dynamic Testing

The reader may again easily apply the method outlined in Section "Dynamic Testing" to the complex representation of the stress or strain responses of a modified Voigt–Kelvin model to oscillating strains or stresses, respectively.

---

[10] Cf. Footnote 8 of Chap. 2.

[11] The initial condition is $t = 0$, $\sigma(t = 0) = \sigma_0 = E'_{0,e}\varepsilon_0$, because only the top spring will be able to deform instantaneously by ε_0, the dashpot then behaving as a rigid body and preventing the second spring to deform.

Study Question 3.4 Prove that the dynamic response of simple Maxwell and Voigt–Kelvin units cannot represent real behavior. [This is the fourth deficiency mentioned in Footnotes 1 and 2].

Study Question 3.5 As in Question 3.2, obtain all dynamic functions of the modified Voigt–Kelvin model as

$$
E'(\omega\theta) = \frac{E'_{0,e}E'_{1,e}}{E'_{0,e} + E'_{1,e}} + \frac{E'_{0,e}{}^2}{E'_{0,e} + E'_{1,e}} \frac{(\omega\theta)^2}{1 + (\omega\theta)^2} = \frac{E_\infty + E_0(\omega\theta)^2}{1 + (\omega\theta)^2}, \qquad (3.13)
$$

$$
E''(\omega\theta) = \frac{\frac{E'_{0,e}{}^2}{E'_{0,e}+E'_{1,e}}\omega\theta}{1 + (\omega\theta)^2} = \frac{(E_0 - E_\infty)\omega\theta}{1 + (\omega\theta)^2}, \qquad (3.14)
$$

$$
\tan\delta_E(\omega\theta) = \frac{E''}{E'} = \frac{(E_0 - E_\infty)\omega\theta}{E_\infty + E_0(\omega\theta)^2}, \qquad (3.15)
$$

$$
D'(\omega\tau) = \frac{\left(E'_{0,e}{}^{-1} + E'_{1,e}{}^{-1}\right) + E'_{0,e}{}^{-1}(\omega\tau)^2}{1 + (\omega\tau)^2} = \frac{D_\infty + D_0(\omega\tau)^2}{1 + (\omega\tau)^2}, \qquad (3.16)
$$

$$
D''(\omega\tau) = \frac{E'_{0,e}{}^{-1}\omega\tau}{1 + (\omega\tau)^2} = \frac{(D_\infty - D_0)\omega\tau}{1 + (\omega\tau)^2}, \qquad (3.17)
$$

$$
\tan\delta_D(\omega\tau) = \frac{(D_\infty - D_0)\omega\tau}{D_\infty + D_0(\omega\tau)^2}. \qquad (3.18)
$$

Study Question 3.6 From the above relationships and that previously established between the relaxation ant retardation times, show that for both the modified Maxwell and Voigt–Kelvin models, we have $\tan\delta_E(\omega) = \tan\delta_D(\omega)$.[12] [That is what is documented in Fig. 3.4 by the bold solid curve].

3.1.3 Equivalence of the Modified Maxwell and Voigt–Kelvin Models—The Standard Linear Solid Model

The reader will notice (or have noticed after working through the preceding Study Questions) that the results for all linear viscoelastic responses are coincident for both the modified Maxwell and Voigt–Kelvin models, which must mean that (3.1) and (3.10) are equivalent and that the elastic and viscous parameters of both models must be related. So, any of the modified models will be representative of what has been named the *standard linear solid*.

---

[12] [But recall that this equality is generally not valid for real materials, and why].

Study Question 3.7 (1) From the mentioned equality of all response functions, show that the elastic and viscous parameters of bot linear models are related by

$$E'_{0,e} = E_{0,e} + E_{1,e}, \quad E'_{1,e} = \frac{E_{0,e}\left(E_{0,e} + E_{1,e}\right)}{E_{1,e}}$$

$$\text{and} \quad \eta'_1 = \eta_1 \left(\frac{E_{0,e} + E_{1,e}}{E_{1,e}}\right)^2.$$

(2)   Obtain the same relationships by equating the three coefficients of the 1st order linear differential constitutive equations of the models, $\sigma + a_1 \frac{d\sigma}{dt} = a_2 \varepsilon + a_3 \frac{d\varepsilon}{dt}$ and find $a_1 = \theta$, $a_2 = E_\infty$ and $a_3 = E_0 \theta$.

As a result, we may use either of the above modified models as the simplest but realistic representation of linear viscoelastic behavior, excluding viscous flow, as well as the simplest limiting representation of non-linear viscoelasticity at very low strains and stresses.

Reminder The reader will now be able and is encouraged to consider Study Questions 2.5 and 2.6.

3.1.4   Physical Qualitative Description of the Models' Behavior

Uniaxial Tensile Stress Relaxation

In the case of a modified Maxwell model, both springs are initially deformed and sustain the initial stress, but then the viscous dashpot starts deforming and thus (at constant total strain) reduces the strain and relaxes the stress on its associated spring, and therefore the overall stress. At very long times, only the left spring will support the limiting stress, σ_∞.

Assuming now a modified Voigt–Kelvin model, the applied strain and initial stress is concentrated on the top spring, but then the viscous dashpot starts deforming, as well as the second spring in parallel with it, and therefore (at constant overall strain) the strain and stress on the top spring is reduced, the latter being equal to the total stress. At very long times, however, the limiting stress σ_∞ will be distributed between the springs, depending on the elastic properties of both springs.

Uniaxial Tensile Creep

In a modified Maxwell model, both springs support the initial equal strain, but then the viscous dashpot starts deforming, dragging with it the left spring, whose stress reduces the one acting on the dashpot at constant overall stress, leading to a continuously decreasing creep rate. At very long times, the ultimate strain is reached when the stress on the dashpot and right spring has reduced to zero.

In a modified Voigt–Kelvin model, the initial strain is fully supported by the top spring, but then the viscous dashpot starts deforming, dragging with it the left spring, whose stress again reduces the one acting on the dashpot at constant overall stress, leading to a continuously decreasing creep rate. At very long times, the ultimate strain is also reached when the stress on the dashpot has reduced to zero.

Dynamic Testing

In dynamic testing, there are no significant *qualitative* differences between the way both models respond, with the viscous dashpot dampening, and therefore retarding, the strain variations, which will always lag the stress.

3.2 Accounting for Irreversible, Viscous, Flow—The Standard Linear Liquid Model

The possibility of irreversible viscous flow, simultaneous with the viscoelastic response, cannot be totally ignored, depending on the nature of the materials. As the behavior of melts (melt rheology) is outside the scope of this text, we will only consider the ideal case of linear, constant viscosity (Newtonian), viscous flow in trying to identify the nature of the effects it adds to the viscoelastic response. This is a very difficult subject even in the linear viscoelastic domain,[13] the only one considered in this section. Future developments of the subject matter of Part II of this book will eventually have to also take viscous effects into account.

The simplest way of adding viscous flow to the behavior of a standard linear solid, changing it to a *standard linear liquid*, is by adding a viscous dashpot in series. Figures 3.6 and 3.7 show the two possibilities for the representation of a standard linear liquid.

Before giving this subsection the same sequence of treatment adopted for the standard linear solid, in addition to recalling that a single viscous dashpot does not allow stress relaxation and surely increases creep, it is instructive to also identify its expected effects on the dynamic response. Using the complex representation of the excitation and response, from a deformation $\varepsilon^* = \varepsilon_0 e^{i\omega t}$ and stress $\sigma^* = \sigma_0 e^{i(\omega t+\delta)}$ the reader may easily obtain the dynamic modulus components of the dashpot $E' = 0$

---

[13] Their treatment is not widespread in texts of similar nature and, in the authors' opinion, even in such a comprehensive work on linear viscoelasticity as Tschoegl's [3], viscous flow is not given the emphasis that condensed matter theoretical developments may require.

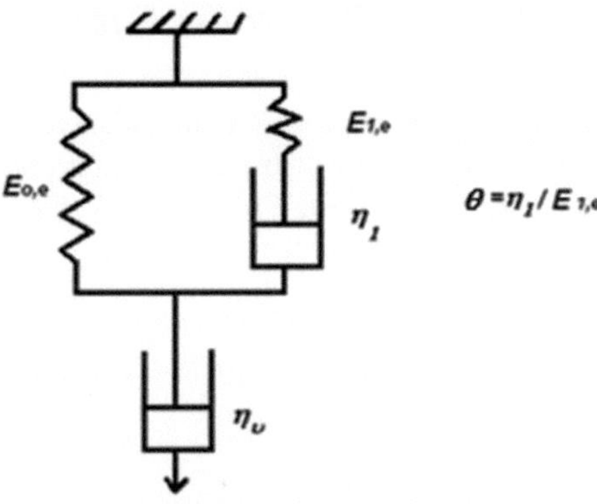

Fig. 3.6 One representation of the standard linear liquid

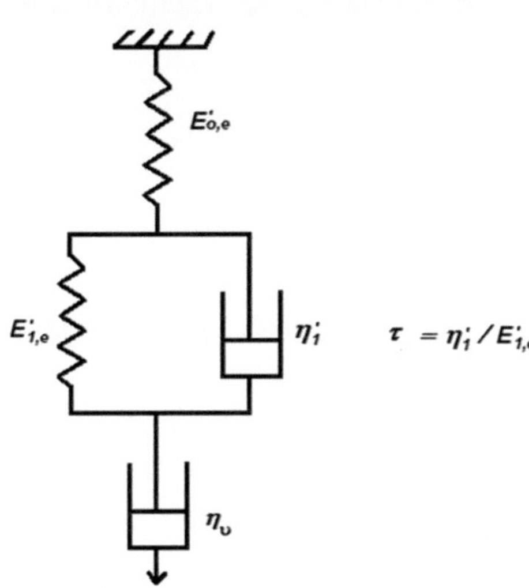

Fig. 3.7 An alternative representation of the standard linear liquid

and $E'' = \omega\eta_v$ and the dynamic components of the compliance $D' = 0$ and $D'' = 1/\omega\eta_v$. So, viscous flow (decreasing viscosity) is obviously expected to change the loss properties,[14] increasing the compliance and decreasing the modulus, but these results would suggest that all storage properties are not affected.

   In the following subsections we will see that, while D' remains unchanged, viscous flow does strongly interfere with the storage modulus, E', of an otherwise standard linear solid, by accelerating stress relaxation and therefore decreasing the modulus. The effect is of such magnitude that it may yield much reduced *effective relaxation times* relative to the nominal ones.[15]

---

[14] [Because additional energy is indeed dissipated. We will see that its influence on the time-dependent and loss compliances are the easiest to obtain].

[15] The effect is also of more difficult accurate formulation.

3.2.1 Basic Constitutive Equation

Given the equivalence or identical behavior of the modified Maxwell and Voigt–Kelvin models, any of the representations of Figs. 3.6 and 3.7 may be adopted to formulate the behavior. From Fig. 3.7, the basic constitutive equation may be written as

$$\sigma = \eta_v \frac{d}{dt}\left[\varepsilon - \frac{\sigma}{E'_{0,e}} - \frac{\sigma - \eta'_1 \frac{d}{dt}\left(\varepsilon - \varepsilon_v - \frac{\sigma}{E'_{0,e}}\right)}{E'_{1,e}}\right], \tag{3.19}$$

where ε_v is the strain on the dashpot of viscosity η_v, responsible for the effect being studied. Taking $\frac{d\varepsilon_v}{dt} = \frac{\sigma}{\eta_v}$, the equation may be simplified to

$$\sigma + \left(\eta_v \frac{E'_{0,e} + E'_{1,e}}{E'_{0,e} E'_{1,e}} + \frac{\eta'_1}{E'_{1,e}}\right)\frac{d\sigma}{dt} + \frac{\eta'_1 \eta_v}{E'_{0,e} E'_{1,e}}\frac{d^2\sigma}{dt^2} = \eta_v \frac{d\varepsilon}{dt} + \frac{\eta'_1 \eta_v}{E'_{1,e}}\frac{d^2\varepsilon}{dt^2}, \tag{3.20}$$

which is now a second-order linear differential equation.

3.2.2 Uniaxial Tensile Stress Relaxation

After imposing and kept constant a strain ε_0 at $t = 0$,[16] (3.20) yields for the stress relaxation modulus a solution of the form $D(t) = C_1 e^{\lambda_1 t} + C_2 e^{\lambda_2 t}$, with λ_1 and λ_2 obtained by

$$\lambda_{1,2} = \frac{-\left(\frac{1}{\theta} + \frac{E'_{0,e}}{\eta_v}\right) \pm \sqrt{\left(\frac{1}{\theta} + \frac{E'_{0,e}}{\eta_v}\right)^2 - 4\frac{E'_{0,e} E'_{1,e}}{\eta'_1 \eta_v}}}{2},$$

where θ is the same relaxation modulus of the corresponding standard linear solid (excluding viscous flow), and the final stress relaxation modulus may then be worked out to give

$$E_r(t) = e^{-\frac{1}{2}A\left(1 - \sqrt{1 - \frac{B}{A^2}}\right)\left(\frac{t}{\theta}\right)}\left[E'_\infty + (E_0 - E'_\infty)e^{-A\sqrt{1 - B/A^2}\left(\frac{t}{\theta}\right)}\right], \tag{3.21}$$

with $A = 1 + \frac{E_0 - E'_\infty}{E'_\infty}\frac{\theta}{\tau_{vis}}$ and $B = 4\frac{E_0 - E'_\infty}{E_0}\frac{\theta}{\tau_{vis}}$, if we define τ_{vis} (a characteristic *viscous flow time*) as the time that viscous flow alone would need to result in a

---

[16] Cf. Footnote 8 of Chap. 2.

strain (and compliance), in creep under the initial stress $[\sigma(t = 0) = \sigma_0]$, equal to that corresponding to the final $(t = \infty)$ creep compliance plateau in the absence of viscous flow,[17] D'_∞, and so the defining equation of τ_{vis} is[18]

$$\tau_{vis} = \eta_v\left(D'_\infty - D_0\right) = \eta_v\frac{E_0 - E'_\infty}{E_0 E'_\infty}. \tag{3.22}$$

It may also be recognized from $E'_\infty = \frac{E_0 E'_{1,e}}{E_0 + E'_{1,e}}$, that $\tau_{vis} = \frac{\eta_v}{E'_{1,e}} = \frac{\eta_v}{\eta'_1}\tau$, where τ is the retardation time of the standard linear solid, a very simple and meaningful relationship.

To obtain the final relaxation modulus of (3.21), we had of course to consider the initial condition $(t = 0, E_0 = E'_{0,e} = C_1 + C_2)$ and also the long-time condition $(t = \infty)$ in the absence of viscous flow, corresponding to also making $\eta_v = \infty$ in (3.21), which may be seen to give $E_r(\eta_v = \infty, t) = C_1 + C_2 e^{-t/\theta}$ and, making $t = \infty$, we obtain $C_1 = E'_\infty$ and, from the mentioned relaxation modulus in the absence of viscous flow (that of a standard linear solid), $C_2 = E_0 - E'_\infty$. As it should, $\lim_{\tau_{vis}=\infty} E_r(t) = E'_\infty + \left(E_0 - E'_\infty\right)e^{-t/\theta}$.

Study Question 3.8 Work through the physical and analytical detail of the derivation of (3.21).

Figure 3.8 plots the relaxation modulus given by (3.21) as a function of time for various extents of viscous flow, as defined by (3.22), using $\frac{\theta}{\tau_{vis}}$ as a parameter. The effect of viscous flow turns out twofold: (1) the reduction and possible near disappearance followed by reappearance and growth of the fully relaxed modulus plateau typical of a standard linear solid, and (2) significant reductions of the *effective relaxation time* relative to that of the corresponding standard linear solid, $\theta_{eff} \sim \frac{\theta}{A\sqrt{1-B^2/A}}$. The variations mentioned in (1) above can only be seen in a log-modulus scale and are very surprising, thus warranting future detailed physical justification! Note that, as expected, the inflection of the modulus curve for the standard linear solid $(\tau_{vis} \rightarrow \infty)$ occurs at $t/\theta = 1$ (cf. Fig. 3.9).

The very significant effect of viscous flow on the stress relaxation times shows that, in theoretical interpretations of experimental data, and given the practical impossibility of ensuring the absence of viscous flow in experiments of finite duration with most real materials, it is of the utmost future relevance to generalize the formulation and evaluation of the effect beyond the above idealized case of the standard linear liquid.

---

[17] This somewhat convoluted definition that we propose for τ_{vis} will be seen to enable to quantify the importance of viscous flow in all types of materials' responses.

[18] Similarly, $E'_\infty = 1/D'_\infty$ is the fully relaxed modulus in the absence of viscous flow.

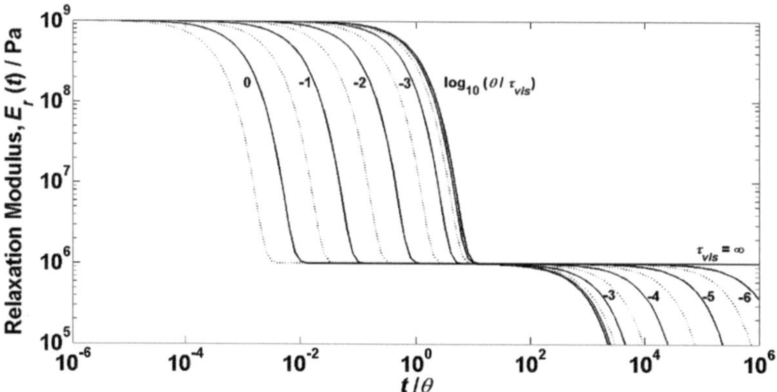

Fig. 3.8 Time-dependent uniaxial tensile relaxation modulus of a standard linear liquid with $E_0 = 10^9$ Pa and $E'_\infty = 10^6$ Pa for varying extents of viscous flow[19]

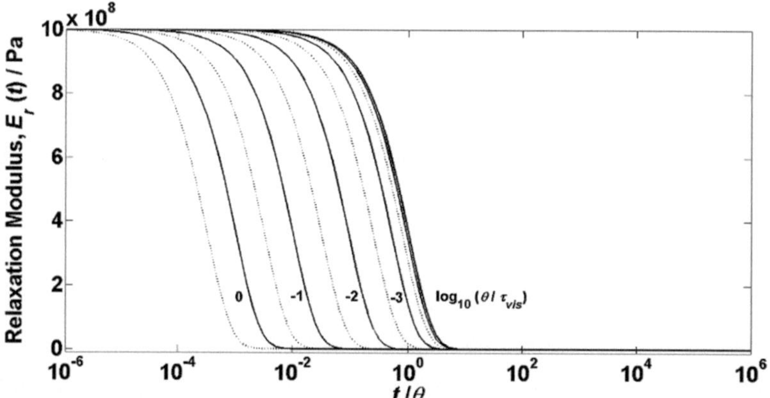

Fig. 3.9 The same data of Fig. 3.8 in linear modulus scale

3.2.3   Uniaxial Tensile Creep

We do not need to go back to (3.20) and integrate it from scratch. To obtain the overall creep compliance we just need to add the viscous flow compliance, t/η_v, to that of the standard linear solid, leading to

$$D(t) = D_0 + \left(D'_\infty - D_0\right)\left[1 - e^{-\frac{t}{\tau}} + \left(\frac{\tau}{\tau_{vis}}\right)\left(\frac{t}{\tau}\right)\right], \qquad (3.23)$$

whose results are plotted in Fig. 3.10, showing the known behavior of viscoelastic materials under the influence of varying extents of viscous flow. The effect is visible

---

[19] Remember that θ is the relaxation time of the corresponding standard linear solid.

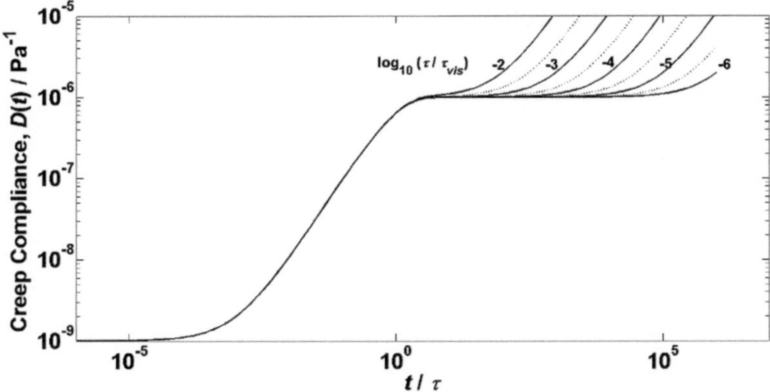

Fig. 3.10 Time-dependent uniaxial tensile creep compliance of a standard linear liquid with $D_0 = 10^{-9}\,\mathrm{Pa}^{-1}$ and $D'_\infty = 10^{-6}\,\mathrm{Pa}^{-1}$ for varying extents of viscous flow[20]

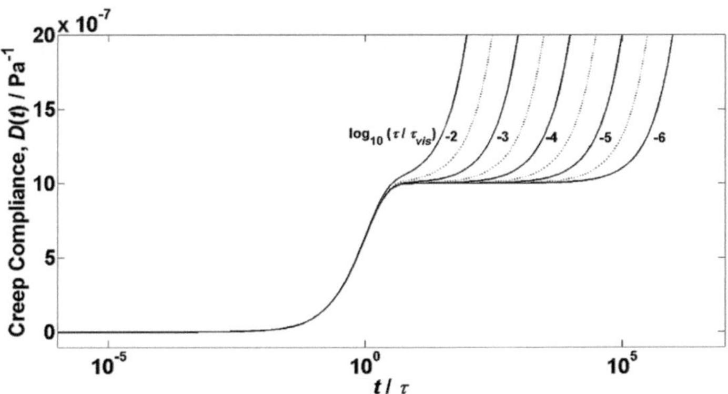

Fig. 3.11 The same data of Fig. 3.10 in linear compliance scale

for long times, but, contrary to what happens in stress relaxation, the viscous flow does not affect the *effective retardation time*, with the inflection at $t/\tau = 1$ (cf. Fig. 3.11).

Study Question 3.9 Discuss qualitatively whether and in what measure the curves of Figs. 3.10 and 3.11 could represent the behavior of an amorphous non-crosslinked polymer. Which polymer parameter would be related with τ/τ_{vis}? Which specific characteristic of the plotted behavior is *not* representative of such material?

---

[20] τ is the same retardation time of the corresponding standard linear solid.

3.2.4 Dynamic Testing

Considering as before a complex strain $\varepsilon^* = \varepsilon_0 e^{i\omega t}$ and the corresponding complex stress $\sigma^* = \sigma_0 e^{i(\omega t + \delta)}$, and substituting them in (3.20), the separation of its real and imaginary components leads to the following two equations,

$$\left(\frac{1}{\theta} + \frac{E_0}{\eta_v}\right) E'' \omega - \left(\frac{E'_\infty}{\theta \eta_v} - \omega^2\right) E' = E_0 \omega^2, \tag{3.24}$$

$$\left(\frac{E'_\infty}{\theta \eta_v} - \omega^2\right) E'' + \left(\frac{1}{\theta} + \frac{E_0}{\eta_v}\right) E' \omega = \frac{E'_\infty}{\theta} \omega, \tag{3.25}$$

which may be solved and give

$$E'(\omega\theta) = \frac{E'_\infty + \frac{1}{A}\left[1 - \frac{B}{4(\omega\theta)^2}\right] E_0 (\omega\theta)^2}{A + \frac{1}{A}\left[1 - \frac{B}{4(\omega\theta)^2}\right]^2 (\omega\theta)^2}, \tag{3.26}$$

$$E''(\omega\theta) = \frac{E_0 - \frac{1}{A}\left[1 - \frac{B}{4(\omega\theta)^2}\right] E'_\infty}{A + \frac{1}{A}\left[1 - \frac{B}{4(\omega\theta)^2}\right]^2 (\omega\theta)^2} \omega\theta, \tag{3.27}$$

$$\tan \delta_E = \frac{E_0 - \frac{1}{A}\left[1 - \frac{B}{4(\omega\theta)^2}\right] E'_\infty}{E'_\infty + \frac{1}{A}\left[1 - \frac{B}{4(\omega\theta)^2}\right] E_0 (\omega\theta)^2} \omega\theta, \tag{3.28}$$

with A and B the same given just below (3.21). It may be checked that, as η_v and τ_{vis} go to infinity (and A and B approach 1 and 0, respectively), the above dynamic property values reduce to those of (3.13), (3.14) and (3.15), respectively, that characterize a standard linear solid.

The storage and loss modulus of a standard linear liquid with the specified elastic properties are plotted in Figs. 3.12 and 3.13. Two very significant effects of viscous flow become apparent in the first figure: (1) the increase of the storage modulus with frequency is severely "retarded", occurring at increased frequencies, because of the already mentioned acceleration of stress relaxation process ($\theta_{eff} < \theta$), and (2) the low frequency plateau is reduced and will tend to disappear as η_v and τ_{vis} decrease. As to the loss modulus, the viscous flow reveals the emergence of additional energy absorption at low frequencies, the lower the greater η_v and τ_{vis}.

Given that the compliance of the second dashpot simply adds to that of the standard linear solid, and the viscous flow does not affect the storage compliance, the storage and loss compliances, and the loss factor, of the standard linear liquid may be written as

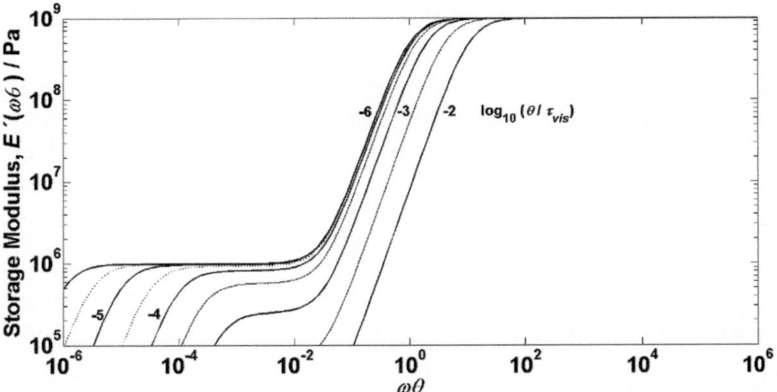

Fig. 3.12 Uniaxial tensile storage modulus of a standard linear liquid with $E_0 = 10^9$ Pa and $E'_\infty = 10^6$ Pa for varying extents of viscous flow[21]

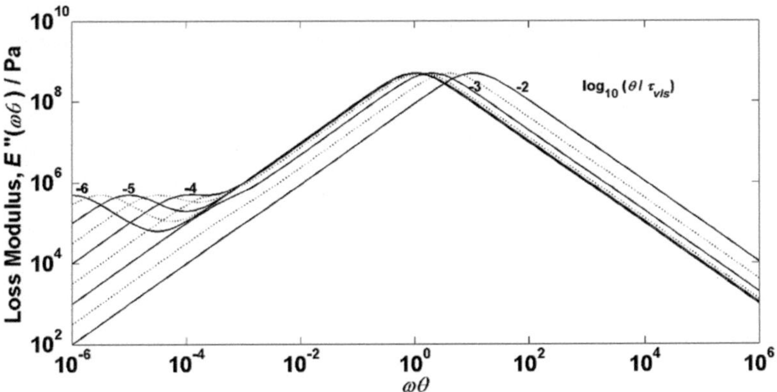

Fig. 3.13 Uniaxial tensile loss modulus of a standard linear liquid with $E_0 = 10^9$ Pa and $E'_\infty = 10^6$ Pa for varying extents of viscous flow[22]

$$D'(\omega\tau) = D_0 + \frac{D'_\infty - D_0}{1 + (\omega\tau)^2} = \frac{D'_\infty + D_0(\omega\tau)^2}{1 + (\omega\tau)^2}, \tag{3.29}$$

$$D''(\omega\tau) = \left(D'_\infty - D_0\right)\left[\frac{\omega\tau}{1 + (\omega\tau)^2} + \frac{1}{\omega\tau}\frac{\tau}{\tau_{vis}}\right], \tag{3.30}$$

$$\tan\delta_D = \frac{\left(D'_\infty - D_0\right)\left[\omega\tau + \frac{1+(\omega\tau)^2}{\omega\tau}\frac{\tau}{\tau_{vis}}\right]}{D'_\infty + D_0(\omega\tau)^2}, \tag{3.31}$$

---

[21] See Footnote 19.

[22] See Footnote 19.

which are plotted in Figs. 3.14, 3.15 and 3.16 as functions of $\omega\tau$, $\frac{\tau}{\tau_{vis}}$ being a parameter, with D' and D'' plotted as variations relative to $D'_\infty - D_0$.

Other than the absence of viscous flow effects on the storage compliance, the loss compliance and loss factor show a clear maximum (at the vicinity $\omega\tau = 1$ in the case of D'', as physically expected) and a minimum at lower and lower frequencies (before steep increases) as the viscosity and τ_{vis} are increased. Nevertheless, in the vicinity of the characteristic response frequency of the corresponding standard linear solid ($\omega\tau = 1$) the effect of viscous flow is present but very small.

Study Question 3.10 For the case of the above loss compliance, show that two stationary points are indeed expected at $\omega\tau =$

$$\left\{ \frac{1-2\frac{\tau}{\tau_{vis}}}{2\left(1+\frac{\tau}{\tau_{vis}}\right)} \left[1 \pm \sqrt{1 - 4\frac{\tau}{\tau_{vis}} \frac{1+\frac{\tau}{\tau_{vis}}}{\left(1-2\frac{\tau}{\tau_{vis}}\right)^2}} \right] \right\}^{1/2}, \text{ the first one (a maximum) at the}$$

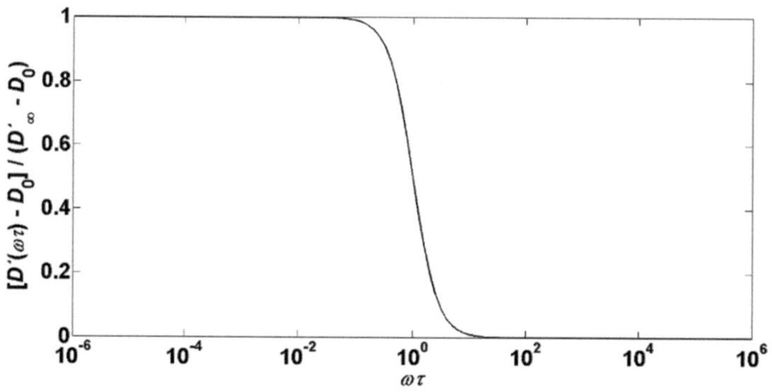

Fig. 3.14 Relative uniaxial tensile storage compliance of a standard linear liquid, showing its independence of viscous flow

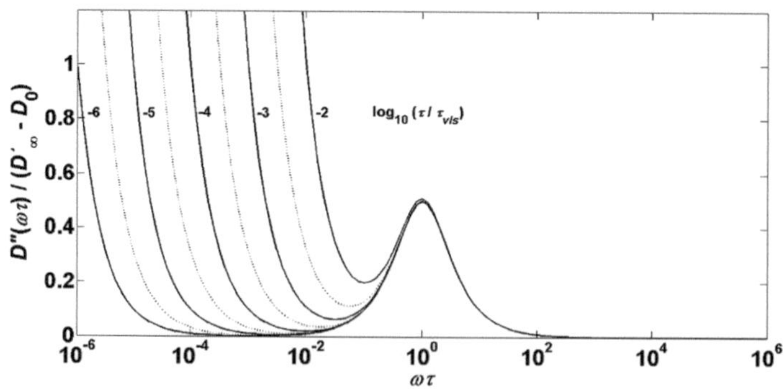

Fig. 3.15 Relative uniaxial tensile loss compliance of a standard linear liquid for varying extents of viscous flow

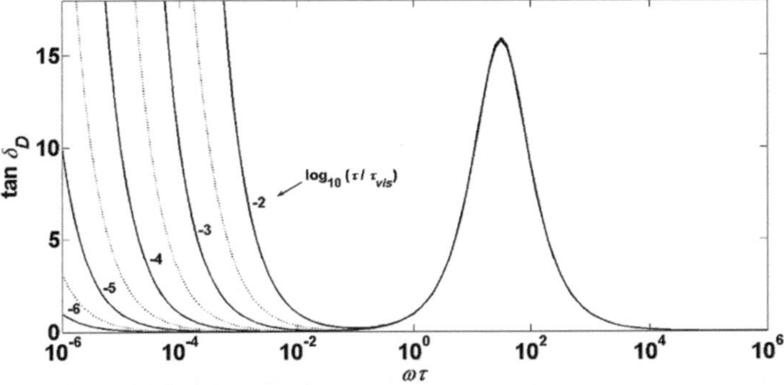

Fig. 3.16 Uniaxial tensile loss factor of a standard linear liquid for varying extents of viscous flow

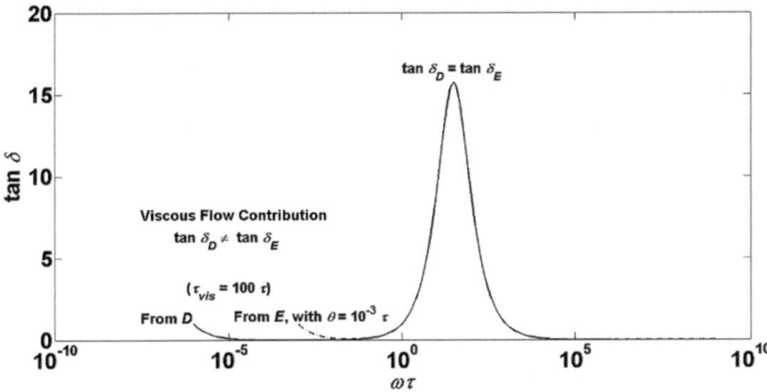

Fig. 3.17 Uniaxial tensile loss factors of a standard linear liquid calculated from dynamic moduli and compliances for the indicated extent of viscous flow

vicinity of $\omega\tau = 1$ (exactly there for infinite viscosity—a standard linear solid), and the other (a minimum) at lower frequencies.

Finally, Fig. 3.17 shows that, where viscous flow effects are negligible (for very high viscosity or at the vicinity of $\omega\tau = 1$), tan δ_D = tan δ_E, which is the known result for a standard linear solid.

3.3 Relaxation and Retardation Spectra

As the reader may have noticed in the qualitative Figs. 2.9 and 2.12, from available knowledge of materials' experimental behavior, the transition region between the stress relaxation modulus and creep compliance short and long-time plateaus covers more than the two or three decades of log t that characterize single response time systems such as the standard linear solid or liquid (cf. Figs. 3.2 and 3.3). This has rightly been interpreted as indicating that a range of response times determine

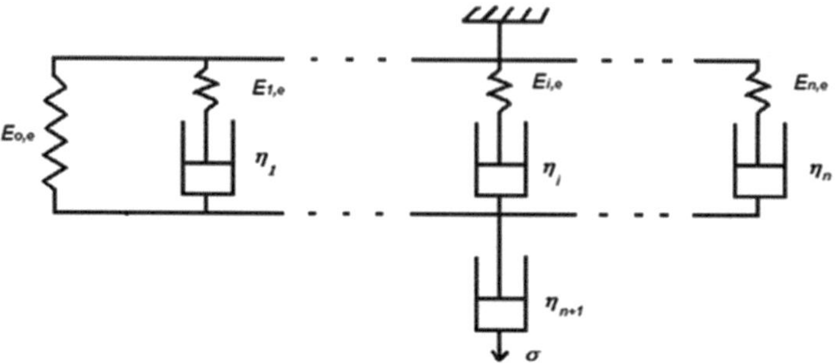

Fig. 3.18 One simplified[25] representation of a generalized standard linear liquid

the behavior of materials and justifies the attempts to relate that with the existence of a range of different structures at the molecular scale contributing to the overall response.[23]

The obvious classical solution of imagining adjustable parallel or series associations of Maxwell, Voigt–Kelvin and standard linear solid units, such as those represented in Figs. 3.18 and 3.19, became popular in books and articles[24] and is the simplest way to understand and, say, "visualize" the need for, and role of, the *response* (relaxation and retardation) *time spectra*.

For the simplified model of Fig. 3.18, the stress relaxation modulus excluding viscous flow would thus be formulated as

$$E_r(t) = E_{0,e} + \sum_{i \geq 1} E_{i,e} e^{-t/\theta_i}, \tag{3.32}$$

---

[23] The identification of such structures is, of course, a very difficult endeavor, which stands and remain an important challenge in the field. Following Ngai [5], we adopt in Part II of the book the designation of *primitive relaxor* for the smallest of those responsive (moving) structures.

[24] However, the widespread attempts that keep being published at describing the behavior of complex materials by mere numerical fitting of the elastic and viscous constants of such adjustable associations to experimental data does *not*, in our opinion (and also Ferry's [6]), serve any useful scientific purpose whatsoever. At most, they may find use as a means of correlating single specific sets of data for *one* single material, at *one* single temperature and *one* stress and/or strain level.

[25] The simplification consists in assuming one single viscous "dashpot" in series with the parallel association of Maxwell or modified Maxwell units. An arguably more reasonable representation would consider a range of different "dashpots", each associated with its own Maxwell or modified Maxwell unit, because one may guess that, at the molecular scale, each elementary moving structure may be immersed in its own variable local environment, and so be affected by a different local viscosity, but that would only increase the complexity of the above representation and ensuing formulation. It is however important to recognize that the above unique "dashpot" at the bottom of the figure should perhaps be assigned the role of a parallel association of "dashpots", each associated with its own Maxwell or modified Maxwell unit. Likewise, the first spring on the left of the figure should better stand for a parallel association of possibly different springs, such that $E_{0,e} = \sum_{i \geq 1} E_{0,e,i}$.

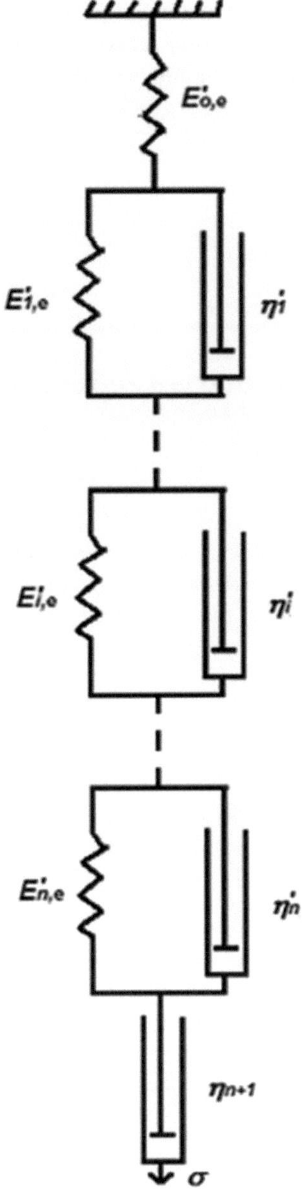

Fig. 3.19 Another simplified representation of a generalized standard linear liquid

where $E_{0,e} = E_\infty$, $\theta_i = \eta_i/E_{i,e}$ and $\sum_{i\geq 1} E_{i,e} = E_0 - E_\infty$. To our knowledge, the literature did not yet consider how to include the effect of viscous flow in real materials, but we may guess that a future solution might require the modification of (3.32) by generalizing (3.21) to the above parallel association of simple structures, each with its θ_i and $\tau_{vis,i}$.

Generalizing a bit more, and imagining that we may not have a finite number of characteristic relaxation times but a continuum of infinitely close times,[26] making up a continuous *relaxation time spectrum*, we may replace (3.32) by

$$E_r(t) = E_\infty + \int_0^\infty \mathfrak{E}(\theta)e^{-t/\theta}d\theta, \tag{3.33}$$

$$E_r(t) = E_\infty + \int_{-\infty}^{+\infty} H(\theta)e^{-t/\theta}d\ln\theta, \tag{3.33a}$$

with $H(\theta) = \theta\,\mathfrak{E}(\theta)$ or, as often done and adopted in Part II of the book,

$$E_r(t) = E_\infty + (E_0 - E_\infty)\int_{-\infty}^{+\infty} H'(\theta)e^{-t/\theta}d\ln\theta, \tag{3.33b}$$

in which case $\int_{-\infty}^{+\infty} H'(\theta)d\ln\theta = 1$. Either $H(\theta)$ or, as we prefer, $H'(\theta)$ is what is named the *relaxation spectrum*. Likewise, the dynamic properties in terms of that spectrum may be obtained as

$$E'(\omega) = E_\infty + (E_0 - E_\infty)\int_{-\infty}^{+\infty} H'(\theta)\frac{(\omega\theta)^2}{1+(\omega\theta)^2}d\ln\theta, \tag{3.34}$$

$$E''(\omega) = (E_0 - E_\infty)\int_{-\infty}^{+\infty} H'(\theta)\frac{\omega\theta}{1+(\omega\theta)^2}d\ln\theta, \tag{3.35}$$

which generalize (3.4) and (3.5). A corresponding generalization of (3.26) and (3.27) may also be anticipated in future inclusions of viscous flow effects.

With reference now to the model of Fig. 3.19, the reader may easily conclude that the creep compliance (excluding viscous flow[27]) and all relationships analogous to (3.33)–(3.35) may be written as in the list that follows:

$$D(t) = D_{0,e} + \sum_{i\geq 1} D_{i,e}\left(1 - e^{-t/\tau_i}\right), \tag{3.36}$$

---

[26] [To include all possible variations in the local environment of whatever elementary moving structures that may be present within the material].

[27] [Which was already considered in Sect. 3.2.3, but only for a standard linear liquid (with a unique response time), not a real material].

with $D_{j\geq0,e} = 1/E_{j\geq0,e}$, $\sum_{i\geq1} D_{i,e} = D_\infty - D_{0,e}$, $D_{0,e} = D_0$ and $\tau_{i\geq1} = \eta_{i\geq1}/E_{i\geq1,e}$,

$$D(t) = D_0 + \int_0^\infty \mathfrak{D}(\tau)\left(1 - e^{-t/\tau}\right)d\tau, \tag{3.37}$$

$$D(t) = D_0 + \int_{-\infty}^\infty L(\tau)\left(1 - e^{-t/\tau}\right)d\ln\tau, \tag{3.37a}$$

with $L(\tau) = \tau\mathfrak{D}(\tau)$, or

$$D(t) = D_0 + (D_\infty - D_0)\int_{-\infty}^{+\infty} L'(\tau)\left(1 - e^{-t/\tau}\right)d\ln\tau, \tag{3.37b}$$

with $\int_{-\infty}^{+\infty} L'(\tau)d\ln\tau = 1$,

$$D'(\omega) = D_0 + (D_\infty - D_0)\int_{-\infty}^{+\infty} \frac{L'(\tau)}{1 + (\omega\tau)^2}d\ln\tau, \tag{3.38}$$

$$D''(\omega) = (D_\infty - D_0)\int_{-\infty}^{+\infty} \frac{L'(\tau)\omega\tau}{1 + (\omega\tau)^2}d\ln\tau. \tag{3.39}$$

Estimate of $H'(\theta)$ from Stress Relaxation and Dynamic Data

There are very few comprehensive studies on reliable calculations of characteristic time spectra from experimental response data. The classical study by Marvin [7], who proposed the "wedge and box" (bimodal) shape of the relaxation and retardation time spectrum for poly(isobutylene), keeps being referenced as one of few examples, but it is not certain that such shape might be representative of a wide range of materials (despite its inclusion of the glassy, viscoelastic, rubbery and flow regions). In addition, the link between calculated or estimated spectra and structural details is even farther from reliable clarification. Within a narrower range of timescales such as to exclude viscous flow, one however expects a unimodal shape[28] skewed towards long timescales, corresponding to the "wedge" part of Marvin's proposal.

---

[28] [At least for materials having one single type of "moving" structures at the molecular scale].

From reasonably accurate stress relaxation data defining $E_r(t)$, it would in principle be possible to calculate $H'(\theta)$ by Laplace inversion [8].[29] This possibility may be understood if one makes the substitution $\theta = 1/s$ in (3.33), giving $E_r(t) - E_\infty = \int_0^\infty \frac{\mathfrak{E}(\frac{1}{s})}{s^2} e^{-ts} ds = \mathcal{L}_t\left[\frac{\mathfrak{E}(\frac{1}{s})}{s^2}\right]$, where t now stands as the Laplace complex parameter or frequency and, by Laplace inversion of $\{[E_r(t) - E_\infty]\}_{s=1/\theta}$,

$$H'(\theta) = \frac{\theta \, \mathfrak{E}(\theta)}{E_0 - E_\infty} = \frac{\frac{1}{s} \, \mathfrak{E}(\frac{1}{s})}{E_0 - E_\infty} = \frac{s\left\{\mathcal{L}_t^{-1}[E_r(t) - E_\infty]\right\}_{s=1/\theta}}{E_0 - E_\infty}. \tag{3.40}$$

A less accurate estimate may be obtained by applying Alfrey's approximation [9][30] to (3.33b), whereby for $\theta < t$ one may make $e^{-t/\theta} \sim 0$ (instead of only for $\theta \ll t$) and for $\theta > t$ one may take $e^{-t/\theta} \sim 1$ (instead of only for $\theta \gg t$), giving for $E_r(t)$ and $H'(\theta)$

$$E_r(t) \sim E_\infty + (E_0 - E_\infty) \int_{\ln t}^{\infty} H'(\theta) d \ln \theta, \tag{3.41}$$

$$H'(\theta) \sim - \frac{\left[\frac{dE_r(t)}{d \ln t}\right]_{t=\theta}}{E_0 - E_\infty}. \tag{3.42}$$

With similar approximations of θ relative to ω^{-1}, from (3.34) we obtain

$$H'(\theta) \sim \frac{\left[\frac{dE'(\omega)}{d \ln \omega}\right]_{\omega=\theta^{-1}}}{E_0 - E_\infty}. \tag{3.43}$$

An important note should however be made that, in view of the significant reductions of the effective relaxation times by the possible presence of viscous flow (highlighted in Sect. 3.2.2 and 3.2.4), drawing conclusions on response times and spectra from stress relaxation and/or mechanical dynamical experimental data is problematic and require extreme precautions.

Study Question 3.11 Obtain the equivalent of (3.40)–(3.43) for the retardation time spectrum, $L'(\tau)$, by generalizing the representation of Fig. 3.19.

---

[29] [Which may require the use of specialized numerical methods, currently implemented in various computing packages such as MATLAB®].

[30] In Chap. 4 (Sect. 4.4.3) we offer an appraisal of the type of errors incurred with this approximation and a more comprehensive one in Chap. 11 (Sect. 11.9.3).

Closing Note

Interested and senior readers may find comprehensive but more complex treatments of these spectra in the linear viscoelastic domain, such as the one by Tschoegl [3]; in the non-linear domain, however, venturing into the subject is a very challenging task even in a conjectural and approximate way [10–12].

As a result of the wide availability of finite element software packages that use the mentioned superpositions of viscoelastic units (also known as Prony series) in the linear viscoelastic domain, some authors neglect the importance of the use of response time spectra [13],[31] but they are considered essential when trying to associate the adopted formulations with the response of fine structural features of the materials.

References

1. I.M. Ward, *Mechanical Properties of Solid Polymers*, 2nd edn. (Wiley, 1985)
2. I.M. Ward, J. Sweeney, *An Introduction to the Mechanical Properties of Solid Polymers*, 2nd edn. (Wiley, 2004)
3. N.W. Tschoegl, *The Phenomenological Theory of Linear Viscoelastic Behavior—An Introduction* (Springer-Verlag, Berlin, Heidelberg, New York, London, Paris, Tokyo, 1989)
4. J. Jäckle, R. Richert, Phys. Rev. E **77**, 031201 (2008)
5. K.L. Ngai, J. Non-Cryst. Solids **353**, 709 (2007)
6. J.D. Ferry, *Viscoelastic Properties of Polymers* (Wiley, New York, 1980)
7. R.S. Marvin, *Proceedings of the 2nd International Congress of Rheology* (Butterworth, London, 1954)
8. D. Fleisch, *A Student's Guide to Laplace Transforms* (Cambridge University Press, Cambridge, UK, 2022)
9. T. Alfrey, *Mechanical Behavior of High Polymers* (Interscience, New York, 1948)
10. J.R.S. André, Fluência de Polímeros—Fenomenologia e Modelação Dinâmica Molecular (Polymer Creep—Phenomenology and Modeling of Molecular Dynamics), Doctoral Thesis, Universidade de Aveiro, 2004
11. J.R.S. André, J.J.C. Cruz Pinto, Polym. Eng. Sci. **54**, 404 (2014)
12. J.J.C. Cruz Pinto, J.R.S. André, Polym. Eng. Sci. **56**, 348 (2016)
13. J. Bergström, *Mechanics of Solid Polymers—Theory and Computational Modeling*, Chap. 6 (Elsevier, 2015)

---

[31] This, however, does not in our opinion justify somewhat confusing statements such as "… even if a relaxation spectrum is chosen, the Boltzmann's superposition principle *does not allow* for a sigmoidal shaped stress–strain response that is typical for polymers at intermediate to large strains", cited from [13] (our emphasis). Whether the author is strictly meaning stress–strain curves, or also including stress relaxation modulus or compliance curves, Chaps. 5 and 11 and [10–12], the latter in the non-linear viscoelastic domain by the present authors, show that such sigmoidal curves will result by using spectra, either from constitutive equations, or by applying the above principle. (1) Spectra may be continuous or discontinuous, (2) the result of using discontinuous spectra is analogous to that of using Prony series, and (3) the smoothing of discontinuous spectra, when one associates each term of the series to a specific type (even if unspecified) of moving structure with variable local properties, will make them continuous.

Chapter 4
Non-linear Viscoelastic Behavior: Approximate Approaches, Simplified Models and the Basis for Improved Versions

4.1 Introduction

This chapter does not intend to provide a comprehensive and detailed coverage of all types of approach to the difficult and still immature subject of the non-linear viscoelastic behavior of materials. Those may be found in various other sources [1–8] and references thereof. After a very brief summary of some of the classical and approximate formulations, this discussion will limit itself to introduce the basic concepts that suggested and played a prominent role in the developments of Part II of the book, where it is highlighted that, particularly near and below the glass transition temperature, thermal activation holds the key to understand and quantitatively formulate the thermal and other forced responses of amorphous condensed matter.

4.2 Summary of Early Approximate Approaches

Before this brief presentation of the early approximate approaches to non-linear viscoelasticity, one should point out that a general and accurate formulation would in principle be possible through a generalization of Boltzmann's superposition principle, by considering that the $E_r(t - t_i')$ or $D(t - t_i')$ of (2.1) and (2.2) would also be functions of all previous (or differential) changes in the strain or stress that could be expressed by series expansions of those changes. Mathematically, that would lead to expressions of the type

$$\sigma(t) = \sum_i (\Delta \varepsilon)_i E_1(t - t_i') + \sum_i \sum_j (\Delta \varepsilon)_i (\Delta \varepsilon)_j E_2(t - t_i', t - t_j') + \cdots, \quad (4.1)$$

© The Author(s), under exclusive license to Springer Nature Switzerland AG 2024
J. J. Cruz Pinto and J. R. dos Santos André, *Analytical Molecular Dynamics of Amorphous Condensed Matter*, Springer Series in Materials Science 342,
https://doi.org/10.1007/978-3-031-56517-5_4

$$\varepsilon(t) = \sum_i (\Delta\sigma)_i D_1\big(t - t_i'\big) + \sum_i \sum_j (\Delta\sigma)_i (\Delta\sigma)_j D_2\big(t - t_i', t - t_j'\big) + \cdots ,$$

$$(4.2)$$

or the corresponding multiple integral versions, but this more exact representation of non-linear viscoelasticity is impractical and nearly always inaccessible for lack of knowledge of the functions $E_{l\geq2}$ and $D_{l\geq2}$. Some simplifications of the above formulas would be possible, like for example by neglecting all terms of order higher than 3 in the strain or stress, as there are creep situations where the compliance appears proportional to the stress squared, but one simplification that has found some use was Leaderman's [1], whereby in both the stress and strain the dependence on time may be separated from that of the other variable,

$$\sigma(t, \varepsilon) = h(\varepsilon)p(t) \quad \text{or} \quad \varepsilon(t, \sigma) = f(\sigma)g(t), \tag{4.3}$$

which, for a series of excitations, would yield

$$\sigma(t) = \sum_i p\big(t - t_i'\big)[\Delta h(\varepsilon)]_i , \tag{4.4}$$

or

$$\varepsilon(t) = \sum_i g\big(t - t_i'\big)[\Delta f(\sigma)]_i , \tag{4.5}$$

with $[\Delta h(\varepsilon)]_i = h(\varepsilon_i) - h(\varepsilon_{i-1})$ or $[\Delta f(\sigma)]_i = f(\sigma_i) - f(\sigma_{i-1})$.

Study Question 4.1 Non-linear viscoelastic materials are observed to recover faster than they creep. Show that Leaderman's modification of Boltzmann's superposition principle does not predict such behavior, but that recovery turns out the exact opposite of creep, as in the linear viscoelastic domain.

Non-linear creep has been much more extensively studied than stress relaxation, and an obvious alternative for the treatment of non-linearity was the direct use of limited experimental creep data in interpolation procedures. From a limited set of creep curves, one may obtain *isometric* (constant strain) and *isochronous* (constant time) *curves*. Isometric curves are similar to stress relaxation ones, which is useful given the easier execution of creep measurements, and so such curves provide approximate estimates of the stress relaxation modulus, $E_{r,c}(t, \varepsilon_0) \sim \left[\frac{\sigma(t)}{\varepsilon}\right]_{isom}$. One should however note that such estimates differ from those of a *creep modulus*, directly obtainable from creep curves by $E_c = \frac{\sigma_0}{\varepsilon(t)}$, and some sources may lead to confusion by using the latter name either to E_c or to $E_{r,c}$. As to the isochronous curves (which should not be confused with stress–strain curves—cf. Sect. 2.3.5), they may easily disclose the non-linearity of the behavior by a slope different from 1 if plotted on log–log scales. A practical observation by Turner [2] was that the creep behavior could be approximately evaluated with ease for a wide range of conditions from only two creep curves for an extended range of times, combined with various isochronous creep tests for a range of applied stresses, or even only one in case of approximate separability between time and stress.

Alternative treatments of creep non-linearity have involved the use of simplified correlations, such as

$$\varepsilon(t, \sigma) = A(\sigma)t^n,\tag{4.6}$$

with $A(\sigma)$ a non-linear function of the stress or, as suggested mainly by experiments with metals, Nutting's correlation [3],

$$\varepsilon(t, \sigma) = C\sigma^m t^n.\tag{4.7}$$

Another type of empirical correlation involves expressing $A(\sigma)$ as a sinh,

$$\varepsilon(t, \sigma) = C't^n \sinh\left(\frac{\sigma}{\sigma_c}\right),\tag{4.8}$$

all the above examples of possible separability of the time and stress dependencies, where σ_c is a "critical" stress value above which the non-linearity intensifies, marking an approximate limit of linear viscoelasticity, σ_l (cf. Fig. 2.16).

The above approximate stress-time separability allowed Turner [2] to observe that, by applying Boltzmann's modified superposition principle (cf. 4.5) to creep and recovery, one could approximately correlate the fraction of recovered strain, $FR = [\varepsilon(t_c) - \varepsilon(t - t_c)]/\varepsilon(t_c)$, to the fractional recovery time, $t_R = (t - t_c)/t_c$, where t_c is the creep time, by $FR = f(t_R) \approx 1 - (1 + t_R)^n + t_R^n$.

4.3 Simplified Models with Thermal Activation

The first models of this nature were proposed for creep, and so we will treat it first in this very brief review.

4.3.1 Uniaxial Tensile Creep

Various authors [1, 9–18] rightly considered that creep would be driven by local *activated transitions* at the molecular scale, whose dynamics could be expressed by calculating the difference between the frequency of the direct and reverse transitions relative to the increase in strain, in which the applied stress was assumed to impose symmetrical changes to the potential energy barriers, $E_{a,0}$, to be surmounted (for simplicity, assumed in both cases to be identical to those at equilibrium with zero

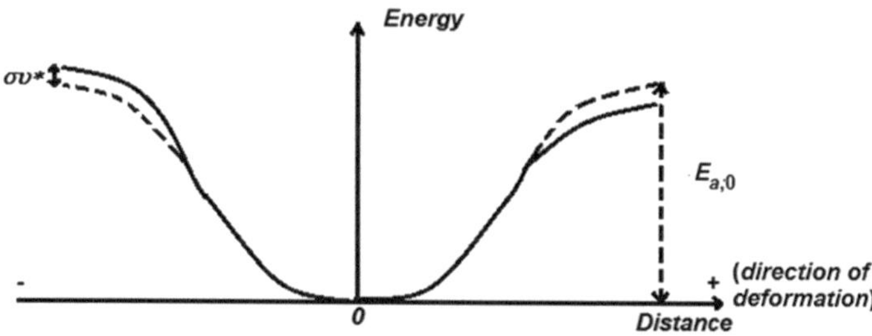

Fig. 4.1   Qualitative energy diagram assumed in the early models of creep

applied stress). With reference to the qualitative energy diagram of Fig. 4.1 (conceptually and functionally equivalent to its more usual representation, as in Figs. 9.28
and 10.7 of [6, 7], respectively), the creep rate was formulated as

$$\frac{d\varepsilon}{dt} \propto \underbrace{\exp\left(-\frac{E_{a,0} - \sigma\upsilon^{\#}}{k_B T}\right)}_{positive\ transitions} - \overbrace{\exp\left(-\frac{E_{a,0} + \sigma\upsilon^{\#}}{k_B T}\right)}^{negative\ transitions}, \quad \text{or}$$

$$\frac{d\varepsilon}{dt} \propto e^{-\frac{E_{a,0}}{k_B T}} \sinh\left(\frac{\sigma\upsilon^{\#}}{k_B T}\right),$$

(4.9)

also know as the Eyring model, where $\upsilon^{\#}$ is an activation volume.

The rationale of the effective activation energies specified was that $\sigma\upsilon^{\#}$, the work
done by the applied stress per unit resisting area up to the activated state in each
of the direct transitions, allowed an identical reduction of the thermal fluctuation
necessary to reach it, while a higher fluctuation would be needed to counteract the
applied stress in each of the reverse transitions. It is the resulting imbalance of the
direct and reverse transitions that leads to the increase in strain.

So, in its simplest formulation, creep could be modeled by incorporating an *activated "dashpot"* in the classical standard linear solid obeying (4.9) (changing it to
a non-linear one). Most authors did not go into great details, but rightly assumed
that the proportionality "constant" in (4.9) ought to depend on the strain reached
up to any given time (as in a simple standard linear solid), and so easily concluded
that, at high stress, the behavior would approximately follow an equation of the form
$\dot{\varepsilon} e^{\frac{E_0 - \sigma\upsilon^{\#}}{k_B T}} = f(\varepsilon)$ or $\dot{\varepsilon} = f_1(T) f_2\left(\frac{\sigma}{T}\right) f_3(\varepsilon)$, in addition to some other generalizations
about the effects of the shear and hydrostatic components of the stress [14], and
the possible need to consider two coupled activated "dashpots" at high stresses with
different activation energies and pre-exponential factors. However, the subject did
not since then seem to advance much farther, apart from the recognition that one
needed to account for a spectrum of contributing elementary processes with specific
retardation times.

4.3.2 Uniaxial Tensile Stress Relaxation

The activated "dashpot" was later [19–21] applied in the formulation of stress relaxation by means of a simple Maxwell-type activated series association with an elastic spring. The viscous strain rate was written as the high stress approximation of (4.9),

$$\dot{\varepsilon}_v \propto e^{-\frac{E_0}{k_B T}} e^{\frac{\sigma v^{\#}}{k_B T}} = A e^{B\sigma}, \tag{4.10}$$

and, adding the elastic component of the creep rate, $\dot{\varepsilon}_l = \frac{\dot{\sigma}}{E}$, with $\dot{\varepsilon} = \dot{\varepsilon}_l + \dot{\varepsilon}_v = 0$, it leads to the highly simplified final result for the stress

$$\sigma(t) = \sigma_0 - \frac{1}{B} \ln\left(D + \frac{t}{C} \right), \tag{4.11}$$

where B, C and D are constants, with $C = (ABE)^{-1}$ and $D = B\sigma_0 + e^{-B\sigma_0} \approx B\sigma_0$. This differs from (10.38) quoted in [7], because their authors used a modification of (4.10) and (4.11) to account for strain hardening, justified by the type of step-stress relaxation experiments they used. Notwithstanding the claimed reasonable agreement obtained (by numerical fitting) between their slightly different correlation and some specific polymer stress relaxation data [19–21], one should recognize the oversimplified nature and lack of physical quantitative detail of the above formulation.

4.4 The Basis for Improved Versions

As mentioned in the preceding section and abundantly described in the literature, the attempts at formulating non-linear stress relaxation and creep with an explicit (albeit approximate) link to the microscopic (molecular) behavior did not go much beyond modifications of the simplest linear viscoelastic model(s) by one (or maximum two) activated dashpots [5–24], thus oversimplifying the problem. The next two subsections describe the first recent attempts by the authors to develop comprehensive formulations of non-linear viscoelasticity intended to better portray the behavior at the molecular scale. Two main improvements of the mentioned formulations will be seen to involve (1) allowing for a spectrum of response (relaxation or retardation) times and (2) a more realistic (or phenomenologically acceptable) description of the processes involved at the molecular scale.

4.4.1 The Need to Account for Relaxation and Retardation Spectra

A wide range of localized motions (and transitions) at the molecular scale may be anticipated to be involved in any of the physical responses of materials, and it is only logical to associate to such motions and transitions specific response times, adding up to the characteristic relaxation and retardation spectra, $H'(\theta)$ and $L'(\tau)$, presented in Sect. 3.3. Those spectra are thus expected to depend and shed light on some of the materials' structural details and their role on the responses under study. Assuming that many of those transitions may be *activated*,[1] Feltham [25] suggested that the simplest relaxation and retardation spectra[2] would be approximately log-normal, because the local activation energies, E_a, would likely be normally distributed due to randomly varying local environments, and $\ln \tau \propto E_a$.

In stress relaxation we would then have

$$H'(\theta) = \frac{H(\theta)}{E_0 - E_\infty} = \frac{b}{\sqrt{\pi}} e^{-\left[b \ln \frac{\theta}{\theta^*}\right]^2}, \qquad (4.12)$$

where θ^* is an average relaxation time and b is proportional to the reciprocal of the standard deviation, $\sigma_{\ln \theta}$ $\left(b = \frac{1}{\sigma_{\ln \theta}\sqrt{2}}\right)$. The resulting stress relaxation modulus of a material would then be

$$E_r(t) = E_\infty + (E_0 - E_\infty)\frac{b}{\sqrt{\pi}} \int_{-\infty}^{+\infty} e^{-t/\theta} e^{-\left[b \ln \frac{\theta}{\theta^*}\right]^2} d \ln \theta, \qquad (4.13)$$

or, adopting Alfrey's approximation [26] (cf. Sect. 3.3),

$$E_r(t) \approx E_\infty + (E_0 - E_\infty)\frac{b}{\sqrt{\pi}} \int_{\ln t}^{+\infty} e^{-\left[b \ln \frac{\theta}{\theta^*}\right]^2} d \ln \theta, \qquad (4.14)$$

where the integral may be evaluated to yield

$$E_r(t) \approx E_\infty + \frac{1}{2}(E_0 - E_\infty)\left\{1 + \mathrm{erf}\left[b \ln \frac{\theta^*}{t}\right]\right\}. \qquad (4.15)$$

---

[1] The existence and importance or not of *activated processes* in materials physics became a standing issue in the materials research community. No matter how obvious they might be to so many, there seems to be some (or perhaps as many) saying the exact opposite. Ironical is that (1) hopping-like (activated) processes had to come to fix deficiencies of an advanced theory such as the MCT—Mode Coupling Theory (cf. Chaps. 7 and 10) and, (2) in the opinion of some of their own proponents, the free volume theories (cf. Sect. 6.10) will also need similar fix(es), particularly at low temperatures. [Much more on *activation*, though, in Chap. 8].

[2] [When one single type of *"moving structures"* or *"relaxors"* participate in the material's response].

Likewise, in creep, a similar (but not identical) spectrum would be

$$L'(\tau) = \frac{L(\tau)}{D_\infty - D_0} = \frac{b'}{\sqrt{\pi}} e^{-[b' \ln \frac{\tau}{\tau^*}]^2}, \tag{4.16}$$

and the resulting creep compliance in the absence of viscous flow would be

$$D_r(t) \approx D_0 + \frac{1}{2}(D_\infty - D_0)\left\{1 + \mathrm{erf}\left[b' \ln \frac{t}{\tau^*}\right]\right\}, \tag{4.17}$$

where τ^* is the average retardation time. In the non-linear viscoelastic domain, b and θ^* are expected to depend on the applied strain, ε_0, while b' and τ^* will depend on the applied stress, σ_0.

However, it is reasonable to consider that there may be a minimum relaxation and retardation time, θ_1 and τ_1, respectively, corresponding to the smallest structural contributor[3] to the process being considered, and so a truncated distribution might be assumed. In the case of creep, for example, $L'(\tau \geq \tau_1)$ would be divided by $\int_{\ln \tau_1}^{+\infty} L'(\tau)d \ln \tau$ and give for the creep compliance

$$D(t) = D_0 + (D_\infty - D_0)\frac{\int_{b \ln \frac{\tau_1}{\tau^*}}^{b \ln \frac{t}{\tau^*}} e^{-[b \ln \frac{\tau}{\tau^*}]^2} d\left(b \ln \frac{\tau}{\tau^*}\right)}{\int_{b \ln \frac{\tau_1}{\tau^*}}^{+\infty} e^{-[b \ln \frac{\tau}{\tau^*}]^2} d\left(b \ln \frac{\tau}{\tau^*}\right)}, \tag{4.18}$$

or

$$D(t) \approx D_0 + (D_\infty - D_0)\frac{\mathrm{erf}(b_0) + \mathrm{erf}\left(b \ln \frac{t}{\tau^*}\right)}{1 + \mathrm{erf}(b_0)}, \tag{4.19}$$

with $b_0 = b \ln \frac{\tau^*}{\tau_1}$, where b_0 may reasonably be expected to vary very little with temperature and stress, as $\ln \frac{\tau^*}{\tau_1}$ and $1/b$ are both proportional to the distribution's standard deviation. The latter, however, when confronted with real data, will be expected to decrease with stress and temperature.

Study Question 4.2 Obtain the analogous to (4.19) for the stress relaxation modulus, $E_r(t)$.

---

[3] [Which we first mentioned in Sect. 3.3 and, following Ngai [27], we name "*primitive relaxor*" throughout Part II of the book].

4.4.2 *First New Definitions and Behavior of Standard and Real Non-linear Solids. An Idealized Case of Polymer Creep by Chain Unfolding*

With the objective of representing and formulating the behavior of amorphous materials at the molecular scale, we considered first [28–32] the specific case of creep of an initially equilibrated polymer, assuming that the process could (very artificially) be reduced to the uncoiling of its chains under the applied stress σ_0. In these conditions, we then arbitrarily assumed that the creep rate could be formulated as the time-dependent difference between the number of favorable ("*gauche*"-to-"*trans*") and unfavorable ("*trans*"-to-"*gauche*") conformational transitions per unit time. Depending on the exact local topology of the polymer chains, "*gauche*"-to-"*trans*" transitions could, in other instances, contribute negatively and "*trans*"-to-"*gauche*" positively to the overall strain, but we kept this idealized problem as simple as possible. We recognize that this picture of the process cannot be generalized to other types of materials, but it may help us in "bootstrapping" a possible more general future solution.[4]

From an initial equilibrated set of the mentioned chain conformers at a given temperature, the strain may then be expected to grow as

$$\frac{d(\varepsilon - \varepsilon_0)}{dt} \propto f_{g^-} e^{-\frac{E_{gt}}{k_B T}} e^{\alpha \sigma_0} + f_{g^+} e^{-\frac{E_{gt}}{k_B T}} e^{\alpha \sigma_0} - 2 f_t e^{-\frac{E_{tg}}{k_B T}} e^{-\alpha \sigma_0}, \tag{4.20}$$

where ε_0 is the instantaneous elastic strain, E_{gt} and E_{tg} the activation energies of the "*gauche*"-to-"*trans*" and "*trans*"-to-"*gauche*" individual transitions, respectively, f_{g^-}, f_{g^+} and f_t are the instantaneous population numbers of each type of conformer which, at $t = 0$, are assumed to obey Boltzmann's distribution, and α is proportional to an activation volume, $\upsilon^{\#}$, and to the reciprocal of temperature, T, $\alpha = \frac{\upsilon^{\#}}{k_B T}$, as in the early thermally activated proposals [1, 7–23] (cf. Fig. 4.1), k_B being Boltzmann's constant. Equation (4.20) takes into account that there are of course two possible paths for the "*trans*"-to-"*gauche*" transitions.

The number of "*gauche*" conformers will be expected to decrease and that of the "*trans*" ones increase proportionally to the strain increase, according to $f_{g^-} = f_{g^+} = f_{g,0}[1 - a(\varepsilon - \varepsilon_0)]$ and $f_t = f_{t,0}\big[1 + a'(\varepsilon - \varepsilon_0)\big]$, with $f_{t,0} = f_{g,0} \dfrac{e^{-\frac{E_t}{k_B T}}}{e^{-\frac{E_{g\pm}}{k_B T}}} = f_{g,0} e^{\frac{E_{tg} - E_{gt}}{k_B T}}$ from Boltzmann's distribution. The conservation of the total number of conformers provides the relationship between a' and a, $a' = 2 f_{g,0} a / f_{t,0}$ and (4.20), with the remaining a parameter determined from the final mechanical equilibrium of zero strain growth at $t = \infty$,[5] may then be rearranged to yield [28, 29] $\frac{d(\varepsilon - \varepsilon_0)}{dt} = c_0 \sinh(\alpha \sigma_0)\Big[1 - \frac{\varepsilon - \varepsilon_0}{\varepsilon_\infty - \varepsilon_0}\Big]$, where c_0 is proportional to some fundamental frequency

---

[4] In Part II of the book, we venture into possible strategies aiming at more "universal" and physically realistic solutions.

[5] [Assuming no viscous flow, or a very lightly crosslinked polymer].

identified in Chap. 9, a relationship that specifies the $f(\varepsilon)$ function of Sect. 4.3.1 and, for the creep compliance,

$$D(t) = D_0 + (D_\infty - D_0)(1 - e^{t/\tau}),\tag{4.21}$$

the response of a *standard non-linear solid*, not of a mere activated dashpot of the earlier works [1, 7–23], with retardation time

$$\tau = \frac{\sigma_0}{c_0' \sinh(\alpha\sigma_0)},\tag{4.21a}$$

where $c_0' = \frac{c_0}{(D_\infty - D_0)} \propto \frac{4 f_{g,0}}{(D_\infty - D_0)} e^{-\frac{E_{gt}}{k_B T}}$. The limiting domain of linear viscoelasticity at low stresses, with a stress-independent retardation time, is also predicted with $\tau_l = 1/(c_0'\alpha)$. The classical proportionality of the creep rate to a sinh of an argument proportional to the stress is recovered, but now the instantaneous and ultimate elasticity of the system are also accounted for. In addition, this formulation predicts that the time and stress dependencies are *not* separable.

Study Question 4.3 From (4.20) and the foregoing discussion, work out the results of (4.21) and (4.21a).

As pointed out from the outset (cf. the title of this subsection), this formulation assumes an extremely idealized and unrealistic picture of the creep process, as no single (uncorrelated) conformational transitions will be possible in a bulk polymer, but the result of (4.21) and (4.21a) is (or would be) compatible with a time-stress equivalence or superposition (TSE/TSS). This is because to the assumed set of single $g^\pm \rightleftharpoons t$ transitions would correspond a stress-dependent decaying function $\chi(t, \sigma_0) = \frac{D_\infty - D(t)}{D_\infty - D_0}$ which is a stress-dependent power of its value in the linear limit, $\chi_l(t) = e^{-\frac{t}{\tau_l}}$ with $\tau_l = 1/(c_0'\alpha)$, $\chi(t, \sigma_0) = [\chi_l(t)]^{\frac{\sinh(\alpha\sigma_0)}{\alpha\sigma_0}}$, which means that the mentioned equivalence or superposition is (or would be) valid, with a stress shift factor $a_{\sigma_0} = \frac{\sigma_0 \sinh(\alpha\sigma_{0,ref})}{\sigma_{0,ref} \sinh(\alpha\sigma_0)}$.

However, for strict overall stress-time equivalence or superposition to be valid for a real system (not the idealized set of single $g^\pm \rightleftharpoons t$ transitions), that same shift factor would need to be valid for all types of motions at the molecular scale contributing to the response. This would result in retardation time spectra of invariant shape, being just shifted to shorter times at increased stresses, which is not what is known to occur in experiments like those reported in [28–32].

Single uncorrelated $g^\pm \rightleftharpoons t$ transitions being impossible except at chain ends, peculiar combinations (like one or *clusters* of several chain crankshafts) [33] may nevertheless realistically represent some of the contributors to the response. Therefore, we may conjecture that, combining elementary responses of the type of (4.21) and (4.21a) with the truncated log-normal retardation spectrum of the preceding section, (4.19) might be an approximate estimate of a *"real"*[6] *standard non-linear*

---

[6] The "real" qualifier is only justified by the fact that a retardation time spectrum is taken into account.

solid's creep compliance. References [28–32] showed that such conjecture is not unreasonable, as the abundant experimental and calculated creep data of [28] shows for two semi-crystalline (PP and UHMWPE[7]) and two amorphous (PMMA and PC[8]) polymers, for which estimates were obtained of the materials' τ_1, τ^* and b_0 values, in addition to the average activation energies (controlling τ^*) and the initial and final equilibrium compliances, b_0 being confirmed as stress- and temperature-independent constants for each of the materials (cf. Sect. 4.4.1). The optimized retardation spectra shifted and narrowed with increasing temperature and stress [28, 29], as physically expected, due to growing activation energies with the size of the participating structures, thereby leading to individual retardation time shifts that grow with that size.

In the calculations, τ^* was taken as $\tau^* = \frac{\sigma_0}{c_0'^* \sinh(\alpha^* \sigma_0)}$ and $c_0'^*$ and α^* were optimized against the experimental creep data, both parameters referring to the average microscopic structure taking part in the process. The minimum retardation times, τ_1 (likely associated with the smallest of those microstructures, whatever their nature) were estimated from the optimized stress- and temperature-independent values of b_0 and the stress- and temperature-dependent values of b and τ^*. But, in more recent works [34, 35], dealing with stress relaxation, the τ_1 (or θ_1) were formulated as the τ^* (or θ^*) with scaled-down activation energies, $E_{a,1} = E_{a,n}/n^*$, and $\alpha_1 = \alpha^*/n^*$, n^* being the size of the average relaxor.[9] These results are presented and further discussed in Sects. 11.9.1 and 11.9.2 in the light of the developments of Part II of the book.

Notwithstanding the idealized and incomplete character of the foregoing formulation of creep, it becomes clear the importance of the definition and use of specific, physically meaningful, fundamental parameters such as (in the above simplified example) one or two basic activation energies and the above mentioned fundamental frequency.

4.4.3 Physical and Practical Evaluation of These Models. Possible Way(s) Forward

There is no doubt that the foregoing $g^{\pm} \rightleftharpoons t$ transitions model is not an accurate picture of a creep (or relaxation) process, not even in polymers. Nevertheless, the fact that an obvious association may be presumed between τ^* and τ_1 with the *sizes of the participating structures* (whatever they might be) strongly suggests that the growing awareness of materials scientists to the suspected or even firmly established

---

[7] PP—polypropylene.

   UHMWPE—ultra-high molecular weight polyethylene.

[8] PMMA—poly(methylmethacrylate).

   PC—bisphenol-A polycarbonate.

[9] The reader will see how the latter two works "bootstrapped" the developments described in Part II of the book, where *clustering* and *cooperativity* are the main themes.

role of *cooperativity* of local structures (subject to *clustering*) in the responses holds the key to a successful understanding and formulation of the behavior of amorphous condensed matter. The microscopic details will not be those of the simplified transitions assumed in the above model, but one expects that a more realistic (and desirably universal) microscopic dynamic model might be worked out. Such an attempt underlies the conjectures and the entire contents of Part II of the book.

In that endeavor, however, a few conditions, steps and milestones should be borne in mind:

(1)   the formulations should concentrate on the *dynamics at the molecular scale*, but *microscopic details* should be the bare *minimum* compatible with the desirable *universality*;

(2)   a range of different local structures should be involved and determine all physical behavior, and so *spectra of response times* will have to be taken into account and explicitly obtained;

(3)   random and/or topologically driven *clusters* of those structures should be allowed and their weights obtained;

(4)   to objectively validate response predictions, approximations such as Alfrey's [26] should be evaluated and desirably avoided;

[Figure 4.2 illustrates the type of errors resulting from that approximation in stress relaxation calculations for a standard linear solid at a single value of time (cf. the difference between the shaded areas in the diagram). In Chap. 11 (Fig. 11.33) a graphical illustration is provided of the errors that result for full relaxation modulus curves].

(5)   allowance for the effect of the strain or stress ramps up to the applied strain or stress.

[A presentation of what is involved and how to solve this problem is the subject of Chap. 5 for stress relaxation].

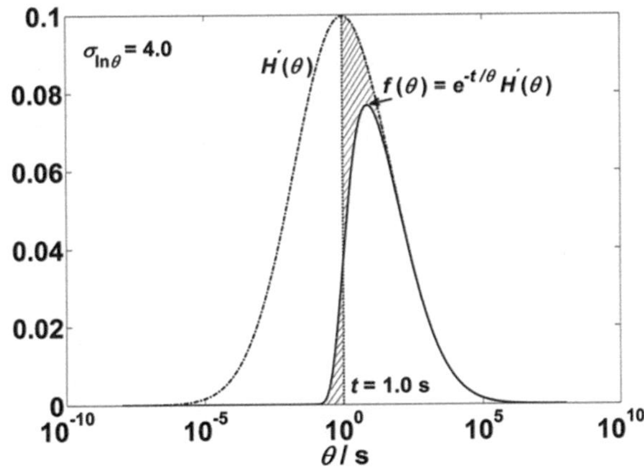

Fig. 4.2 Graphical representation of the type of errors resulting from Alfrey's approximation [26] on the time-dependent relaxation modulus of a "real" linear solid characterized by normal relaxation spectra [cf. (3.41) vs. (3.33b)]

References

1. H. Leaderman, *Elastic and Creep Properties of Filamentous Materials and Other High Polymers* (Textile Foundation, Washington, DC, 1943)
2. S. Turner, Polym. Eng. Sci. **6**, 306 (1966)
3. P.G. Nutting, J. Franklin Inst. **235**, 513 (1943)
4. F.J. Lockett, *Nonlinear Viscoelastic Solids* (Academic Press Inc., London, 1972)
5. W.N. Findley, J.S. Lai, K. Onaran, *Creep and Relaxation of Nonlinear Viscoelastic Materials (With an Introduction to Linear Viscoelasticity)* (Dover Publications Inc., New York, 1976)
6. I.M. Ward, *Mechanical Properties of Solid Polymers*, 2nd edn., Chap. 9 (Wiley, 1985)
7. I.M. Ward, J. Sweeney, *An Introduction to the Mechanical Properties of Solid Polymers*, 2nd edn., Chap. 10 (Wiley, 2004)
8. R. Lakes, *Viscoelastic Materials*, Sect. 2.12 (Cambridge University Press, 2009)
9. G. Halsey, H.J. White, H. Eyring, Text. Res. J. **15**, 295 (1945)
10. O.D. Sherby, J.B. Dorn, J. Mech. Phys. Solids **6**, 145 (1958)
11. R.E. Robertson, J. Appl. Polym. Sci. **7**, 443 (1963)
12. J.A. Roetling, Polymer **6**, 311 (1965)
13. C. Bauwens-Crowet, J.C. Bauwens, G. Holmes, J. Polym. Sci. **A2**(7), 735 (1969)
14. M.J. Mindel, N. Brown, J. Mater. Sci. **8**, 863 (1973)
15. A.S. Krausz, H. Eyring, *Deformation Kinetics* (Wiley-Interscience, New York, 1975)
16. M.A. Wilding, I.M. Ward, Polymer **19**, 969 (1978)
17. M.A. Wilding, I.M. Ward, Polymer **22**, 870 (1981)
18. S. Houshyar, R.A. Shanks, A. Hodzic, Polym. Test. **24**, 257 (2005)
19. F. Guiu, P.L. Pratt, Phys. Status Solidi **6**, 111 (1964)
20. B. Escaig, in *Plastic Deformation of Amorphous and Semi-Crystalline Materials*, ed. by B. Escaig, C. G'Sell (Les Éditions de Physique, Les Ulis, France, 1982), pp. 187–225
21. J. Sweeney, I.M. Ward, J. Mater. Sci. **25**, 697 (1990)
22. R.G. Larson, Rheol. Acta **24**, 327 (1985)
23. J. Sweeney, H. Shirataki, A.P. Unwin et al., J. Appl. Polym. Sci. **74**, 3331 (1999)
24. F.X. Kromm, T. Lorriot, B. Coutand, R. Harry, J.M. Quenisset, Polym. Test. **22**, 463 (2003)
25. P. Feltham, Br. J. Appl. Phys. **6**, 26 (1955)
26. T. Alfrey, *Mechanical Behavior of High Polymers* (Interscience, New York, 1948)
27. K.L. Ngai, J. Non-Cryst. Solids **353**, 709 (2007)
28. J.R.S. André, Fluência de Polímeros—Fenomenologia e Modelação Dinâmica Molecular (Polymer Creep—Phenomenology and Modeling of Molecular Dynamics), Doctoral Thesis, Universidade de Aveiro, 2004
29. J.R.S. André, J.J.C. Cruz Pinto, *Time-Temperature and Time-Stress Correspondence in Non-Linear Creep. Experimental Behavior of Amorphous Polymers and Quantitative Modeling Approaches*, No. 079, e-Polymers 2004 (12th Annual POLYCHAR World Forum on Advanced Materials, Guimarães, Portugal, 2004)
30. J.R.S. André, J.J.C. Cruz Pinto, Mater. Sci. Forum **455–456**, 759 (2004)
31. J.R.S. André, J.J.C. Cruz Pinto, Mater. Sci. Forum **480–481**, 759 (2005)
32. J.R.S. André, J.J.C. Cruz Pinto, Macromol. Symp. **247**, 21 (2007)
33. T.F. Schatzki, J. Polym. Sci. **57**, 496 (1962)
34. J.R.S. André, J.J.C. Cruz Pinto, Polym. Eng. Sci. **54**, 404 (2014)
35. J.J.C. Cruz Pinto, J.R.S. André, Polym. Eng. Sci. **56**, 348 (2016)

Chapter 5
Treatment of Experimental Data: The Problem of the Initial Strain (or Stress) Ramps

5.1 Introduction

Strains and stresses cannot be applied instantaneously, whatever the instrumentation used, and therefore there is always a short period, generally up to a few seconds, during which both strain and stress increase before the final nominal applied strain or stress is accurately reached. A typical variation for the case of stress relaxation is the one shown in Fig. 5.1 for a PMMA, corresponding to an approximate applied strain ramp up to a nominal strain of 3%. As the stress reached may be high and modulus approaches the instantaneous value, E_0, errors may result if corrections are not made, before extracting values of the stress relaxation modulus.

Authors Flory and Mckenna [1] and various other workers, by means of numerical computations, proposed several step rate corrections, but the relatively general conclusion was that the effect was only important in materials with significant contribution of relaxation times of the order of (or shorter than) the ramp time. In addition, relative errors affecting the relaxation modulus, $E_r(t)$, were generally small, though the same would not necessarily apply to best-fit KWW parameters,[1] given their extreme sensitivity to only slight changes in the data.

We do not use the above empirical expressions to interpret experimental data, but short times may be important in the evaluation of any specific physical model. The parameters of the theory developed from Chap. 8 onwards[2] have very precise physical meanings, and so the experimental data used in Chap. 11 in a first assessment of the theory ought to be purged of possible significant errors resulting from the rapidly varying applied strain during the initial ramp. We could adopt one of the already

---

[1] [The values of θ or τ and β in expressions such as $E_r(t) = E_\infty + (E_0 - E_\infty)e^{-(t/\theta)^\beta}$ or $D(t) = D_0 + (D_\infty - D_0)[1 - e^{-(t/\tau)^\beta}]$].

[2] CTMD—Cooperative Theory of Materials Dynamics.

© The Author(s), under exclusive license to Springer Nature Switzerland AG 2024
J. J. Cruz Pinto and J. R. dos Santos André, *Analytical Molecular Dynamics of Amorphous Condensed Matter*, Springer Series in Materials Science 342,
https://doi.org/10.1007/978-3-031-56517-5_5

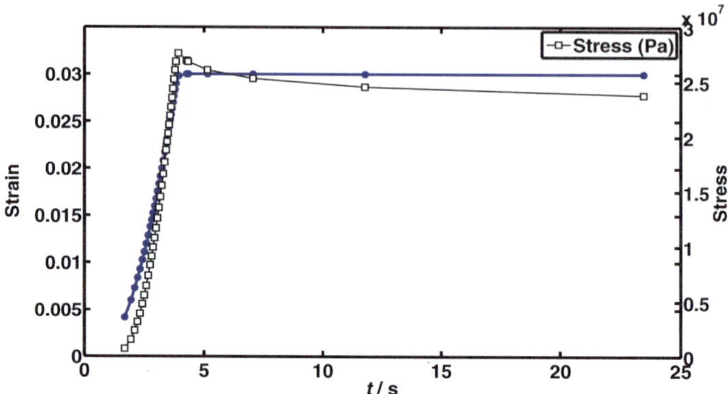

Fig. 5.1 Example of initial strain and stress variations in a real uniaxial tensile stress relaxation experiment with a PMMA

published solutions [1] of only registering stress data at time values longer than the ramp one[3] (or, better, ten times longer, according to the so-called "factor of ten rule of thumb", for materials with sufficiently long relaxation times), subtracted of some fraction of the ramp time, but we deemed relevant to revisit the correction procedure for best results.

The objective is to obtain from experimental measurements the best estimate of the stress relaxation moduli that would result from a corresponding pure, instantaneous, strain step change. Rather than using Boltzmann superposition principle (cf. Chap. 3) in the linear viscoelastic domain to deduce the response to an idealized strain ramp followed by a period of constant strain[4] [1], we directly integrate the constitutive equation of a standard linear solid, and then of a generalized linear solid as predicted by CTMD for a real PMMA material, for both ideal and more realistic (delayed) strain ramps, to compare the responses to those of ideal strain step changes. The effects within the non-linear viscoelastic domain will also warrant a brief comment, but its future accurate treatment will require the use or modification of the developments formulated in Chap. 9.

---

[3] [Taken as the time at which the strain exactly reaches its intended nominal value].

[4] [Equivalent to the superposition of one infinite or unbound positive ramp to a second but negative one starting at the end of the first ramp].

5.2   Response of a Standard Linear Solid to an Ideal Linear Strain Ramp from $t = 0$ to t_r, Followed by Stress Relaxation at $\epsilon = Constant = kt_r$

For a *standard linear solid* represented by a spring in parallel with a Maxwell unit (cf. Fig. 3.1), with the indicated parameters, $E_{0,e}$, $E_{1,e}$ and η_1, we have seen in Chap. 3 that its instantaneous and final relaxed modulus are $E_0 = E_{0,e} + E_{1,e}$ and $E_\infty = E_{0,e}$, respectively, and its relaxation time is $\theta = \frac{\eta_1}{E_{1,e}} = \eta_1/(E_0 - E_\infty)$, and so its constitutive equation may be written in terms of the latter parameters as $\sigma = E_\infty \epsilon + \eta_1 \frac{d}{dt}\left[\epsilon - \frac{\sigma - E_\infty \epsilon}{E_0 - E_\infty}\right]$ or, during an ideal ramp, $\epsilon = kt$,

$$\frac{d\sigma}{dt} = -\frac{\sigma}{\theta} + kE_0 + kE_\infty \frac{t}{\theta}, \tag{5.1}$$

whose solution during the ramp for the corresponding modulus, $E_r(t \leq t_r)$, after introducing the initial condition, $(t = 0, \sigma = 0)$, and referring it to the final intended strain (kt_r), becomes

$$E_r(t \leq t_r) = E_\infty \left(\frac{t}{t_r}\right) + (E_0 - E_\infty)\left(\frac{\theta}{t_r}\right)(1 - e^{-t/\theta}), \tag{5.2}$$

while, for $t \geq t_r$, with $\epsilon = kt_r = const$, we will have (by integration between (t_r, t) and $[E_r(t_r), E_r(t)]$)

$$E_r(t \geq t_r) = E_\infty + (E_0 - E_\infty)\left(\frac{\theta}{t_r}\right)(1 - e^{-t_r/\theta})e^{-(t-t_r)/\theta}. \tag{5.3}$$

5.3   Response of a Standard Linear Solid to a Delayed Strain Ramp from $t = 0$ to t_r, Followed by Stress Relaxation at $\epsilon = Constant = k(t_r - \tau')$

A real strain ramp (cf. Fig. 5.1) does not instantaneously start with a fixed slope, due to the delayed response of the instrumentation, and its more realistic representation will be as in Fig. 5.2 (bold line), where one recognizes that the system will not show a strict delay (τ'), followed by an ideal accurate ramp, but a gradually (though fast) increasing strain rate to a final value. As the intended final strain is reached, conventional instrumentation reacts much faster than at the ramp's start, and very little error is then incurred by assuming a nearly instantaneous drop to zero strain rate.

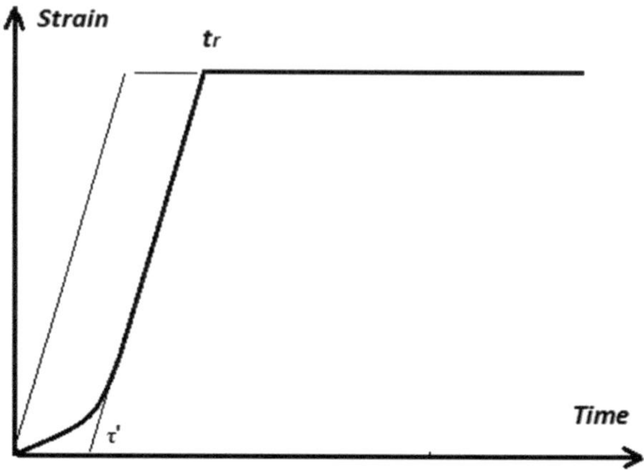

Fig. 5.2 Illustration of ideal linear and delayed strain ramps up to a given constant strain

A reasonable representation of such delayed ramps would be as $\epsilon(t) = k[t - \tau(1 - e^{-t/\tau})]$, where τ (smaller than the extrapolated τ' shown) depends on the instrumentation's response, which may only be approximated by $\epsilon(t) = k(t - \tau)$ or $\epsilon(t) = k(t - \tau')$ for very fast instrumentation. Using the most accurate of the previous expressions, one may find that, for an intended final strain, $\epsilon_0 = \epsilon(t_r)$, $k = \epsilon_0/[t_r - \tau(1 - e^{-t_r/\tau})]$, and the equation giving the ramp moduli referred to that final strain will be (from 5.1)

$$\frac{dE_r}{dt} = -\frac{E_r}{\theta} + \frac{1}{t_r - \tau\left(1 - e^{-\frac{t_r}{\tau}}\right)}\left[\frac{E_\infty}{\theta}t + \left(E_0 - \frac{E_\infty \tau}{\theta}\right)\left(1 - e^{-\frac{t}{\tau}}\right)\right], \quad (5.4)$$

whose solution under the same obvious initial condition ($t = 0, \sigma = 0$) may be obtained as

$$E_r(t \leq t_r) = \frac{1}{t_r - \tau\left(1 - e^{-\frac{t_r}{\tau}}\right)}\left\{E_\infty t + [(E_0 - E_\infty)\theta - E_\infty \tau](1 - e^{-t/\theta}) \right.$$
$$\left. -(E_0\theta - E_\infty \tau)\frac{\tau}{\theta - \tau}(e^{-t/\theta} - e^{-t/\tau})\right\}. \quad (5.5)$$

With τ infinitely small, we recover the expression for the ideal linear ramp, (5.2). The resulting stress relaxation modulus for $t \geq t_r$ will then be

$$E_r(t \geq t_r) = E_\infty + [E_r(t_r) - E_\infty]e^{-(t-t_r)/\theta}. \quad (5.6)$$

Study Question 5.1 Show that the stress relaxation modulus for a perfectly linear strain ramp subject to a simple delay of τ', as sketched in Fig. 5.2: $\epsilon = k(t - \tau')$, is

$$E_r(t \le t_r) = E_\infty \frac{t - \tau'}{t_r - \tau'} + (E_0 - E_\infty)\frac{\theta}{t_r - \tau'}\left[1 - e^{-(t-\tau')/\theta}\right]. \qquad (5.7)$$

5.4   Responses of a Generalized Standard Linear Solid[5] to the Above Types of Strain Ramps

In Chaps. 8 and 9, a new *cooperative theory of materials dynamics* (CTMD) is developed, whereby each set of clusters of primitive relaxors (as defined by Ngai [2, 3] and in Chap. 8) is proved to behave mechanically as a standard linear or non-linear solid depending on the level of strain and stress. Their combination allows the formulation of the behavior of an amorphous material (PMMA in this case) by adequately weighing the contributions of the various clusters. The resulting moduli for the strain ramps of (5.2) and (5.7),[6] in the linear viscoelastic domain, are

$$E_r(t \le t_r) = E_\infty \left(\frac{t}{t_r}\right) + (E_0 - E_\infty)\sum_{k \ge 1} F_k'\left(\frac{\theta_k}{t_r}\right)\left(1 - e^{-t/\theta_k}\right) \qquad (5.2a)$$

and

$$E_r(t \le t_r) = E_\infty \frac{t - \tau'}{t_r - \tau'} + (E_0 - E_\infty)\sum_{k \ge 1} F_k'\frac{\theta_k}{t_r - \tau'}\left[1 - e^{-(t-\tau')/\theta_k}\right], \qquad (5.7a)$$

respectively, where the F_k' are the weights of each set of clusters of k primitive relaxors, as formulated in Chap. 9 or by some future better alternative.[7]

In a similar way to the standard linear solid cases, the corresponding stress relaxation modulus expressions for the period under constant strain will be, respectively,

---

[5] [According to the CTMD theory developed in Chaps. 8 and 9].

[6] From Fig. 5.2, one recognizes that the most accurate delayed ramp of (5.5) is intermediate between the two simple limiting cases of (5.2) and (5.7), and so these two should suffice to assess the magnitude of the ramp effects.

[7] Depending on future detailed analysis, it might prove possible to extend or modify the above expressions to the non-linear viscoelastic domain, in the light of the non-linear formulations of the θ_k and F_k' developed in Chap. 9.

$$E_r(t \geq t_r) = E_\infty + (E_0 - E_\infty) \sum_{k \geq 1} F_k' \left(\frac{\theta_k}{t_r}\right)\left(1 - e^{-t_r/\theta_k}\right)e^{-(t-t_r)/\theta_k} \qquad (5.2b)$$

and

$$E_r(t \geq t_r) = E_\infty + (E_0 - E_\infty) \sum_{k \geq 1} F_k' \frac{\theta_k}{t_r - \tau'}\left[1 - e^{-(t_r - \tau')/\theta_k}\right]e^{-(t-t_r)/\theta_k}. \quad (5.7b)$$

Study Question 5.2 Obtain the expressions for the creep compliance during ideal linear and delayed linear stress ramps in creep experiments on a hypothetical standard linear solid.

5.5   Corrections to the Measured Relaxation Modulus Curves

The above calculations were performed for a hypothetical standard linear solid with an average relaxation time of 100 s (Figs. 5.3 and 5.4) and the PMMA experimentally[8] and theoretically studied in Chap. 11 within its linear viscoelastic domain (Figs. 5.5 and 5.6), for an ideal linear and a delayed linear ramp (with $\tau' = 1$ s) up to a time of $t_r = 5$ s (approximately representative of the instrumentation used in the experiments there described), and then onwards under constant strain until each system reached its full relaxed modulus.

It is confirmed that the so-called "factor of ten rule of thumb" for accurate data is reasonably verified, but immediately after the ramp the stress and modulus values are overestimated. The best correction to the data to obtain accurate estimates of the stress relaxation modulus since the start of the simulated experiment (including the ramp) turns out to be a simple negative shift of $2t_r/3$ to the entire set of measured times, where t_r is the total ramp time. The zoomed-in plots of Fig. 5.4 also illustrate the quality of the correction, which turns out identical for the ideal and delayed linear strain ramps.

The same series of calculations were performed for a PMMA material (Figs. 5.5 and 5.6), with the same conclusions as for the hypothetical standard linear solid. The figures also illustrate how the best correction may be found by varying the shift as a fraction of the total ramp time.

---

[8] The stress relaxation experiments served to obtain first estimates of the theory's physical parameters (cf. Chaps. 8, 9 and 11), which were then used in the present ramp and full relaxation curve calculations.

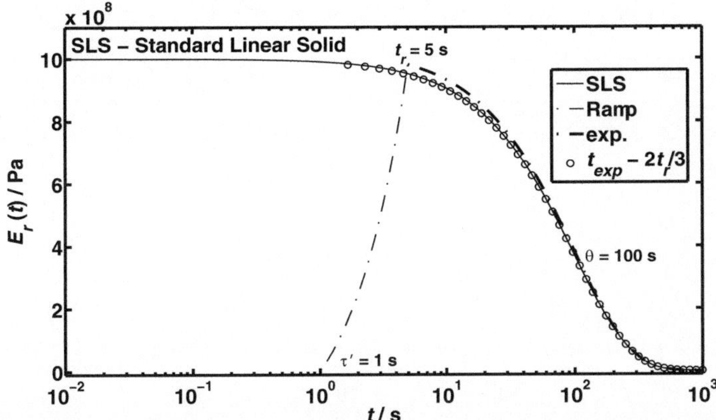

Fig. 5.3 Calculated uniaxial tensile stress relaxation modulus of a standard linear solid during and after (exp.) a delayed ideal strain ramp (broken lines) and under the corresponding ideal strain step change (solid curve + markers of the corrected data)

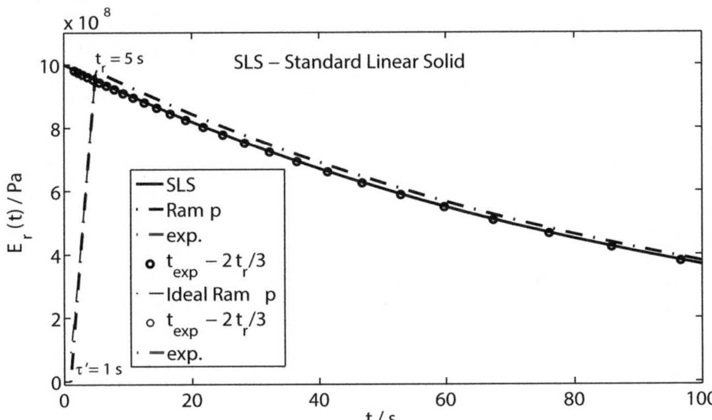

Fig. 5.4 Zoomed-in short time calculated uniaxial tensile stress relaxation modulus of a standard linear solid during and after (exp.) a delayed ideal strain ramp (broken lines) and under the corresponding ideal strain step change (solid curve + markers of the corrected data)

It may however be justified, for safer conclusions, to investigate the influence of varying ratios τ'/t_r on the best correction, and use in each real case the values of t_r and τ' that best fit the response of the instrument being used.

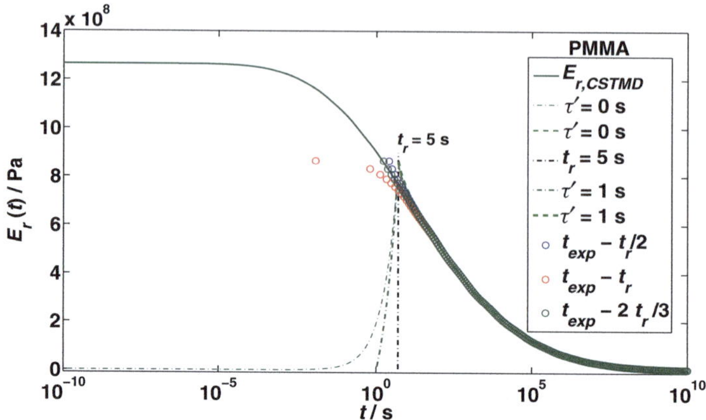

Fig. 5.5 Calculated uniaxial tensile stress relaxation modulus of a PMMA during and after (exp.) an ideal and delayed ideal strain ramp (broken lines) and under the same ideal step change in strain (solid curve + markers of the corrected data)

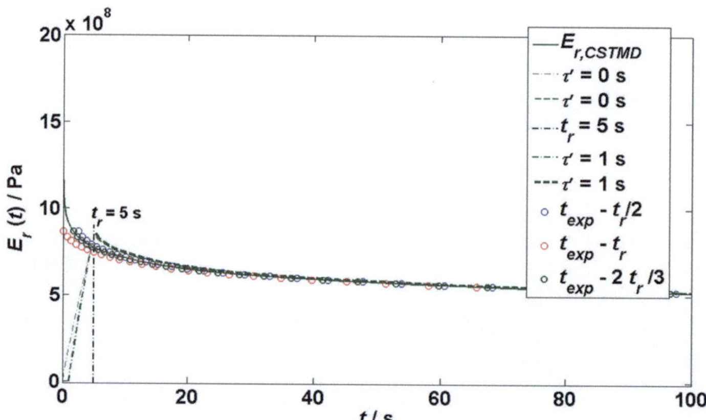

Fig. 5.6 Zoomed-in calculated uniaxial tensile stress relaxation modulus of a PMMA during and after (exp.) an ideal and delayed ideal strain ramp (broken lines) and under the same ideal step change in strain (solid curve + marks of the corrected data)

References

1. A. Flory, G.B. McKenna, Mech. Time-Depend. Mater. **8**, 17 (2004)
2. K.L. Ngai, J. Non-Cryst. Solids **353**, 709 (2007)
3. K.L. Ngai, *Relaxation and Diffusion in Complex Systems* (Springer, New York Dordrecht Heidelberg London, 2011)

Part II
Advanced Theories of the Dynamics of Amorphous Condensed Matter

Chapter 6
Physical Characterization of the Dynamics of Amorphous Condensed Matter

6.1 Range of Response Times and Approximate Stretched Exponential Nature of the Behavior

As we have seen in Chaps. 1–4, the viscoelastic nature of the behavior implies that the transient response to forced excitations will extend from very short to very long times,[1] the longer the lower the temperature, as abundantly illustrated for many materials, namely polymers, and/or to significant and extended dependencies on the frequency of cyclic excitations [1–5]. Further, the time-extended (or non-first order) nature of behavior has been classically described (though only approximately) by fitting it to two-parameter KWW[2] [6, 7] relationships of the type $\sim e^{-(t/\theta)^{\beta}}$ or $\sim \left[1 - e^{-(t/\tau)^{\beta}}\right]$ [6–8], in the cases of time-dependent stress relaxation or creep compliance, respectively, where θ and τ are average response (relaxation and retardation) times and $\beta < 1$ is an extension parameter, known to be lower at lower temperature, whose exact physical meaning and nature is still under active discussion [8].

6.2 Non-linearity with Respect to the Excitation Intensity

Within viscoelasticity, the non-linearity consists in a strain dependence of the relaxation modulus and a stress dependence of the creep compliance, whose effects were graphically illustrated in Chap. 2 and given their earliest formulations in Chap. 4.

---

[1] [And that is why logarithmic time and frequency scales are invariably used in graphical representations of the behavior, to expand the scale for very short times and low frequencies and compress it for very long times and high frequencies].

[2] KWW: Kohlrausch–Williams–Watts.

© The Author(s), under exclusive license to Springer Nature Switzerland AG 2024
J. J. Cruz Pinto and J. R. dos Santos André, *Analytical Molecular Dynamics of Amorphous Condensed Matter*, Springer Series in Materials Science 342,
https://doi.org/10.1007/978-3-031-56517-5_6

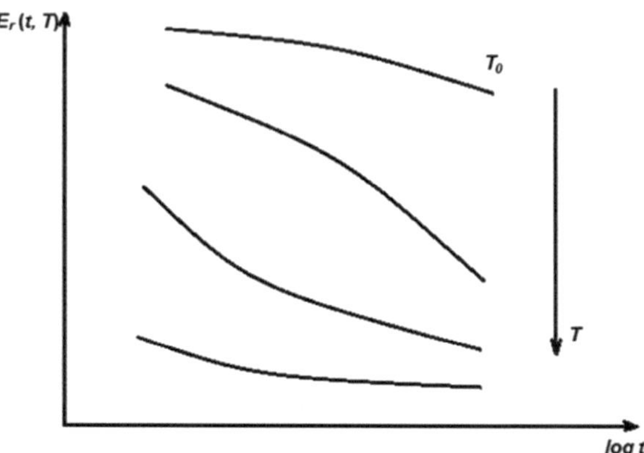

Fig. 6.1 General and qualitative plot of the effect of temperature on the uniaxial tensile stress relaxation modulus

6.3   Sensitivity to Temperature

6.3.1   Description of the Effect

Given the significant time dependence of the behavior of all viscoelastic materials and the temperature dependence of the rate of all structural transitions and motions within their molecular structure, it is easily understood that temperature will be a variable of paramount importance. So, an increase in temperature will lead to an acceleration of the materials' responses—faster stress relaxation $(-d\sigma/dt)$, creep $(d\varepsilon/dt)$, and higher frequencies of energy absorption peaks—resulting in lower relaxation moduli, $E_r(t, T)$ or storage moduli, $E'(\omega, T)$, and higher creep or storage compliances, $D(t, T)$ or $D'(\omega, T)$, for the same values of time or frequency, where $\omega = 2\pi\nu$ (with ν in Hz) is the angular frequency in s^{-1}.

Those effects are qualitatively illustrated in Figs. 6.1, 6.2, 6.3 and 6.4. As to the loss peaks, $E''(\omega, T)$, $D''(\omega, T)$ and $\tan \delta_{E \, or \, D}(\omega, T)$, they will be essentially displaced to higher frequencies, but may also show changes in their detailed shape, because the various motions at the molecular scale responsible for energy absorption will not accelerate uniformly when temperature increases.

In other words, an increase in temperature shifts to shorter times (or higher frequencies) the same approximate behavior that may be detected at longer times (or lower frequencies). The outstanding still unanswered question, however, is whether these effects result mostly from the changes in temperature, or from the associated changes in volume.[3]

---

[3] For example, what will happen when temperature is changed at constant overall volume or constant *free volume* (defined in Sect. 6.10)? Nothing?

Fig. 6.2 General and qualitative plot of the effect of temperature on the uniaxial tensile creep compliance

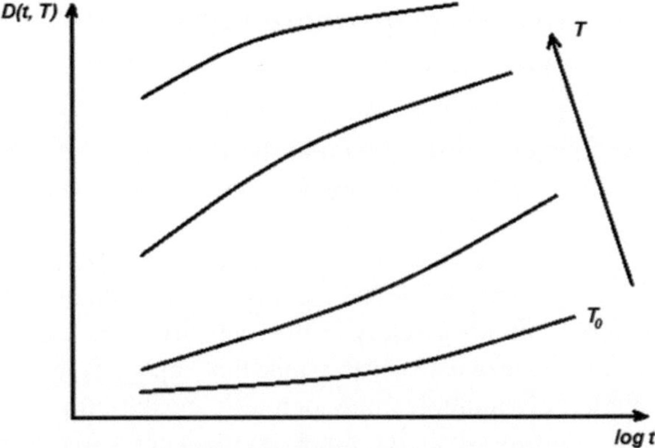

Fig. 6.3 General and qualitative plot of the effect of temperature on the uniaxial tensile storage modulus

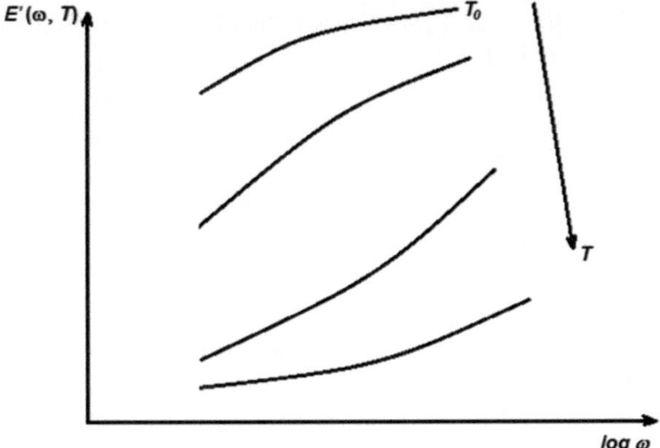

Fig. 6.4 General and qualitative plot of the effect of temperature on the uniaxial tensile loss peaks (ignoring possible multimodality and changes in peak width or shape with temperature)

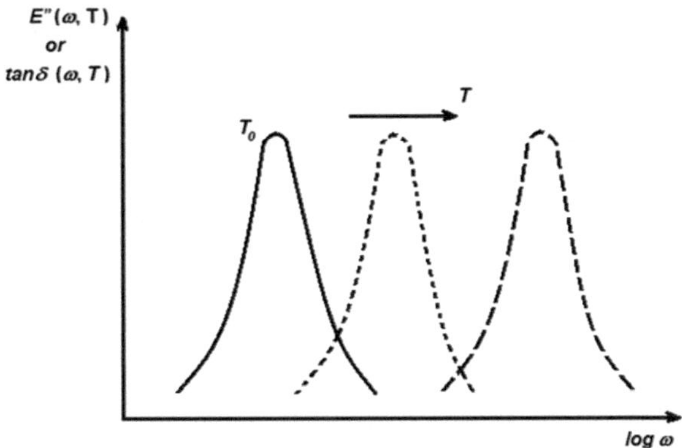

6.3.2   Time–Temperature Equivalence or Superposition (TTE/TTS)

The foregoing discussion and the mentioned known results [1–5] led to the early acceptance of the following (at least) approximate "principle" of *time–temperature equivalence or superposition*:

An increase/decrease in temperature relative to any given reference value, $\Delta T = T - T_{ref}$, changes the behavior (namely mechanical) of an amorphous viscoelastic material, for any given type and intensity of the excitation within the linear domain, such as to make it nearly the same that applies at the reference temperature but now at submultiple/multiple times, or multiple/submultiple frequencies, of the original ones. The factor by which the times or frequencies will be changed, $a_{T/T_{ref}} = \frac{t}{t_{ref}} = \frac{\omega_{ref}}{\omega}$, will be a univocal function of ΔT, and is the so-called *temperature shift factor*.

In simpler and more practical terms, what happens is that all curves sketched in Figs. 6.1, 6.2, and 6.3 may approximately be reduced to a single curve spanning a much wider time or frequency range (a so-called *reduced* or *master curve*) by horizontal shifts or displacements along their $\log(t)$ or $\log(\omega)$ scale, univocally related to ΔT. Identical shifts would roughly apply to the stress relaxation modulus, creep compliances curves, and the various dynamic (storage and loss) viscoelastic functions, $E'(\omega, T)$, $D'(\omega, T)$, $E''(\omega, T)$, $D''(\omega, T)$ and, to some measure, even $\tan \delta_{E or D}(\omega, T)$. Figure 6.5 qualitatively illustrates two sketches of the mentioned reduced or master curves for the stress relaxation and storage moduli, where $t/a_{T/T_0}$ and $\omega a_{T/T_0}$, with $T_0 = T_{ref}$, stand for the time and angular frequency values for the corresponding full curves at the reference temperature, T_{ref}. In the literature, including classical textbooks, many analogous plots have been published from shifts of actual experimental data at a range of temperatures [1–4].

6.3.3   WLF (Williams-Landel-Ferry) Equation

Williams et al. [5, 9], after much experimentation, have shown that the relationship $a_{T/T_{ref}} = f(T - T_{ref})$ may be represented by

$$\log_{10} a_{T/T_{ref}} = -\frac{C_1(T - T_{ref})}{C_2 + (T - T_{ref})}, \tag{6.1}$$

where C_1 and C_2 are constants that only depend on the material and the chosen reference temperature, or

$$-1/\log_{10} a_{T/T_{ref}} = 1/C_1 + (C_2/C_1) \cdot \left[1/(T - T_{ref})\right]. \tag{6.1a}$$

They also verified that, if for any amorphous polymer T_{ref} were selected as its most usually measured (volumetric or calorimetric) glass transition temperature, T_g

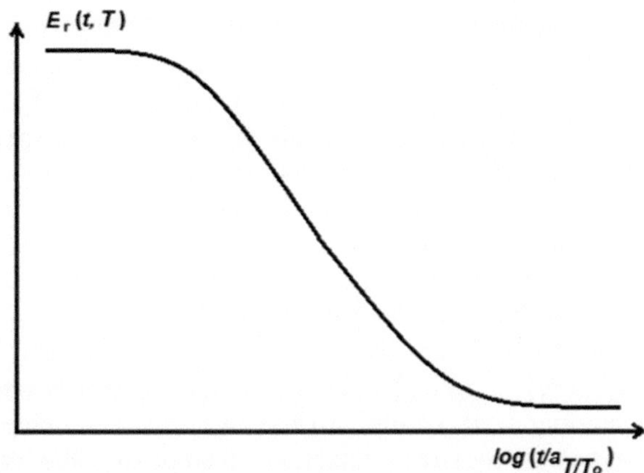

Fig. 6.5 Qualitative reduced or master curves for uniaxial tensile stress relaxation modulus (top) and storage modulus (bottom)

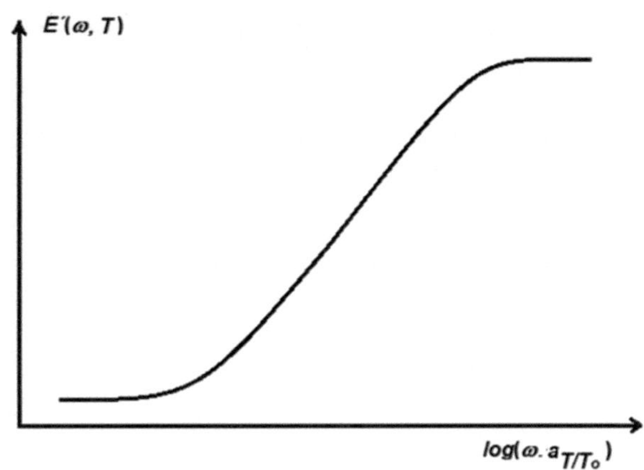

(cf. 6.5), those constants, although showing variations among different polymers, would be distributed around the average values of $C_1 = 17.44$ and $C_2 = 51.6K$, *i.e.*

$$\log_{10} a_{T/T_{ref}} \sim -\frac{17.44(T - T_g)}{51.6 + (T - T_g)}. \tag{6.1b}$$

Equations 6.1 are different forms of the so-called *WLF equation*, where the minus sign gives the right direction of the shift that must be given to the *reduced* or *master curves* at the reference temperatures to obtain the approximate material's behavior at any other temperature, T, as the reader may easily verify, and this is consistent with the indicated abscissae of the curves in Fig. 6.5.

It must be stressed that the above values of the constants are not universal, but mere approximate average values for a range of amorphous polymers; amorphous materials of widely different nature may be characterized by different values. As to semi-crystalline materials, it will only be reasonable to expect a similar (though not strictly identical) behavior in temperature on condition that it will be mostly influenced by the amorphous zones of the material, namely around its glass transition temperature. For materials with significant crystallinity, or amorphous but significantly crosslinked, WLF correlations are not applicable, not even as crude approximations, and the behavior appears to depend on specific energy or activation barriers. It should be noted that a similar interpretation may be extended to the WLF case, except that the energy barriers that appear to be in control turn out to depend on temperature, according to (if referred to 1 mol of whatever moving structures within the material responsible for stress relaxation, creep or any other type of response)

$$E_{a,WLF} = (\ln 10) R \frac{C_1 T T_{ref}}{C_2 + (T - T_{ref})}, \tag{6.2}$$

where $R = 8.31451 \, \mathrm{JK^{-1} mol^{-1}}$ is the gas constant, showing that such conjectured (and still somewhat disputed) activation barriers will have to increase as temperature is reduced. This is so because $T T_{ref}$ varies very much less than $(T - T_{ref})$, $E_{a,WLF}$ and the response times being "predicted" to diverge at $T = T_{ref} - C_2$, a much referred (but difficult to understand and validate) possibility,[4] which will warrant detailed discussion in later chapters of this book.

Study Question 6.1 Obtain 6.2 from WLF and Arrhenius equations. Discuss what may physically justify such *temperature-dependent average activation energies* in the context of the dynamic behavior of amorphous materials. [Be prepared to later relate and revise your discussion as you work through Chaps. 8–10 and 11].

An important issue is the temperature range of applicability of the WLF correlation. It turns out that it is only approximately valid within $T_{ref} \pm 50K$ and generally always above volumetric or calorimetric T_g,[5] i.e. within $T_g \leq T \leq T_g + 100K$. The likely physical reason for the mentioned lower limit is that, as T_g is approached, there is a very significant freeze of most of the motions within the structure of the materials, which will then always find themselves far out of equilibrium conditions. Further, as discussed in Chap. 11, in addition to the fact that WLF correlation is approximate at most, the above limitations strongly suggest that its physical basis is doubtful (and does happen to be very conjectural at present), and so may be questioned.

To be accurate, it is often found that the horizontal shifts of the curves need to be combined with small vertical shifts to obtain satisfactory superposition, due at least to a small influence of temperature on the plateau values of the relaxation

---

[4] Response times much longer than whatever timescale of an observation or experiment *do not mean* divergence.

[5] Cf. Sect. 6.5.

modulus and creep compliances, in particular at very long times (where $E_{r,\infty} \propto T$ and $D_{\infty} \propto 1/T$), but these effects are most often neglected.

6.3.4 Practical Usefulness of the WLF Equation

The mentioned possibility of approximate superposition of the response curves by mere horizontal shifts upon temperature changes results in important practical applications of the WLF Equation. Two examples are:

1. Within the design of a plastics part, one wants to estimate an adequate elasticity modulus for a given amorphous polymer, for which one knows its master relaxation modulus curve $E_r(t_0, T_0)$ at a temperature higher than its future service temperature and its shift factor curve a_{T/T_0} (for example, by a WLF type of equation), and one needs to ensure a safe service of the part during 10 years at an average temperature of 20 °C.

    A service life of 10 years at $T = 20\,°C$ corresponds to a time in years of $t_0 = 10/a_{T/T_0}$ at the reference temperature T_0, and so the part should be designed assuming a modulus of $E_r(t_0, T_0)$, not counting any safety factor to allow for data and material uncertainties.

2. One needs to conduct experiments not longer than 6 months to evaluate the creep behavior of a material up to 10 years at a given intended service temperature, assuming known or easy to estimate (by a WLF type of equation) its shift factors, the question being at which temperature should the experiments be conducted.

    What we then need is a reduction in time by a factor of 20, corresponding to $\log a_{T/T_0} = -1.301$, from which the intended temperature will be easily calculated.

6.4 Non-affinity of Local Stresses and Strains to the Overall Average Ones

Even ignoring any non-uniform initial residual stresses within any given material, during stress relaxation, creep, or any other situation it is only logical to expect that the processes involved within the material's molecular structure will generate non-uniform local stresses and strains; for example, local stresses may relax at different times at different locations. So, despite the difficulty, this should be allowed and accounted for as much as possible in any realistic description of the materials' responses.

6.5   Onset (and Definition) of a Glass Transition and Its Dependence on the Timescale

The reader is expected to be familiar with the general concept and meaning of the *glass transition* and the *glass transition temperature*, T_g [10]. For example, when heating an amorphous material, above that temperature the thermal expansion coefficient (cf. Sect. 6.10) changes very significantly, as well as its heat capacity, the opposite happening in cooling, though generally not in a reversible way, thus showing hysteresis (cf. Sect. 6.11).

The fact that the mentioned changes are so significant (appearing as near discontinuities) and, being thermal expansivities and heat capacities (among other properties) second derivatives of the free energy, led to the idea that the glass transition could ideally be considered as a second-order thermodynamic phase transition subject to, or somehow blurred by, kinetic effects, namely motion freezing at the molecular scale as temperature is lowered, or delayed motions' activation as temperature is increased at growing rates. Earlier literature tended to support that view, while the existence of such thermodynamic glass transition has progressively been questioned [11–19], meaning that the mentioned relatively sudden changes in expansivities and heat capacities might disappear, being replaced by continuous smooth changes, as one would decrease further and further the rate of cooling or heating. However, this is still an open question.

The above behavior results in small but measurable increases in T_g as the rate of cooling or heating increases, typically a few degrees Kelvin (K) per power of 10 of scanning rate (K/s). Understanding and theoretically quantifying this experimentally well-known dependence on the *timescale* of the observation or experiment is one of the outstanding challenges of amorphous condensed matter theories.

6.6   Cooperativity and Dynamic Heterogeneity of the Response

The *cooperativity* of the response of amorphous materials at the molecular scale has since long [20] been recognized as one of the features of their behavior and is at present unanimously invoked by all researchers in the field [8, 21–23]. The concept was first proposed by Adam and Gibbs [20], who coined and defined the term "cooperatively rearranging regions", within which the relevant "particles" move cooperatively, whose size grows as temperature is reduced.

That concept was soon coupled with the idea that such cooperative regions are, by nature, highly intermittent, frequently changing in morphology and size at different locations within the material, leading to what is called *dynamic heterogeneity*, another major outstanding challenge in the study and modelling of amorphous materials [8, 21–23].

6.7   Universality of the Behavior

Widely different amorphous materials display many common characteristics of their dynamic behavior, and this substantial *universality* must of course play a role and be apparent in the theoretical formulations devised to quantitatively portray their behavior. A wide range of concepts, equations and graphical plots (relaxation maps being just one example), are universally applicable [8, 22, 23].

6.8   Critical Dynamic Properties

As widely recognized [22], the universality mentioned above entails that the same type and number of specific physical properties are involved in any comprehensive formulation of the behavior, such as the *crossover temperature*, T_c, and *frequency*, v_c or $\omega_c = 2\pi v_c$, and arguably some characteristic or minimum activation energy.

6.9   Effect of Partial Crosslinking or Crystallization

Crosslinks and crystalline zones within the structure of an otherwise amorphous material will hinder the motions of whatever adjacent relaxing structures may be involved in the response to a thermal or other physical excitation. As experimentally found, the effects will be the lengthening of all response times and the reduction of the maximum levels of stress relaxation and creep at very long times.

6.10   Influence of Molecular Packing and Free Volume

It is obvious and experimentally known that increased molecular packing will decelerate materials' responses. How exactly, and whether nothing else (e.g. temperature) will also directly determine the behavior, are the questions. For example, which other variable would control the behavior if molecular packing could be forced to be kept constant, and what is its real weight at ordinary constant pressure conditions? These are the questions that ultimately need to be answered.

Figure 6.6 very schematically shows the typical distribution of the volume of amorphous materials at constant pressure within a wide range of temperatures, and how it changes as the material is cooled at two different rates. To the volume of the molecular skeleton, v_s, one must add the volume of the voids, v_{void}, to obtain the total occupied volume, v_o, and the so-called *free volume*, v_f,[6] will be $v_f = v - v_o =$

---

[6] For example, in a *random-close-packed* (*rcp*) arrangement of hard spheres, the voids amount to about 0.363 of the total volume, but the free volume is zero, because the particles will be touching

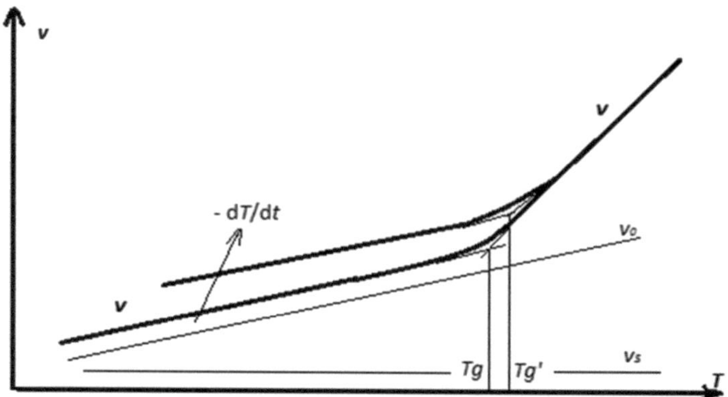

Fig. 6.6 Typical variations of the specific volume of an amorphous material at different scanning rates

$v - (v_s + v_{void})$, where v is the total volume at each temperature. It is within the free volume that the materials' relaxing structures will be able to move in response to thermal or other physical excitations.

The Figure shows the more widely assumed typical traces of the total volume, v, when the amorphous (non-crystallizable) material is cooled from equilibrium in the liquid state (high T) at two different rates, as indicated. At some low temperature, defining a *glass transition temperature*, T_g, the material falls out of equilibrium, as the motions at the molecular scale decelerate in relation to the rate of cooling, and the total volume cannot then keep the original packing rate. Below T_g, the volume contracts as an in ordinary hard elastic solid, keeping the free volume nearly constant, $v_f = v - v_o$, because it is (or has been) generally assumed that no wide-scale motions have time to develop to allow further reductions in v_f[7]; most of what is believed and claimed to happen below T_g is a reduction in the occupied volume, v_o, as a result of the reduced amplitudes of vibration of the molecular skeleton.

Repeating the experiment from the same initial high temperature but at a higher cooling rate, the material will fall out of equilibrium at a higher T_g, as indicated. Stopping the cooling at some temperature well below T_g and letting the material *age* at constant temperature during significant time, the point representing the material will gradually drop along a vertical line (not shown) and, if then heated at some constant rate, another trace will result though not at all similar to the one in cooling, likely showing a delayed but more sudden rise in volume above the T_g obtained in cooling,

---

each other and cannot move. However, in a real material, we must include in v_o also the volume spanned by the mere vibrations of the relevant "particles", which is also inaccessible to large-scale motions.

[7] However, this is questionable, because large scale motions are possible after very long times at lower temperatures (and the v line may gradually approach the v_0 one, corresponding to zero free volume at extremely low temperatures), but current textbook "wisdom" is not far from the one implied in Fig. 6.6, except in recent percolation-based approaches. These will warrant discussion towards the end of this section.

before joining the equilibrium line.[8] The T_g values in cooling and in heating will thus generally not coincide,[9] showing that ordinary T_g's, being dependent on aging time and, above all, temperature scanning rate, are not equilibrium thermodynamic properties.

The behavior described in the preceding paragraph is one of the manifestations of *hysteresis*, which will be further discussed in the next section.

The type of plots in Fig. 6.6 has in the past justified the following heuristics and approximate analysis, subject to the restriction that the material will not be too far from equilibrium, from high temperature down to near T_g:

1. It is known from linear viscoelasticity (Chap. 3) that materials' response times are proportional to a viscosity, η, divided by a modulus, E, or a sum of moduli and, as a viscosity is much more variable with temperature than moduli, the shift factors defined in Sect. 6.3.2 may be formulated as $a_{T'/T} = \theta(T')/\theta(T) \sim \eta(T')/\eta(T)$.

2. In 1951, A. K. Doolittle [24] obtained an empirical correlation[10] between the viscosity of liquid n-paraffins and free volume—his free-space equation—$\eta = Ae^{Bv_0/v_f}$, valid at high enough temperatures, where the fraction of free volume is *high*, and B was found to be ~ 1.

3. In a somewhat questionable extension of Doolittle's equation to amorphous solids (assumed supercooled enough to have *low* free volume[11]), Williams, Landel and Ferry [9] argued that, v_f being low, $v_o \sim v$, and Doolittle's equation would become $\eta = Ae^{B/f}$, where f is the fraction of free volume, and therefore the shift factor would be expressed as

$$\ln a_{T/T_g} = B\left(\frac{1}{f} - \frac{1}{f_g}\right). \tag{6.3}$$

4. Considering and formulating the volume relationships implicit in the approximate (but questionable) traces of Fig. 6.6, one would write

$$v_f \cong v_g f_g + v_g \Delta\alpha(T - T_g), \quad f \cong \frac{v_g}{v} f_g + \frac{v_g}{v} \Delta\alpha(T - T_g) \text{ and}$$

$$v \cong v_g[1 + \alpha(T - T_g)], \text{ and so}$$

$$f \cong \frac{f_g + \Delta\alpha(T - T_g)}{[1 + \alpha(T - T_g)]} < f_g + \Delta\alpha(T - T_g), \tag{6.4}$$

---

[8] To obtain in heating a trace similar (though not coincident) with those in cooling, the scanning rate would need to be infinitely slow, or the material allowed to equilibrate (for a sufficiently long time) after each very small heating step.

[9] The measurements are also affected by temperature lags, whose correction requires careful temperature calibration procedures at finite scanning rates, which must of course be different in heating and cooling (because, in general, $T_{meas} > T_{sample}$ in heating, and $T_{meas} < T_{sample}$ in cooling).

[10] Further below in this same section, a statistical mechanical justification of Doolittle's equation will also be shown, due to Cohen and Turbull (cf. [25–27]).

[11] Note the obvious inconsistency of items 3 and 2.

not the usual expression we see in the original papers and many textbooks (just the numerator of the fraction in 6.4), where α and $\Delta\alpha$ are the thermal expansion coefficients of the total and free volumes (cf. Fig. 6.6), respectively. From 6.3 and 6.4, one would then obtain

$$\ln a_{T/T_g} \cong -\frac{(B/f_g)(1 - \alpha f_g/\Delta\alpha)(T - T_g)}{\frac{f_g}{\Delta\alpha} + (T - T_g)}, \tag{6.5}$$

with the form of WLF equation but whose constants (for $T_{ref} = T_g$, using decimal logarithms of the shift factors) would appear as

$$C_{1,g} = -\frac{B}{(\ln 10)\, f_g}\left(1 - \frac{\alpha}{\Delta\alpha} f_g\right) \quad \text{and} \quad C_{2,g} = \frac{f_g}{\Delta\alpha}, \tag{6.5a}$$

and where B, contrary to what happens with liquids, is expected to be lower than 1 and, in addition, dependent on temperature, as shown further below.

The above discussion and much simplified formulation appears to suggest (and many authors conveyed the view) that the effect of changes in temperature only appears via their resulting changes in molecular packing or volume! Can that be? Would materials entirely miss a glass transition or any other less abrupt change of behavior if they were heated and compressed such as to keep free volume constant? Later chapters will examine the possible relative roles of temperature and molecular packing.

Before closing this section, it is relevant to follow Cohen and Turnbull [25–27] in a statistical mechanical justification of Doolittle's equation. Assuming that the material may be assumed close to equilibrium at any temperatures above its T_g, and considering that each of its N elementary relaxors (or dynamically relevant "particles") will be surrounded at any given temperature by a free volume $\upsilon_{f,i}$, υ_f^* being the minimum free volume per "particle" necessary to allow long range motions or molecular transport, we may conjecture that an equilibrium distribution of the particles among different possible individual free volumes, $n_i(\upsilon_{f,i})$, such that $N = \sum_i n_i$ and $V_f = \sum_i n_i(\upsilon_{f,i})\upsilon_{f,i}$, might be obtained by maximizing $W = N!/\prod_i n_i!$ at constant N and V_f, the total free volume. The reader may easily apply the same reasoning and Lagrange multiplier's method used to derive Boltzmann's energy distribution in a canonical ensemble (with energies replaced by volumes), to obtain

$$\frac{n_i(\upsilon_{f,i})}{N} = p(\upsilon_{f,i}) = \frac{e^{-\beta\upsilon_{f,i}}}{\sum_j e^{-\beta\upsilon_{f,j}}} \quad \text{and} \quad \frac{V_f}{N} = \frac{\sum_i \upsilon_{f,i} e^{-\beta\upsilon_{f,i}}}{\sum_j e^{-\beta\upsilon_{f,j}}}.$$

or, in the continuum limit, as $\int_0^\infty p(\upsilon_f)d\upsilon_f = 1$, $p(\upsilon_f) = \beta e^{-\beta\upsilon_f}$ and $\frac{V_f}{N} = \int_0^\infty \upsilon_f p(\upsilon_f)d\upsilon_f$, yielding $\beta = N/V_f$, the reciprocal of the average free volume per "particle". Therefore, one obtains

$$p(\upsilon_f) = \frac{N}{V_f} e^{-\frac{N}{V_f}\upsilon_f}$$

for the probability distribution of "particle" free volumes.

Finally, relating the "fluidity" or the reciprocal of the viscosity to the cumulative probability of "particle" free volumes above the minimum necessary for molecular transport, υ_f^*, one obtains

$$\theta \propto \eta \propto \left(\int_{\upsilon_f^*}^{\infty} p(\upsilon_f) d\upsilon_f \right)^{-1} \propto e^{N\upsilon_f^*/V_f} \propto e^{B/f}, \tag{6.6}$$

i.e. Doolittle's equation, which will also yield the WLF one (cf. two paragraphs below) with $B = f^* = N\upsilon_f^*/V$, the minimum fractional free volume for molecular transport, much lower than (*not* equal to) 1, perhaps of the order of f_g. With f given by 6.4, one obtains

$$\theta \propto e^{B\frac{1+\alpha(T-T_g)}{f_g+\Delta\alpha(T-T_g)}} \sim e^{\left[\frac{f_g}{\Delta\alpha}\frac{1+\alpha(T-T_g)}{T-T_0}\right]} \text{ with } T_0 = T_g - f_g/\Delta\alpha,$$

where T_0 becomes $\sim 50K$, assuming the typical values of f_g and $\Delta\alpha$ for amorphous polymers. So, this formulation predicts the divergence of response times at $T = T_0$,[12] as the WLF equation and the empirical VTF (Vogel-Tamman-Fulcher) equation [28–30], $\theta \propto e^{A/(T-T_0)}$.

According to the above formulation, T_0 is also the ideal temperature at which the extrapolated free volume becomes zero, if in Fig. 6.6 we extend the equilibrium liquid trace down to the occupied volume (υ_o) one. As $\alpha(T - T_g) \ll 1$ (down from ~0.05 at the accepted limiting temperature of $T = T_g + 100\,\mathrm{K}$), the relationship obtained above is virtually the same as VTF equation.

Like the WLF equation, the fact that the VTF one predicts that it is *volume* alone, and not *temperature* directly, that determines materials' dynamics warrants closer analysis and possible criticism.

Cohen and Grest subsequently extended and combined the foregoing treatment of free volume with percolation ideas [31–35]. These developments predicted that free volume will only reduce to zero at 0 K (which is physically very reasonable indeed) and average response times such that[13]

$$\log_{10}\theta(T) = A + \frac{B}{T - T_0 + \left[(T - T_0)^2 + CT\right]^{1/2}}, \tag{6.7}$$

---

[12] [Which we will be questioning in Chap. 11].

[13] In fact, 6.7 was obtained to correlate (and was evaluated by) liquid viscosity data of very simple systems, considering that the viscosity should be proportional to some unspecified average response time θ. It should be observed that no distribution of response times are accessible to this theory.

where A, B, C and T_0 are adjustable material constants (A being dimensionless and B and C with dimensions of temperature), of which T_0 is more easily assigned definite and clear physical meaning, as the temperature at which supercooled liquids and amorphous materials in general exhibit a significant change in their dynamics, which may be suspected to coincide with the threshold for percolation of liquid-like domains. It could also correspond to the so-called *crossover temperature*, as defined in the literature [22], also mentioned in Sect. 6.8, and given a redefinition and abundant discussion and use in later chapters. However, the correlation shows limited predictive potential.

This percolation approach has more recently warranted detailed discussion and comparison with experimental data [36], and the main conclusion was that a direct role of temperature, by itself, cannot be neglected, given that Cohen and Grest's approach predicts that, below T_0, free volume fluctuations, in addition to always being temperature-dependent, require energy expenditure. Will not then *thermal activation* be suspected and even required to play a measurable role?[14] That is exactly what other workers have already suggested, via *volume-dependent* effective *activation energies* or, more precisely, by combining the requirements of a minimum energy fluctuation *and* a minimum available free volume to *activate* large-scale motions within the structure of an amorphous material [37, 38]. The present text also considers in Sect. 9.11 a practical solution of equivalent nature, followed by a tentative extension of CTMD to the formulation of *free volume relaxation* in Chap. 15.

The reported identified need to make A, B and C (the latter two being measured in units of temperature) themselves temperature-dependent for an acceptable general description of supercooled dynamics [36–38] may raise reasonable doubts about the physical basis of the functional form and even practical utility of 6.7.

6.11 Hysteresis and Aging Effects

Following the discussion of the behavior illustrated in Fig. 6.6, another similar example of hysteresis is the one sketched in Fig. 6.7 [39]. If an amorphous polymer is cooled at a given finite rate from equilibrium in the liquid state, at some temperature the material will fall out of equilibrium, because the indicated thermodynamic functions do not have time to relax such as to exactly follow the equilibrium line.[15] If the cooling is stopped, and immediately started a heating scan, the low temperature will substantially freeze the motions necessary to change H or U, and so they will not vary much until the point representing the material crosses the equilibrium line.

---

[14] In fact, the same may (at least in principle) apply even without energy expenditure! Are not *a-thermal* chemical reactions (or can they not be) controlled by their respective *activation energies*? Cohen and Grest themselves write in [31] that "there surely exist *activated processes* by which solid-like and liquid-like cells can exchange volume" and that "such thermally activated motion should be important at low temperatures". They also state in the same reference that there are also time-dependent and relaxation effects that were neglected.

[15] [Which of course does not have to be straight as in the qualitative diagram shown].

Fig. 6.7 Typical qualitative traces of enthalpy or internal energy in non-equilibrium cooling and heating experiments, showing the deviations relative to the extrapolated equilibrium line, where the indicated T_g refers to the heating branch of the cycle

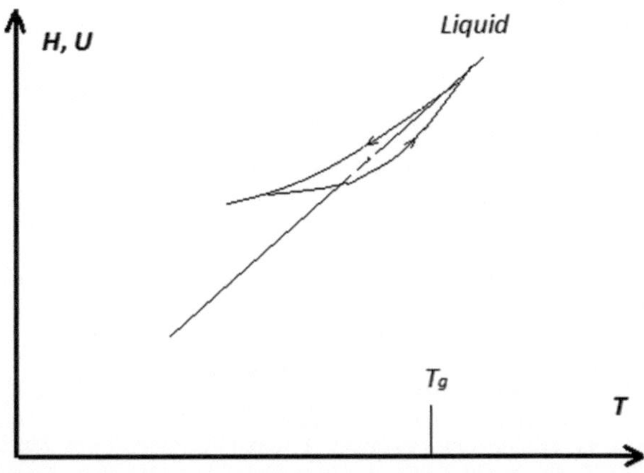

After that, the large thermodynamic *driving forces*, e.g. $(H^* - H)$, and the increasing temperature then allow accelerated motions that drive the material towards equilibrium.[16] If a similar cooling and heating cycle is conducted at a slower temperature scanning rate, the trace of the corresponding cycle will appear below the previous one, because the material will then be able to follow the equilibrium line down to a lower temperature before falling out of equilibrium.

Study Question 6.2 Sketch the hysteresis cycle corresponding to a cooling scan from equilibrium followed by an aging period at constant temperature, and then a heating scan until equilibrium is again reached.

A complete *equilibrium* (ETG) and *non-equilibrium* (NETG) *theory of glasses* is recognized as what is still lacking in our understanding of the behavior of these materials, to improve the insight provided by, and predictive ability of, the many available approaches and semi-quantitative formulations [20, 39–52]. The definition and evaluation of a *fictive temperature* to characterize each transient (non-equilibrium) state of an amorphous material is a common objective of most approaches.

A conjecture on a possible strategy towards the above overarching goal is tentatively sketched in the closing sections of Chaps. 8 and 14.

References

1. J.R. McLoughlin, A.V. Tobolsky, J. Colloid Sci. **7**, 555 (1952)
2. E. Castiff, A.V. Tobolsky, J. Colloid Sci. **10**, 375 (1955), J. Polym. Sci. **19**, 111 (1956)
3. A.V. Tobolsky, *Properties and Structure of Polymers* (Wiley, New Tork, 1960)
4. A.V. Tobolsky, J. Polym. Sci. Lett. **2**, 103 (1964)
5. J.D. Ferry, *Viscoelastic Properties of Polymers*, 3rd edn. (Wiley, New York, 1980)

---

[16] At the start of the heating branch of the cycle, the thermodynamic driving force is negative and also large, but the low temperature hinders the molecular motions necessary to approach equilibrium within the timescale of the scan, $(dT/dt)^{-1}$.

6. R. Kohlrausch, Ann. Phys. Chem. **91** (1), 56, 179 (1854)
7. G. Williams, D.C. Watts, Trans. Faraday Soc. **66**, 80 (1970)
8. K.L. Ngai, *Relaxation and Diffusion in Complex Systems* (Springer, New York, 2011)
9. M.L. Williams, R.F. Landel, J.D. Ferry, J. Am. Chem. Soc. **77**, 3701 (1955)
10. L.H. Sperling, *Introduction to Physical Polymer Science* (4th Edn.), Chap. 8 (Wiley-Interscience, Heboken, New Jersey, USA, 2006)
11. M. Wolfgardt, J. Baschnagel, W. Paul, K. Binder, Phys. Rev. E **54**, 1535 (1996)
12. J. Baschnagel, M. Wolfgardt, W. Paul, K. Binder, J. Res. Nat. Inst. Stand. Technol. **102**, 159 (1997)
13. M. Pyda, B. Wunderlich, Macromolecules **32**, 2044 (1999)
14. G.P. Johari, J. Chem. Phys. **265**, 217 (2000)
15. L. Sante, W. Krauth, Nature (London) **405**, 550 (2000)
16. G.P. Johari, J. Non-Cryst, Solids **288**, 148 (2001)
17. F.H. Stillinger, P.G. Debenedetti, T.M. Truskett, J. Phys. Chem. B **105**, 11809 (2001)
18. M. Pyda, B. Wunderlich, J. Polym. Sci. Phys. Ed. **40**, 1245 (2002)
19. S.L. Simon, G.B. Mckenna, J. Non-Cryst, Solids **355**, 672 (2009)
20. G. Adam, J.H. Gibbs, J. Chem. Phys. **43**, 139 (1965)
21. F. Blackburn, M.T. Cicerone, G. Hietpas, P.A. Wagner, M. Ediger, J. Non-Cryst, Solids **172**, 256 (1994)
22. E.-J. Donth, *The Glass Transition: Relaxation Dynamics in Liquids and Disordered Materials* (Springer-Verlag, Berlin, 2001)
23. L. Berthier et al. (eds.), *Dynamical Heterogeneities in Glasses, Colloids, and Granular Media* (OUP, Oxford, 2011)
24. A.K. Doolittle, J. Appl. Phys. **22**, 1471 (1951)
25. M.H. Cohen, D. Turnbull, J. Chem. Phys. **31**, 1164 (1959)
26. D. Turnbull, M.H. Cohen, J. Chem. Phys. **34**, 120 (1961)
27. D. Turnbull, M.H. Cohen, J. Chem. Phys. **52**, 3038 (1970)
28. H. Vogel, Phys. Z. **222**, 645 (1921)
29. G.S. Fulcher, J. Am. Ceram. Soc. **8**, 339 (1923)
30. V.G. Tammann, W.Z. Hesse, Anorg. Allg. Chem. **156**, 245 (1926)
31. M.H. Cohen, G.S. Grest, Phys. Rev. B **20**, 1077 (1979)
32. G.S. Grest, M.H. Cohen, Phys. Rev. B **21**, 4113 (1980)
33. G.S. Grest, M.H. Cohen, Adv. Chem. Phys. **48**, 455 (1981)
34. M.H. Chen, G.S. Grest, Ann. N. Y. Acad. Sci. **371**, 199 (1981)
35. M.H. Cohen, G.S. Grest, J. Non-Cryst. Solids **61/62**, 749 (1984)
36. M. Paluch, R. Casalini, C.M. Roland, Phys. Rev. E **67**, 021508 (2003)
37. P.B. Macedo, T.A. Litovitz, J. Chem. Phys. **42**, 245 (1965)
38. T. Paluka, J. Mol. Liq. **86**, 109 (2000)
39. C.T. Moynihan, S.-K. Lee, M. Tatsumisago, T. Minami, Thermochim. Acta **280/281**, 153 (1996)
40. A.Q. Tool, J. Am. Ceram. Soc. **29**, 240 (1946)
41. J.H. Gibbs, E.A. Di Marzio, J. Chem. Phys. **28**, 373 (1958)
42. J.H. Gibbs, E.A. Di Marzio, J. Chem. Phys. **28**, 807 (1958)
43. O.S. Narayanaswamy, J. Am. Ceram. Soc. **54**, 491 (1971)
44. C.T. Moynihan et al., Ann. N. Y. Acad. Sci. **279**, 15 (1976)
45. L.C. Struik, *Physical Aging in Amorphous Polymers and Other Materials* (Elsevier, Amsterdam, 1978)
46. A.J. Kovacs, J.J. Aklonis, J.M. Hutchinson, A.R. Ramos, J. Polym. Sci. Polym. Phys. Ed. Polym. Phys. Ed. **17**, 1097 (1979)
47. E.A. Di Marzio, Ann. N. Y. Acad. Sci. **371**, 1 (1981)
48. G.W. Scherer, J. Am. Ceram. Soc. **674**, 504 (1984)
49. L.C. Struik, J. Non-Cryst, Solids **131–133**, 395 (1991)
50. I. Hodge, J. Non-Cryst, Solids **169**, 211 (1994)
51. E.A. Di Marzio, Comput. Mat. Sci. **4**, 317 (1995)
52. C.T. Moynihan, J.-H. Whang, Mater. Res. Soc. Symp. Proc. **235–237**, 1 (1998)

Chapter 7
General Overview and Brief Physical Discussion of Two Recent Dynamic Models of Amorphous Condensed Matter

7.1 Ngai's Coupling Model or Correlation (CM)

Ngai's coupling model or *correlation* [1–6] started from the observation that most materials' responses are significantly extended in time—being well (even if only *approximately*) represented by an extended, KWW [7, 8], exponential—and attempts to interpret them as the result of a transition, at some time, t_c, from the response of single, uncorrelated, *"primitive relaxors"* to the cooperative response of groups of such relaxors, in clusters whose sizes are expected to increase as the temperature is lowered.

Taking the above assumed transition, at $t = t_c$, from a single-exponential (Debye) to an extended exponential (KWW) response, i.e. $e^{-t_c/\theta_{0\alpha}} = e^{-(t_c/\theta_\alpha)^{\beta_\alpha}}$, Ngai obtained the fundamental relationship of his *coupling model* (CM), $\theta_\alpha = \left[t_c^{-(1-\beta_\alpha)} \theta_{0\alpha} \right]^{1/\beta_\alpha}$, where $\theta_{0\alpha} = \theta_1$ is the characteristic response time of the primitive relaxors (the first expected to show their contribution to the measured response), θ_α is the average response time of the material during its cooperative, extended-exponential, overall time-dependent response, and $\beta_\alpha < 1$ is identical to Ngai's $(1 - n)$ parameter.

Ngai is credited to have defined and highlighted the importance of the concept of *primitive relaxors* [5] as true precursors of the cooperative behavior of materials at the microscopic scale and argued about its role in the *Johari–Goldstein β-relaxation*, a universal property of glasses and other disordered materials [6]. He rightly considered and named β_α a *"quintessential parameter"* characterizing the *cooperative nature* of the behavior, calling for the clarification of its exact physical meaning and quantification, and for approaching the description and formulation of the behavior as a true *many-body problem* [6].

© The Author(s), under exclusive license to Springer Nature Switzerland AG 2024   93
J. J. Cruz Pinto and J. R. dos Santos André, *Analytical Molecular Dynamics of Amorphous Condensed Matter*, Springer Series in Materials Science 342,
https://doi.org/10.1007/978-3-031-56517-5_7

7.2 Götze's Mode Coupling Theory (MCT)

In this book, only a very brief qualitative sketch of this advanced theory will be included, given its conceptual and computational complexity, but its formal basis and an itemized summary of its most important predictions are presented, to be used as reference in Chaps. 10 and 11 of this book, where they will be looked at in parallel with those of the new approach suggested in Chaps. 8 and 9. The interested student or junior researcher may try to delve into MCT's abundant (though "hard-going") literature, as for example [9–13].

In very general terms, MCT tries to probe microscopic relaxation dynamics via a time-dependent *density–density correlation function*, $\varphi(k, t)$, which characterizes correlations in particle density fluctuations over space (k being a given wavenumber) and time intervals, t, i.e. the extent to which the instantaneous molecular configurations within a material may resemble the corresponding configurations some time t later.[1] The underlying concepts and formalism are widely used in advanced materials science but may intimidate students and novice researchers. Nevertheless, MCT succeeds in shedding some light into the dynamics of amorphous materials.

MCT is at present considered by many researchers the best first-principles formal route to describing glassy behavior, without any phenomenological assumptions,[2] just from the static properties (at $t = 0$) of a given structure and their fluctuations. This expectation is entirely reasonable and consistent with the so-called *fluctuation– dissipation* theorem or, better, *relationship*, FDR [14–17], which points, though only near equilibrium, to an identity of the timescale of (linear) system's responses to that of property and structural thermal fluctuations. Out of equilibrium, however, violations of FDR have been and are observed, and are indeed expected, particularly at low temperatures.

The theory has an obvious mean-field character, as the material is characterized by the above single function, plus a memory function, itself containing that same function for many different wavenumbers linked by a set of *coupling parameters*. The time dependence of the density–density correlation function is then mathematically developed and followed by means of a single differential equation—analogous to that of a one-dimensional damped (but, in general, otherwise free) harmonic oscillator—whose solution is only possible by a numerical algorithm. Full physical and mathematical accuracy, however, is difficult to be obtained—the former because, among various simplifications, one's focus is on the "slow" variables or collective

---

[1] [Whose evaluation at $t = 0$—the theory's only input—is equivalent to obtaining the so-called *static structure factor*, directly accessible by (e.g. X-ray or neutron) scattering experiments].

[2] *Phenomenology* has often very unfavorable connotations in physical science, more than in engineering, but an undeniable fact is that one goal and sure guide towards any successful physical theory must of course be the ability to closely describe some specific observable macroscopic or microscopic phenomena. Its soundness and quality always depend on the physics behind it, including whether it uses physically relevant, known or measurable, system properties (instead of just "fitting constants"), and how many good (qualitative and quantitative) predictions it yields.

density modes, in such a way as to neglect any variations of the macroscopic density (only local fluctuations being considered), and the latter because the above governing equation cannot generally be solved without approximations.

Despite the formal depth (and even "beauty") of the analysis and the correctness of some specific predictions of MCT (partly listed further below), it is not easy to gain insight into—or, say, "visualize"—the detailed dynamic behavior of a material at the molecular level. Without indulging in any exaggerated or unjustified criticism of such advanced theoretical tool, a question that arises in anyone's mind might well be how, and to what significant extent, it is possible to characterize complex materials' dynamic behavior, especially if far from equilibrium and under forced conditions, by means of such *single function* and *equation*. The aspect of the inaccessibility of significant non-equilibrium behavior by MCT is very clear in the usual initial condition assumed for the rate of change of the density–density correlation function (whose role is analogous to that of the initial velocity condition in classical mechanics)—zero rate of change!

That condition is hardly realistic in situations like non-linear stress relaxation or creep, thermal quenching, etc., and the next obvious question to be asked and answered will be whether the extremely involved recent and ongoing various changes to MCT being tried, including its SCGLE (Self-Consistent Generalized Langevin Equation) modification [18] will bring about significant advances, without disproportionate effort or technical complexity. They might, sooner or later, but let us for the moment just list and (in one case) briefly comment on the main predictions by MCT of practical or engineering relevance, and later (in Chaps. 8, 9, 10 and 11) compare them with what an alternative formulation and experiments may establish or suggest:

1. A *transition* is predicted at some temperature, T_c, where *relaxation times diverge* and $\varphi(k, t)$ fails to decay to zero in whatever long timescale, the response time following a power law, $\theta \sim (T - T_c)^{-\gamma}$, for $T > T_c$, with $\gamma > 0$. This temperature has, at least initially, been wrongly likened to the *glass transition temperature*, T_g, but it is now recognized to be well above T_g. However, should relaxation time(s) truly diverge at some or any temperature, above or even below experimental glass transition temperatures?
2. An easy to understand and expected *cage effect* is also predicted in supercooled liquids, whereby "particles" find themselves temporarily trapped in cages formed by neighboring "particles", corresponding to wavenumbers of the order of the reciprocal of one "particle" diameter, which turn out to be the dominant ones.
3. With a mild extension, MCT may also predict growing and *diverging length scales* at and below T_c.
4. At longer timescales, in the so-called α-relaxation domain, an extended, Kohlrausch, response dynamics is successfully predicted.

A few additional predictions are also made pertaining to the behavior of some very special systems, for example spin-glasses, but those listed above are arguably the most significant with practical relevance.

At this point, other than raising the question above in item 1 and making exhaustive listings and specific comments on MCT's accepted or possible failures (until comprehensive comparisons are made possible by the contents of Chaps. 8, 9, 10 and 11), we instead recognize that MCT is currently being subject to considerable updating, addressed at solving its main outstanding problems, namely, (1) the theory's mean field character, (2) the limitation to (*quasi-*)*equilibrium*, (3) the difficulty in accounting for and explaining *dynamic heterogeneity* and the concept or property of *fragility*, as well as the variability of the latter among different amorphous materials,[3] (4) the mentioned (doubtful or unrealistic) divergence problem at or below T_c, and (5) the apparent *inaccessibility to thermodynamic properties* (internal energies, heat capacities, entropies, etc.).

Various modifications of MCT attempt at solving some or all the above problems in the future, as described in the specialized literature [21–24], among which one may highlight the recognized need to account for some ergodicity-restoring mechanism at low temperatures (to avoid the above divergence problem), inevitably requiring the inclusion of hopping-like *activated processes*.[4]

References

1. K.L. Ngai, Comments Solid State Phys. **9**, 121 (1979)
2. K.L. Ngai, J. Chem. Phys. **109**, 6982 (1998)
3. K.L. Ngai, J. Phys. Condens. Matter **15**, S1107 (2003)
4. K.L. Ngai, M. Paluch, J. Chem. Phys. **120**, 857 (2004)
5. K.L. Ngai, J. Non-Cryst. Solids **353**, 709 (2007)
6. K.L. Ngai, *Relaxation and Diffusion in Complex Systems* (Springer, New York, 2011)
7. R. Kohlrausch, Ann. Phys. Chem. **91**(1), 56, 179 (1854)
8. G. Williams, D.C. Watts, Trans. Faraday Soc. **66**, 80 (1970)
9. U. Bengtzelius, W. Götze, A. Sjölander, J. Phys. C Solid State Phys. **17**, 5915 (1984).
10. E. Leutheusser, Phys. Rev. A **29**, 2765 (1984)
11. W. Götze, L. Sjögren, Transp. Theor. Stat. Phys. **24**, 801 (1995)
12. W. Götze, *Complex Dynamics of Glass-Forming Liquids—A Mode Coupling Theory* (Oxford Science Publications, Oxford, 2009)
13. L.M.C. Janssen, Front. Phys. **6**, Art. 97 (2018)
14. H. Nyquist, Phys. Rev. **32**, 110 (1928)
15. H.B. Callen, T.A. Welton, Phys. Rev. **83**(1), 34 (1951)
16. E.-J. Donth, J. Phys. Condens. Matter **12**, 10371 (2000)
17. U.M.B. Marconi et al., Phys. Rep. **461**(4–6), 111 (2008)
18. L. Yeomans-Reyna et al., Phys. Rev. E **76**, 041504 (2007)
19. G. Tarjus, Invited Plenary Lecture (not published), in *5th International Workshop on Complex Systems*, Sendai, Japan, Sept 2007

---

[3] Tarjus in [19, 20] openly calls for the future development (evolved or not from MCT) of a *"theory of tunable fragility"*.

[4] It may surely be stated that this stands in anyone's mind as an example of acceptable and physically sound *phenomenology*, but more on this later in Chaps. 8 and 9.

20. G. Tarjus, in *Dynamical Heterogeneities in Glasses, Colloids, and Granular Media*, ed. by L. Berthier et al. (OUP, Oxford, 2011)
21. G. Szamel, Phys. Rev. Lett. **90**, 228301 (2003)
22. S.P. Das, Rev. Mod. Phys. **76**, 785 (2004)
23. P. Mayer, K. Miyazaki, D.R. Reichman, Phys. Rev. Lett. **97**, 095702 (2006)
24. D. Coslovich, A. Ikeda, K. Miyazaki, Phys. Rev. E **93**, 042602 (2016)

Chapter 8
An Alternative Look at the Dynamics of Amorphous Condensed Matter

8.1　Introduction

A reasonable specification for a sound and comprehensive theory of the dynamics of amorphous condensed matter is that it must necessarily be *kinetic* in nature, otherwise it will not address the whole range of phenomena and features that must be explained, as specified in Chap. 6, where timescale dependence is most obvious and essential. In the case of temperature scanning experiments, as the observation timescale is expanded (or the scanning rate is reduced), the theory should be able to predict in cooling a smooth convergence of the behavior to, either a true, ideal, abrupt thermodynamic glass transition at some specific temperature[1] (will it really exist?), or no transition at all [1–9], i.e. simply show a gradual, well defined, change in the structure's dynamics and physical properties. For all finite timescales, however, that non-equilibrium, gradual evolution with temperature should depend on the actual path and rates of the temperature changes and/or of those of any other imposed physical stimuli, in addition to the structure's detailed initial state, itself the result of all previously imposed stimuli. *Hysteresis* (cf. Sect. 6.11) in cyclic excitations, for example, should therefore also be one of the qualitative and quantitative final outcomes of a future comprehensive theory.

While acknowledging that the challenges and tasks ahead are enormous, one might pause to remember, and then try to follow, Josiah Willard Gibbs' famous statement: *"One of the principal objects of theoretical research in any department of knowledge is to find the point of view from which the subject appears in its greatest simplicity."*[2] Notwithstanding the advances already achieved in this field of knowledge, and its intrinsic and puzzling complexity—very well discussed in advanced reference works such as those of [10–12], one possible though daring question that may be asked is

---

[1] [To which the ordinary *dynamic glass transition* will gradually converge].

[2] From Gibb's letter accepting the Rumford Medal (1881). [Quoted in A. L. Mackay, Dictionary of Quotations (London, 1994)].

© The Author(s), under exclusive license to Springer Nature Switzerland AG 2024　　　99
J. J. Cruz Pinto and J. R. dos Santos André, *Analytical Molecular Dynamics of Amorphous Condensed Matter*, Springer Series in Materials Science 342,
https://doi.org/10.1007/978-3-031-56517-5_8

whether there will not be some simpler, physically clear, and more manageable view of the behavior of amorphous materials, yielding predictions compatible with what we already know, and capable of shedding additional light on what still remains puzzling or obscure. Should we not even consider it? We think we can and indeed should, though remaining fully conscious that progress will not be forthcoming in just one or a few easy steps. Much work will be required beyond the proposals presented in this book.

8.2 The Physics and Cooperativity of Materials' Behavior at the Molecular Scale[3] [13, 14]

The structure of most materials, in particular those of a macromolecular nature, may be so complex that a wide range of contributing structural features or *relaxors* with different location, size (arguably as the result of intramolecular and intermolecular *clustering*, or any form of physical *entanglement*) and dynamics may be assumed to exist within the material, and this stood as the best physical justification for the wide adoption of the concept of relaxation and retardation spectra within early phenomenological models of linear and non-linear viscoelasticity (cf. Chaps. 3 and 4). A simple but useful concept is that a basic *moving structure* or *"primitive relaxor"*, in Ngai's exact terminology [15], may in principle always be defined for any given homogeneous structure,[4] able to undergo single, uncorrelated, transitions (mostly, or including, thermally activated ones), as well as more or less strongly correlated or clustered ones, contributing to the onset and development of local and bulk specific relaxations and transitions, including the most important (and still mysterious[5]) *glass transition*. The mentioned transitions at the local level may either randomly occur due to local temperature and momentum or energy fluctuations (a form of molecular agitation akin to the one affecting typical molecular velocities in a liquid or gas), or be somewhat biased and driven, i.e. forced, by some type of physical excitation (strain, stress, electrical field, etc.)[6] that might modify the local activation energies influencing the local dynamics, depending on the local potential energy landscape.[7]

---

[3] Sections 8.2 and 8.3 greatly expand on parts of [13, 14] by permission of John Wiley and Sons (Copyright © 2016).

[4] For a chemically and structurally non-homogenous or mixed structure, *primitive relaxors* of different nature may also in principle always be defined, as the smallest structural elements able to contribute to local or wide scale motions.

[5] See Footnote 1 of the Preface.

[6] Relaxor's transitions driven or assisted by specific topological constraints (physical entanglements of any kind) might also play a role, which however will not be considered at this stage, except by recognizing that such constraints might blur the definition and the effective minimum size of the "primitive" relaxors that will be found to determine the final behavior.

[7] [Or multidimensional *"energy topographic map"*, now widely recognized as characterizing super-cooled liquids and amorphous solids in general]. For an accessible introduction, see F. H. Still-inger, *Science*, 267, 31st March, 1935–1939 (1995) and, for a detailed and advanced one, see

From here, one will need to proceed with the double objective of physically and mathematically model (1) the *response of sets of* such individual *primitive relaxors* in the instance under consideration (stress relaxation, creep, etc.), and (2) the actual *clustering processes* and their specific effects and relative (size-dependent) weights on the local and overall responses of the material. Both such *responses* and the underlying *clustering* processes will of course have to depend on the temperature and timescale of the observation or experiment, as explicit results of the theory.

The primitive relaxors to be considered may be of a widely varying nature, depending on the type and structure of the material—e.g. small chain crankshafts made of, say, 4 or 6 main chain atoms in the case of polymers[8] [16]. But, for much less complex atomic or molecular materials, a simple pair of neighboring atoms or molecules, or an atom- or molecule-hole pair, capable of interchanging their positions, could be adequate candidates to define the relevant primitive relaxor(s), and so it may be expected that the same or similar type of analysis and modeling might prove of rather general applicability, in line with the significant universality of amorphous condensed matter behavior, already briefly discussed in Sect. 6.7 [10, 17], and revisited later in this chapter and in Chap. 9.

Critically important but still unsolved questions are:

(a)   why and how *cooperativity* sets in;
(b)   why and quantitatively how the *cooperativity level changes with temperature and the timescale* of the observation;
(c)   how the intervening elementary processes undergo *progressive activation/de-activation* as the material is heated/cooled.

8.3   A Radical New Look at Cooperativity—A Transition State Theory (TST) of "Cooperative Materials Dynamics" (CTMD) [13, 14, 18]

It may not be obvious, but we may draw a parallel between the *clustering* and *cooperative motion of* several disjoint or contiguous *primitive relaxors* of a complex molecular structure and a *multi-molecular chemical reaction* (or *multi-center physical process or transition*), without of course any actual chemical change, and tentatively formulate it according to the *transition state theory* (TST)—one of the foundations of chemical physics, physical chemistry and chemical engineering, despite some recognized imperfections in its simplest forms. The only real difference is that the

---

D. Wales, *Energy Landscapes—With Applications to Clusters, Biomolecules and Glasses* (Cambridge University Press, 2003).

[8] In such materials, chain-end contributions will also come from methyl or other terminal groups, with reduced activation energies, which will expectedly result in molar mass and chain branching dependencies of the final behavior and, of course, eventual partial crystallinity or crosslinking will also play a role, by depressing some of the microscopic motions under discussion, as pointed out in Chap. 6 and more specifically discussed in Chap. 9. The emphasis of this text, however, will be on amorphous un-crosslinked materials.

postulated activated or transition state/complex of such cooperative process is now a *cluster* of coupled single-relaxor activated or transition states/complexes, each in its physical environment and location i.e. as truly distinguishable clustered "particles", most probably in adjacent positions, where they collectively—simultaneously or in "quick" succession relative to the average frequencies of the joint transitions—"yield and jump" to the corresponding final states. This feature of the proposed theory should in the end explain the so-called "*dynamic facilitation*" concept referred to or implied by several authors [19–21], to be expanded further below. It will be the frequency of these *clustered transitions* that will have to be formulated by any proposed theory, as functions of temperature, excitation amplitude (when applicable), cluster size, and other relevant structural parameters.

This analogy, though perhaps unexpected at first, should appear obvious, and all we will be doing is to apply *abductive reasoning*[9] (*retroduction* or *inference to the best explanation*) [18, 22] to features (6.1)–(6.11) discussed in Chap. 6. Max Planck and Albert Einstein did something similar more than a century ago, when they conjectured about, and successfully explained, the spectrum of thermal radiation and the photoelectric effect (by the energy quantization and the photon hypotheses, respectively). Also, now and every-day, the same happens when physicians try to figure out what might be wrong with a patient, an IT-technician or programmer troubleshoots a system or a computer code, and detectives or criminal prosecutors try to make sense of a given crime scene. In our case, the "crime scene" is the same set of features (6.1)–(6.11), which remain largely unexplained [10, 11], and the solution that is being proposed here turns out less strange than the radiation quantization appeared to Planck himself, who explicitly expressed his own doubts. Of course, the abducted (assumed/inferred) explanation, "wild" as it might seem (like Planck's), must prove realistic and provide a thorough, *logical and superior interpretation* of all features of the behavior *relative to* any *possible alternatives*, and be subject to *validation or rejection*.

In "troubleshooting" terms, what we should be asking is why none of the known condensed and soft matter theories and formulations satisfactorily explain and quantify each of those eleven features, such as to quantitatively (or at least approximately) predict the correct physical responses at any temperature and for any timescale and, where possible, also identify and yield the numerical values of relevant physical

---

[9] Not of course as logically safe as pure *deduction*—which aims at going from p being true to q (establishing q as a necessary condition to p, i.e. $p \rightarrow$ q), it aims at finding the "*best*" or "*most probable*" $\underline{p}$ (or sufficient condition) that <u>logically implies</u> q, knowing that q <u>is true</u> (like a real, observable, fact). When fully successful (only <u>if and when "all"</u> other logical, necessary, consequences of p <u>are also found true</u>), one would have then reached *logical equivalence* of p and q ($p \leftrightarrow q$). However, <u>one single false logical, necessary, consequence of p will strictly invalidate it.</u> Where such logically invalid consequence(s) of p may prove spurious and irrelevant to what is being analyzed, the p hypothesis (except if nonsensical) may perhaps still survive, as so many other past *provisional* physical theories, but just until a better theory (or p) comes to light. Therefore, every single prediction or necessary condition of p must be scrutinized and dealt with.

properties of the materials. Ngai's insight [11] led him to rightly suggest where the difficulty lies: in his own words, we need to tackle the problem as the *"many-body"* one that it truly is—head-on, and right from the start.

Modeling Thermally Activated Dynamics in the Absence of Forced Excitations [14]

Eyring's transition state theory (TST) was proposed long ago for simple chemical reactions [23] and was also extended to any (chemical/physical) activated processes [24]. As described in Chap. 4, TST has since also been applied to both creep and stress relaxation situations, though within oversimplified models and formulations. TST's recognized limitations, when they really apply, do not warrant the most common criticism—the much referred to, but wrong, "full equilibrium assumption"—even less the occasional, concealed, but sometimes explicit, doubts about the actual role, when not even the existence, of *activated states or complexes*. In fact, as explained in good references on Physical Chemistry [25], TST does *not* require full overall thermodynamic equilibrium between real or would-be "reactants" and activated complexes, but only *local thermal equilibrium*, enabling the use of Boltzmann statistics throughout [25, 26]. This only requires that, locally, thermal relaxation be much faster than the actual activation and ensuing process [25, 26], which will most often apply, even when the temperature is not uniform, and heat may be slowly flowing through the system.[10] Also activated complexes may nowadays be detected and *directly observed* [27] and characterized in some reactions, by *femtosecond* (10–100×10^{-15} s) *transition state spectroscopy* (FTS or TSS [28]), and we do not know of any acceptable physical reasons to doubt or reject the assumption of an active role of possibly much longer-lived activated states in condensed matter non-reactive processes [24, 26]. It may here also be recalled, as in Chap. 7, that even within MCT one had to resort to activated hopping-like processes to avert the divergence problem of response times at and below the crossover temperature, without significant reservations about excessive "phenomenology".

If necessary, even the Boltzmann-like local thermal equilibrium requirement may in principle be lifted in modified, non-equilibrium, power-law statistical theories, at present still under development [29, 30]. Pending deeper evaluation, and to the extent that the actual experimental behavior under analysis might be reasonably described and predicted (as assessed in Chaps. 9, 10 and 11), Boltzmann statistics TST will be adopted.[11] The reader is here assumed familiar with, or willing to revise, the fundamentals of *Statistical Mechanics*, as for example presented in Chap. 22 of [25], among many other useful texts, and then go through the intermediate details of the derivations below.

---

[10] Chemical Reaction Engineering often successfully relies on the applicability of TST's core concepts to formulate, correlate, and generate relevant basic kinetic data needed to quantify the course of many relevant and efficient industrial non-isothermal reactions in complex flow systems.

[11] Cases where tunneling might be important (as when light molecular species are involved) are also excluded from the present treatment.

While in the case of chemical reactions it is easy to formulate the formation of the activated complex through collisions of the reactant molecules, in these cooperative physical transitions, actual collisions do not have to be involved, bearing in mind that any clustering relaxors may and most probably will be contiguous and, anyway, they are all immersed in the common "thermal and mechanical bath" of all other relaxors. The stoichiometry of clustering may thus be represented by

Random Combination of n Primitive Relaxors

$\rightarrow$ *(Activated Cluster of n Primitive Relaxors)$^{\#}$ $\rightarrow$ Final State,*

but, nevertheless, clustering will be viewed and simplified here as a "pseudo-unimolecular" process from some random combination of the system's primitive relaxors onwards, as a single cooperative or collective transition to a different state. How to consider and account for those various combinations of primitive relaxors will be considered in the first few sections of Chap. 9.

Under the above assumed condition of local thermal equilibrium, and for a mass of material containing a total of N_A (Avogadro's number in 1 mol) primitive relaxors, one may write the following *statistical mechanics relationships* between the numbers of distinguishable non-activated and activated ($^{\#}$) *clusters of primitive relaxors of size n* (as previously defined) at their ground state ($_0$), $N_{n,0}$ and $N_{n,0}^{\#}$, or in any possible state, N_n and $N_n^{\#}$, their energies and complete[12] molecular partition functions (z' and z, where the unprimed zz have all their energy levels relative to the corresponding ground levels):

$$N_{n,0} = N_n \frac{e^{-n\varepsilon_{1,0}/(k_B T)}}{z_1'^n} = \frac{N_n}{z_1^n} = N_{n,0}^{\#} \frac{e^{-n\varepsilon_{1,0}/(k_B T)}}{e^{-\varepsilon_{n,0}^{\#}/(k_B T)}}, \qquad (8.1)$$

where one considers that, at any given time, there will be *very small numbers of activated states* among all potential clusters of size n at any given time, such as to allow considering in the first of the above equalities that the various partition functions $z_n' = z_{n,vib+etc.}'$, that should in fact appear in the denominator of the first of (8.1) may be factorized, and thus will not significantly differ from $z_1'^n$ (corresponding to nearly independent primitive relaxors *prior to activation*[13]). From (8.1) one may then obtain

$$N_n^{\#} = N_{n,0}^{\#} z_n^{\#} = N_n \frac{z_n^{\#}}{z_1^n} e^{-\left(\varepsilon_{n,0}^{\#} - n\varepsilon_{1,0}\right)/(k_B T)} = N_n \frac{z_n^{\#}}{z_1^n} e^{-E_{a,n}/(k_B T)}. \qquad (8.2)$$

---

[12] [Meaning that they include all states and energy components, namely all "non-cooperative" ones, like the vibrational and electronic states].

[13] At the end of this chapter, an itemized summary of the physical basis and end results for all partition functions used and obtained in the present formulation will provide full clarification.

The N_n are the total number of distinguishable possible combinations of $N_1 = N_A$ primitive relaxors in groups of n (most probably, or predominantly) adjacent relaxors, the ε's stand here for the various energy levels, and $E_{a,n} = \varepsilon_{n,0}^{\#} - n\varepsilon_{1,0}$ is the activation energy at 0 K for the transition of a cluster of size n (calculated between their respective ground states). The above equations, of course, are only valid for *large systems*, such as to allow Stirling's approximation[14] in the evaluation of factorials, as required to exactly yield the assumed Boltzmann's or canonical distribution of quantum states [25] and, apart from the mentioned *local thermal equilibrium* condition, the only additional assumption so far was the one implied in the approximation $z_n' \approx z_1'^n$ in (8.1).

The formal resemblance of the product $\left(z_n^{\#}/z_1^n\right)e^{-E_{a,n}/(k_B T)}$ in the last of (8.2) to a (pseudo-) equilibrium constant, $K_n = N_n^{\#}/N_n$—also equal to $e^{-\Delta a_n^{\#}/(k_B T)}$ (if one relates the activated cluster and primitive relaxor partition functions $z_n^{\#}$ and z_1 to the corresponding Helmholtz free energies relative to each of the ground level energies[15]) may perhaps, but ought *not*, explain the "equilibrium" misunderstanding in some hasty and incorrect references to TST. However, $\Delta a_n^{\#}$ is *not* an overall standard free energy change (from an initial to a final state of the system as a whole), as it necessarily had to be in an equilibrium situation, but a *local*, individual cluster, *free energy of activation*. Activated states are stationary, first-order saddle, points (with *maximum local free energy* and a *negative local second derivative* along the relevant process coordinate), *not* equilibrium, *minimum overall* or *local free energy* (with *positive local second derivative*), states. Activated states result from significant but infrequent *local energy fluctuations*—all Boltzmann-like under fast thermal relaxation—and are not some stable "chemical species" in equilibrium with others in the bulk system. Instead of equilibrium, overall minimum free energy and stationary number of activated complexes in the whole system, what we have is a *pseudo-stationary* or *dynamic steady state*, where the rate of activation (number of species being activated per unit time) is at all times balanced by the rate of the activated states' jumps to the final state, which thus equals the actual rate of change at any given instant in time. With due allowance for the respective different specific lifetimes, the role of the activated states is, qualitatively, somewhat analogous to that of short-lived, *unstable*, "intermediates" in multi-step chemical reactions (like e.g. macro-radicals in radical polymerizations), whose kinetics is routinely and accurately formulated by invoking a pseudo-"steady-state assumption", on which both the underlying chemistry and the engineering design of many functional and important industrial processes depend.

The overall frequency of transition of clusters of size n is then simply $R_n = N_n^{\#}\nu_1^{\#} = N_n K_n \nu_1^{\#} = N_n \nu_n^*(T)$, where $\nu_1^{\#}$ is the characteristic frequency (in Hz) at which the activated state of each primitive relaxor librates and jumps to the final state. For obvious practical reasons, notwithstanding expected local, microscopic, environment variations, one has of course to take this frequency as approximately uniform and the same for any primitive relaxor (of a given nature within a given uniform

---

[14] $N! \approx \sqrt{2\pi N}\, N^N e^{-N}\left(1 + \frac{1}{12N} + \cdots\right) \approx N^N e^{-N}$ for $N \to \infty$.

[15] From $\Delta a_n^{\#} = a_n^{\#} - na$, with $a - a^0 = -k_B T \ln z_1$ and $a_n^{\#} - a_n^0 = -k_B T \ln z_n^{\#}$, with $a_n^0 = na^0$.

structure) in any of its clusters of whatever size, as clustering may be considered no more than an occasional random, mainly local, coordination, *synchronization or tuning* (or even a very *quick succession*), of the activated primitive relaxors' jumps to their final states, though possibly also assisted by mere proximity[16] and/or local topological constraints or entanglements of whatever nature. As various authors have described very well [31–33], the material behaves as, and may be characterized by, spatial clusters of fast- and slow-moving "particles" (our *primitive relaxors*), where a given "particle" may remain "slow" for a certain time within a large cluster, and then become "faster" by "coordination" within a smaller one or by decoupling itself completely (and vice-versa), thus displaying *intermittent behavior*. This same intermittency will play an essential role in the formulation of the theory's *spectrum of cluster frequencies*, to be considered in Sect. 9.1.

8.3.1 Characteristic Response Frequencies at Thermal and Mechanical (or Other) Equilibrium [14]

From the previous analysis, the resulting specific frequency of transition in Hz (per potential cluster of each given size n), $v_n^*(T)$, then becomes (from 8.2)

$$
\begin{aligned}
v_n^*(T) &= \frac{R_n}{N_n} = v_1^{\#}\frac{z_n^{\#}}{z_1^n}e^{-E_{a,n}/(k_BT)} \\[2ex]
&= v_1^{\#}\left(\frac{z_{1,r}^{\#}}{z_1}\right)^n\left[\frac{e^{-E_{a,1}/(k_BT)}}{1-e^{-hv_1^{\#}/(k_BT)}}\right]^n \\[2ex]
&= v_1^{\#}\left[\left(\frac{z_{1,r}^{\#}}{z_1}\right)\frac{e^{-(E_{a,1}-hv_1^{\#}/2)/(k_BT)}}{2\sinh\left(\frac{hv_1^{\#}}{2k_BT}\right)}\right]^n,
\end{aligned}
\tag{8.3}
$$

where $z_{1,r}^{\#}$ is the primitive relaxor's activated state residual molecular partition function, after the one of the transition-relevant degree of freedom or coordinate, $z_{1,v,t}$, is factored out and expressed in terms of $v_1^{\#}$ and T simply as a libration (as the expression within [] in the 3rd equality of 8.3), and h is Planck's constant.[17] This procedure

---

[16] [Leading to some sort of "vibration-induced", successive, chain-like, transitions].

[17] The above expression refers all energy states of the "decomposing" primitive relaxors' activated states to their zero-point energy, which is $hv_1^{\#}/2$, and that explains its appearance in the numerator of the argument of the exponential of the last equality of (8.3), where $E_{a,1}$ is also referred to the corresponding zero point, $\varepsilon_{1,0}$, whatever it may be. However, should one (perhaps rightly) argue that the above zero-point energy should be included in $E_{a,1}$, it will then disappear from the mentioned argument of the exponential. However, given the very small relative values of $hv_1^{\#}/2$ and $E_{a,1}$ for most materials (cf. [10], only exceeding 1%, e.g. for $E_{a,1} \sim 50$ kJ/mol, if and when $v_1^{\#}$ is well above 10^{12} Hz, and being $< 2 \times 10^{-8}$ for most cases treated in this book), the resulting change may be neglected.

is identical to the well-established one for the reaction coordinate degree of freedom in chemical reaction dynamics, as may be reviewed in Physical Chemistry texts [25].

It has also been assumed that $z_n^{\#} \approx z_{n,r}^{\#} \left(z_{1,v,t}\right)^n \approx \left(z_{1,r}^{\#}\right)^n \left(z_{1,v,t}\right)^n$, because it is not expected that mainly *slow* transitions of the primitive relaxors (e.g. rotations of short-chain crankshafts in the case of polymers), even if synchronized, will significantly change the other vibrational and electronic energy levels not involved in the transitions, or affect the activated primitive relaxors' distinguishability and near independence implied by the factorizations in the above formulation, up to and including the actual synchronization or tuning of their final "decomposition" at $v_1^{\#}$. Once such tuning occurs at $v_1^{\#}$, which must be and is here taken as a property of the system, the cluster's joint partial partition function for the n transition-relevant degrees of freedom, $z_{n,v,t}$, may not significantly differ from $\left(z_{1,v,t}\right)^n$, as if we had sporadic, intermittent, "pseudo-Einstein crystals" of just n activated adjacent primitive relaxors vibrating (and being rapidly dismantled) at $v_1^{\#}$.[18] What makes possible and kinetically significant any of the clustered transitions under discussion is the entropy gained and, depending on the temperature, the corresponding free energy that may be lost *locally* each time n primitive relaxors simultaneously, or just in quick succession, activate and jump to a new state, instead of doing it at distinct random times. Notwithstanding the doubts that might persist, what if, at the end (cf. Sect. 8.3.3), the present formulation will lead to an effective loss of independence of the primitive relaxors within any effective or active cluster (relative to all other primitive relaxors),[19] such that $z_n(T) \neq [z_1(T)]^n$? We might be surprised, but then finally convinced.

In the case of polymer chain crankshaft primitive relaxors, it would be justified to exclude the factor 2 multiplying the "sinh" (equivalent to including a 2^n factor in the *rhs* of the 2nd, 3rd and 4th equalities of 8.3) to account for the two possible rotation directions of each primitive relaxor, but in the above written equations we arbitrarily assumed that, on average, only one of those would be free in polymers. More accurately though, here and in all cases in general (particularly in less complex atomic or molecular materials, where several transition paths or channels may be available for the transitions of each primitive relaxor), there will be a small undetermined, structure-dependent, factor (a number ≥ 1) in the expression of $v_n^*(T)$, also varying exponentially with n. This problem may be circumvented by following the procedure specified later in Chap. 9 to estimate $z_{1,r}^{\#}/z_1$ (lumping it to the above structure-dependent factor), with the indirect help of actual experimental response

---

[18] In an "Einstein crystal" (as in a Debye or a real one), the vibrations along the assumed single normal mode(s), despite their independence (which justifies the factorization of the partition function), refer to motions "carried" or "transmitted" by standing waves that propagate through the whole crystal. We assume that just the same may occur within the bounds of each active cluster during any of its transitions. A concept of "*clustron*", to describe *multifractal cluster structures*, was even coined in [34] to replace that of "*phonon*" in this context. [As in the past update from Einstein's to Debye's (or better) crystal models, one could envisage that, instead of the unique characteristic $v_1^{\#}$ frequency considered here, a distribution of such frequencies would improve the present theory and its formulation, but the ensuing complexity would not play any useful role at this early stage of the development of the present theory].

[19] [Otherwise, clusters would not differ from the corresponding number of independent primitive relaxors].

data. The value of this specific ratio will depend on the particular structure, being lower than 1 if the activated state(s) will make accessible a smaller number of energy states (due, for example, to increased steric repulsion-induced strains in stiffer molecular structures), greater than 1 in the opposite case, and close to 1 in the absence of such effects; it might also be temperature-dependent, but here it will be treated as a constant. Thus, $z_{1,r}^{\#}/z_1$, $E_{a,1}$ and $v_1^{\#}$ are so far three specific physical parameters of this *cooperative or clustering theory*, which will of course depend on the structure's local energy landscape. The latter two parameters may also not significantly depend on temperature (at least, E_{a1}, relating to zero-point energies, is in fact itself defined at 0 K), but may show some change with intermolecular packing, as discussed in Sect. 9.11.

The last equality of (8.3) is the basis of the proposed *cooperative theory of materials' dynamics* (CTMD). Before proceeding, it must be stressed that the $v_n^*(T)$ just calculated are the *thermally driven* relaxors' *frequencies*, in the absence of any other external physical excitation, i.e. with the material in simple *local thermal equilibrium*, without however requiring macroscopically uniform temperature. Under an applied constant strain (as in stress relaxation) or under an applied constant stress (as in creep), the activation energies, $E_{a,n} = nE_{a,1}$, will of course change, as formulated in Sects. 9.2 and 9.3, leading to modified transition frequencies.

So, with CTMD, we might build a clear picture of the clustering process as follows: All primitive relaxors vibrate at frequency $v_1^{\#}$, characterizing the structure being considered (related to its stiffness), which measures how many times per second each one of them "attempts" a transition to the new state (relaxed or unrelaxed, deformed or undeformed, or whatever). In conditions of thermal equilibrium, some of them will randomly succeed, without clustering with any other, corresponding to an average frequency $v_1^*(T) \leq v_1^{\#}$ below the crossover temperature [14] ($T < T_c$) but, at other random times and (maybe predominantly adjacent) locations within the same structure, a group of $n > 1$ other, or some of the same, primitive relaxors may synchronize and "be found" attempting and succeeding at their transitions at exactly or approximately the same time, such as to "benefit" from an increased local activation entropy,[20] despite the increased activation energy (leading to a decreased, temperature-dependent, local activation free energy), and therefore contribute to an effective overall frequency of this joint (clustered) transition $v_n^*(T) < v_1^*(T) < v_1^{\#}$. When, assuming or not thermal equilibrium, the material is being subject to a forced excitation, the behavior will be similar, but with the modified frequencies referred to in the preceding paragraph and dealt with in the following chapter.

Given the critical role of all the individual *cluster frequencies* just formulated (8.3) on the overall dynamic behavior, their actual calculation and plotting will be discussed in the next section and subsections, via a typical numerical example and additional calculations for two specific materials.

---

[20] More on this in Section "New Concepts of a Crossover Temperature and a Crossover or Critical Frequency [14]".

8.3.2   New Concepts of a Crossover Temperature and a Crossover or Critical Frequency [14]

On close analysis, the most relevant feature of (8.3) is that they predict the existence of a crossover or transition region, and even a *crossover temperature*, T_c, and frequency, ν_c, that depend on structural parameters—$E_{a,1}$, $\nu_1^\#$ and $z_{1,r}^\#/z_1$. In fact, though the *crossover* has been rightly defined and named as a (not necessarily sharp) *transition* between different dynamic behaviors [10], we propose that the crossover temperature may here be identified as the temperature at which the expression within brackets in the last equality of (8.3) exactly equals 1, the crossover frequency, ν_c, then turning out to be identical to $\nu_1^\#$, with all individual cluster frequency curves intersecting each other at $T = T_c$ and $\nu = \nu_c$, as graphically shown further below.[21] The ensuing analysis will make clear that, at or near T_c, a significant change in the structure's dynamics does indeed occur, in line with what is experimentally observed and predicted by a range of other theories [10].

The Crossover and the Predicted Universality of the Behavior [14]

Accepting the above crossover equivalent definition, from any experimental estimates of ν_c and T_c (such as those collected in [10], pages 222–223, and [35]), one may obtain the basic structural energy parameter $E_{a,1}$—the minimum activation energy (pertaining to a single primitive relaxor)—from

$$E_{a,1} = -k_B T_c \ln\left[2\sinh\left(\frac{h\nu_c}{2k_B T_c}\right) \bigg/ \left(\frac{z_{1,r}^\#}{z_1}\right) \right] + \frac{1}{2} h\nu_c. \qquad (8.4)$$

Alternatively, T_c values may be determined for any $(E_{a,1}, \nu_c, z_{1,r}^\#/z_1)$ set of values, by direct numerical iteration on

$$T_c = f(T_c) = -\frac{E_{a,1} - h\nu_c/2}{k_B \ln\left[2\sinh\left(\frac{h\nu_c}{2k_B T_c}\right)\big/\left(\frac{z_{1,r}^\#}{z_1}\right)\right]}, \qquad (8.5)$$

where $f(T_c)$ is found to always satisfy the condition $|df(T_c)/dT_c| < 1$ in the vicinity of the solution, as required to ensure convergence. Where T_c, ν_c and $E_{a,1}$ may be experimentally known or easily estimated [10, 35], $z_{1,r}^\#/z_1$ may be calculated from any of the above two equations, as will be adopted in Chap. 9.

---

[21] Given the variability of the above three local structural parameters, relaxation maps showing the various vibration modes may be expected to feature a more diffuse intersection zone of the lines corresponding to the different CTMD cluster sizes, rather than the sharp single point shown in Figs. 8.3 and 8.4. Qualitative examples of the effect of the variability of those parameters among widely different materials on their relaxation map's average α and β lines (as defined further below) are illustrated in Fig. 2.9a of [10].

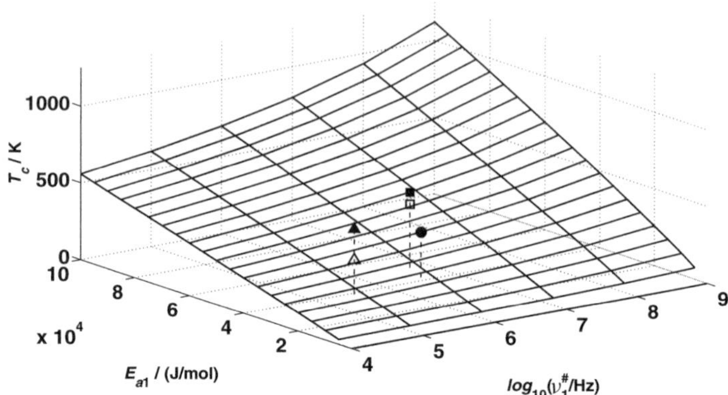

Fig. 8.1 Crossover surface (or solutions map of 8.5): crossover temperature, T_c, as a function of $E_{a,1}$ and $v_1^{\#} = v_c$, for $z_{1,r}^{\#}/z_1 = 1$; •—example of the moderately flexible structure treated in the numerical example that follows; ▲, ■—calculated data for a PMMA (▲) and a PC (■)[22] from stress relaxation data [13]; △, □—projections of ▲, ■ onto the above crossover surface—cf. also Chap. 11. Redrawn with data from [14], by permission of John Wiley and Sons

Equation (8.5) always has a solution and yields a steep, continuous, curved surface plot as a function of $E_{a,1}$ and $v_c = v_1^{\#}$ (keeping $z_{1,r}^{\#}/z_1$ constant), with higher/lower T_c values representing stiffer/softer materials, as shown in Fig. 8.1 for $z_{1,r}^{\#}/z_1 = 1$, including the numerical example specified further below. The plot may be extended to still lower and higher $v_1^{\#}$ values, larger $E_{a,1}$ and higher temperatures, the only effective high temperature or frequency limits being those determined by the actual structure's thermal and mechanical stability. Ceramic materials, for example, would be represented by points significantly high up on a similar surface, for whatever adequate $z_{1,r}^{\#}/z_1$ value.

Once the mentioned association of T_c with a change in dynamics is demonstrated, the striking qualitative similarity of behavior between so many different glass formers, near and across their crossovers, might finally find a physical and mathematical explanation. The crossover temperature has also been associated, within the *mode coupling theory* (MCT) of the glass transition [12, 36–39] {—already presented in Chap. 7 and given additional discussion in Chap. 10, often considered one of the most successful theories of the slow dynamics of soft materials, to the separation between the *ergodic liquid regime* at $T > T_c$ and the *non-ergodic glass regime* at $T < T_c$ [39, 40], a feature that should be highlighted.

The various materials' representations in a CTMD *crossover map*, like the one in Fig. 8.1, still deserve an additional note to provide a more general perspective, bearing in mind that the crossover temperature is a function of three variables, not just two. Figure 8.1 therefore shows just one specific crossover surface, at a constant $z_{1,r}^{\#}/z_1 = 1$. In fact, any amorphous material would require a tetra-dimensional space

---

[22] PMMA—poly(methylmethacrylate); PC—bis-phenol-A polycarbonate.

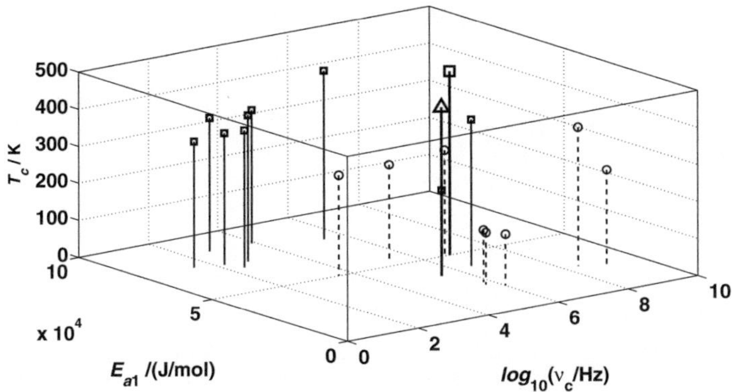

Fig. 8.2 Experimental crossover map data (stem plot) from [10, 35]: □—polymers; ○—small molecules (salol, orthoterphenyl, n-propanol, etc.); △, □—crossover properties for a PMMA (△) and a PC (□) calculated from the experimental stress relaxation data discussed in Chap. 11. Redrawn with data from [13], by permission of John Wiley and Sons

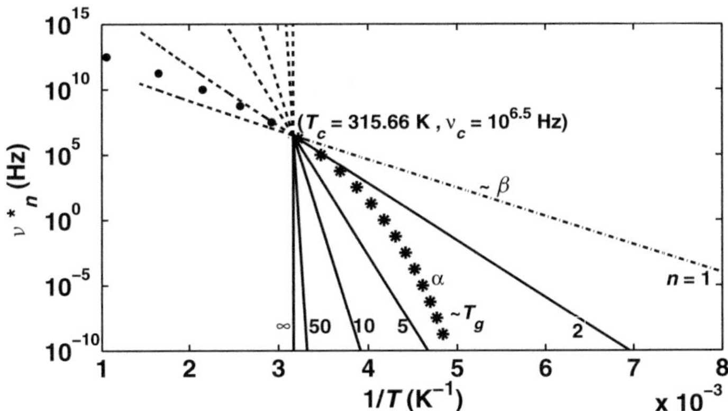

Fig. 8.3 Arrhenius diagram or relaxation map of the cluster transition frequencies, $v_n^*(T)$, at thermal equilibrium, for varying cluster sizes. *, •—rough estimates of transition temperatures (T_g, if below T_c) at varying frequencies, by the method outlined below in Section "Physical Interpretation of the Detailed Dynamics of Amorphous Materials and First Rough Estimates of Glass Transition Temperatures, Tg [14]" for $E_{a,1} = 40$ kJ, $v_1^\# = 10^{6.5}$ Hz, $z_{1,r}^\#/z_1 = 1$. Redrawn with data from [14], by permission of John Wiley and Sons

for complete specification. When the widest possible ranges of published experimental values for different materials [10, 35] are plotted in such a simplified map, what one obtains is the stem plot of Fig. 8.2, where polymers and small molecules are represented by different symbols (cf. caption). No additional data from [10, 35] could be plotted due to unavailable minimum (primitive relaxor) activation energy values.

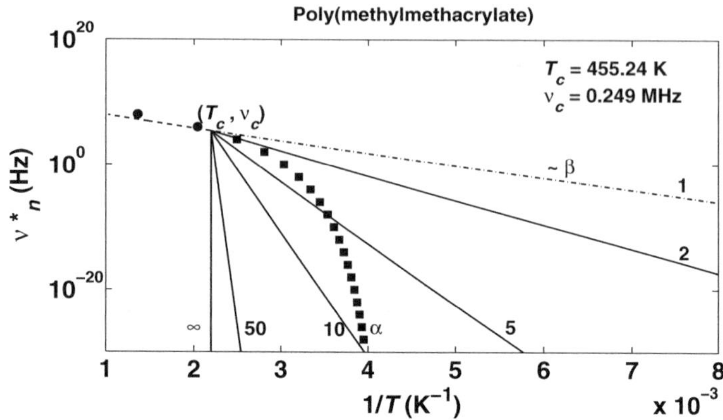

Fig. 8.4 Typical super-Arrhenius relaxation map (■—PMMA), calculated by the method outlined in Section "Physical Interpretation of the Detailed Dynamics of Amorphous Materials and First Rough Estimates of Glass Transition Temperatures, Tg [14]" from experimental stress relaxation data discussed in Chap. 11 [13], showing the cooperative nature of the estimated main α transition ($E_{a,1} = 35.65$ kJ/mol, $v_1^{\#} = 0.248$ MHz, $z_{1,r}^{\#}/z_1 = 0.0003$)

One may recognize that, with perhaps one exception (cf. the only particularly short □-stem datum hidden among the ○-stem ones), associated to a very soft material—poly (1,4-butadiene)—all polymer data, including our own PMMA's (Δ) and PC's (□) data, are clustered near higher crossover surface(s) than the small molecule, ○-stem, data. This seems consistent with likely higher localized stresses (and therefore lower $z_{1,r}^{\#}/z_1$ values) at the activated states of polymer transitions than at those of materials made of small molecules—also a possible indication of some intrinsic connectivity- and entanglement-induced cooperativity among the primitive relaxors of polymer materials.

Specification of a Numerical Example of CTMD's Application [14]

Let us assume, for example, $E_{a,1} = 40$ kJ mol^{-1} (corresponding to what might be expected for some flexible macromolecular materials allowing typical elementary crankshaft chain motions) and $v_c = v_1^{\#} = 10^{6.5}$ Hz, corresponding to an arbitrary middle-range crossover frequency [10, 35]. Equations (8.3) and (8.5) (in this particular example without the factor 2 multiplying the "sinh" function, which only slightly changes the numbers, not the qualitative behavior[23]) yields $T_c = 315.66$ K (cf. filled circle symbol on Fig. 8.1, which also shows the data calculated for the same PMMA and PC materials considered in Chap. 11) and 367.52 K, for $z_{1,r}^{\#}/z_1$ equal to 1 and 0.1, respectively. If we consider the usual bounds of the crossover frequency [10]—10^4 to 10^9 Hz—one would obtain, for the same $z_{1,r}^{\#}/z_1$ values, crossover temperatures

---

[23] [As indicated in the 3rd paragraph of Sect. 8.3.1].

between 232.51 and 260.04 K, and between 485.25 and 613.18 K, respectively. Similarly, higher/lower E_{a1} values will yield higher/lower crossover temperatures. The above variations quantitatively illustrate the effect of the structure's stiffness (through $E_{a,1}$—cf. (8.5), v_c and $z_{1,r}^{\#}/z_1$) on T_c (cf. also Fig. 8.1).

Study Question 8.1 Calculate all the above values of the crossover temperature, T_c, for the conditions specified.

First Theoretical Encounter with Super-Arrhenius Dynamics and Compensation Phenomena [14]

As well known, at the most common timescales of testing, the glass transition temperature, T_g, will normally lie somewhere below the crossover temperature, T_c, with $T_c \sim 1.2T_g$ [40], although slowly varying with the measurement's timescale, as recalled in Chap. 6. The region $T_g < T < T_c$ (where T_g may be as low as its lower limit at infinitely long experimental timescales, *if such limit exists*) is where MCT shows its weakness, and so no uncontroversial microscopic theory is available yet for $T_g < T < T_c$, and even more so for $T < T_g$, where CTMD may still yield valuable data. This will warrant detailed analysis in Chaps. 9 and 11.

Figure 8.3 shows the computed Arrhenius diagram of the cluster frequencies, $v_n^*(T)$ of (8.3), with the above values of $E_{a,1}$ and v_c, and $z_{1,r}^{\#}/z_1 = 1$, for n ranging from 1 to ∞, below (and, conjecturally, also above) the crossover. The individual cluster frequencies, $v_n^*(T)$, are almost, but not exactly, of the Arrhenius type (nearly straight lines), as it may in fact be checked (with a simple ruler) that those lines are slightly curved upwards at sufficiently high temperatures. This is because the associated apparent activation energies, $E_{a,n}$, from their definition, may be seen to increase with temperature according to

$$E_{a,n} = -k_B \frac{d \ln v_n^*(T)}{d(1/T)} = n\left\{ E_{a,1} + \frac{hv_1^{\#}}{2}\left[\frac{1}{\tanh\left(\frac{hv_1^{\#}}{2k_B T}\right)} - 1 \right] \right\}, \qquad (8.6)$$

which turns out approximately proportional to T at high temperatures—$E_{a,n} \approx n\left(E_{a,1} - \frac{hv_1^{\#}}{2} + k_B T \right)$.

Study Question 8.2 Calculate and plot the equilibrium cluster transition frequencies (solid and broken lines) of Fig. 8.3.

The physical justification for the actual occurrence of these highly cooperative (high n) and high activation energy ($\sim nE_{a,1}$) processes is that, as may be deduced from (8.3), and reminding that $\ln\left[v_n^*(T)\right]$ is proportional to the negative of a free energy of activation, the associated activation entropies are also proportional to n and, therefore, to the activation energies. So, processes that are energetically less favored turn out to be *entropically* favored (high $\Delta s_n^{\#} > 0$, resulting in high frequency factors, depending on the temperature), in agreement with the usual interpretations of the so-called *compensation phenomena* [41–45]. This of course means that cooperativity does help energetically less favorable processes, the most obvious manifestation of *dynamic facilitation*. The crossover temperature, T_c, as defined in this theory, therefore appears homologous to a *compensation temperature*.

A relevant note is on the physical likelihood or not of any high n frequencies above the crossover (but not at or below T_c), represented by the broken lines of Fig. 8.3. In fact, it may not be expected that transitions and vibrations might occur at frequencies exceeding those of the structure's uncorrelated transitions (corresponding to $n = 1$). All broken lines at $T > T_c$ in Fig. 8.3 for $n > 1$ would of course then have to be discarded. A detailed discussion of this matter must, however, be postponed until all physical consequences of both alternative hypotheses are derived and evaluated, but the reality of those somewhat "convulsive" cooperative transitions at high temperatures (and frequencies $v_n^* > v_1^{\#}$) may be questioned, inasmuch as T_c might be close to where significant structural changes may occur, such as melting.[24] The existence of multiple paths for activation and transition might be, in our opinion, one possible reason for the occurrence of such high frequency transitions.

Figure 8.4 plots the corresponding cluster frequencies for a PMMA with data derived in Chap. 11 from stress relaxation experiments, in addition to their approximate relaxation maps calculated as for the * curve of Fig. 8.3, following the procedure detailed in the next section.[25]

Physical Interpretation of the Detailed Dynamics of Amorphous Materials and First Rough Estimates of Glass Transition Temperatures, T_g [14]

The first obvious challenge the theory must face is to interpret a super-Arrhenius overall physical behavior from the above individual, nearly Arrhenius, elementary transitions or modes, for each of the values of n—and this has already been graphically illustrated in Figs. 8.3 and 8.4 and is explained now. An exact treatment would require the relative weighing of the above individual cluster modes and the conventional formulation of the specific response dynamics of each and all of them as first-order, Debye, processes, as described later in Chap. 9. A first very crude evaluation of the predictions of the theory for the *dynamics of the glass transition* may nevertheless be easily carried out, without such detail. The problem is to approximately quantify how the measured *glass transition temperature*, T_g, will vary with the dynamic mechanical, dielectric, or other frequency, v, at which the material may be tested.

---

[24] Will such "convulsive" transitions, if real, influence actual rates of melting?

[25] In [18], its Fig. 13.1 shows the corresponding curves for a PC obtained in a similar fashion, from that material's experimental stress relaxation data (cf. Chap. 11).

First of all, for a given test frequency, ν (in Hz) $< \nu_c = \nu_1^{\#}$, the temperature at which $\nu_1^* = \nu$ (calculated using the last of (8.3)—cf. Figs. 8.3 and 8.4) may be associated to Angell's [46] *ergodicity-making temperature*, T_{EM}, when the material is heated, i.e. the temperature at which the various cooperative modes start being activated, one after the other, from $n = 1$ up (towards $n = \infty$). The crossover temperature, T_c, may likewise (also in agreement with MCT interpretations) be associated to the *ergodicity-breaking temperature*, T_{EB}, when the material is cooled, i.e. the temperature at which the same various modes start freezing or being de-activated, but now from $n = \infty$ down to $n = 1$. This agrees with the usual dynamic definition of the crossover temperature. The various activation/de-activation temperatures, $T_n(\nu)$, at which $\nu_n^*(T) = \nu$, may all be accurately calculated (from 8.3) by direct numerical iteration on

$$T_n(\nu) = -\frac{E_{a,1} - h\nu_1^{\#}/2}{k_B \ln\left\{ 2\left(\frac{\nu}{\nu_1^{\#}}\right)^{1/n} \sinh\left[\frac{h\nu_1^{\#}}{2k_B T_n(\nu)}\right] / \left(\frac{z_{1,r}^{\#}}{z_1}\right)\right\}}, \tag{8.7}$$

which may be recognized to obey the same required derivative condition of its right-hand-side as (8.5).

Between the above two temperature limits (T_{EM} and T_{EB}), should lie the material's T_g at the imposed test frequency, and it may be recognized that the width of the transition decreases with ν (cf. Figs. 8.3 and 8.4), becoming very sharp near ν_c and T_c. Pending a much more accurate evaluation, and assuming at this instance that the transition's energy absorption peak would be very roughly symmetrical, T_g (if located at the peak's maximum) could therefore be very crudely estimated as the arithmetic average of T_{EM} and T_{EB}, awaiting the more accurate calculation procedures presented in Chap. 9.

The results of these very crude first estimates are also plotted on Figs. 8.3 and 8.4 (* and ■ symbols), for $\nu \leq \nu_c$. The familiar shape of a *super-Arrhenius relaxation diagram* [10] emerges, as well as the fact that the material is predicted to behave on average (based on the curve's approximate slope) as if clusters of 2 to > 10 (numerical example) or to ~ 50 (PMMA) primitive relaxors were, on average, the main effective contributors to the glass transition. Further, these estimates mean that, as the measurement timescale expands (i.e. ν is lowered), the transition temperatures decrease and the cooperativity levels increase. Figure 8.3 also includes the lines for $n > 1$ above T_c, whose interpretation (as already mentioned) may require close analysis at a later stage or be discarded. Their treatment as done here for $T < T_c$, to estimate a similarly relevant transition temperature, if at all sound and necessary, would yield the • curve data in any of the figures, possibly representative (or *not*, we stress) of the so-called A- and a-regions of the behavior [10]. To unequivocally decide about the validity of the • curve data, one would need to investigate the possibility of experimentally confirming or *not* that, for $T > T_c$, and $\nu > \nu_c$, the progressive activation/de-activation of the various clusters (in heating/cooling) could reverse its order relative to the one below T_c!

Therefore, possible reasonable propositions (based on these first, very crude, estimates for the assumed material structure and dynamics) may thus be that (1) the * and ■ curves might approximately represent the cooperative α-*process*, for which the apparent activation energy and the cooperativity level increase as the temperature is lowered, and (2) the curve for $n = 1$, by its definition, illustrates the behavior of uncorrelated primitive relaxors, but might also be approximately representative of *localized relaxation modes* or of the Johari–Golstein β-relaxation [47, 48], at least speculated to be a true and universal precursor of cooperativity [11, 49, 50], as predicted here by CTMD, which may thus contribute to identify its disputed origin and mechanism. Any other β-*process* relaxation plots[26] may also be more or less close to this line, depending on their exact nature, as determined by the detailed material's structure (e.g. side or in-chain groups vs. primitive relaxors like crankshaft motions, in the case of polymers). These preliminary, only crude, theoretical predictions, including the often-called α-β *bifurcation* at T_c, already resemble what is experimentally found for a wide range of materials [10].

The experimental detection of two absorption peaks, namely by dielectric relaxation [48], in various amorphous (mainly polar) materials, corresponding to two (α and β) relaxation maps, might also be accommodated by CTMD, by considering either (1) specific localized motions of low activation energy, $E_a \leq E_{a,1}$, or (2) that the material might have one set of their primitive relaxors surrounded by larger free volume(s) than the remaining relaxors, resulting in "some islands of mobility in an otherwise frozen matrix", in the exact words of Johari and Goldstein [48]. This possibility might also be explored within CTMD, but only when specific free volume effects will be detailed and included in future analyses.

The still conjectural high frequency behavior and the relative positioning of the proposed α, β, a and A-process traces in the Arrhenius diagram (Figs. 8.3 and 8.4) resemble those experimentally observed by neutron scattering and dielectric spectroscopy (cf. Figs. 2.9, p. 39 and 6.1, p. 379 of [10]). What now appears clear is that a significant change in dynamics (from non-cooperative, Arrhenius, to cooperative, *super-Arrhenius*) is theoretically predicted as the temperature is lowered below the crossover temperature, T_c, as defined here, in qualitative agreement with known experimental behavior, as discussed in Chap. 11. Another important observation is that CTMD succeeds in predicting that *slowness* (cf. the low values of ν_n^{*}[27]) and *heterogeneity* (cf. the wide range of n values) manifest themselves even at *equilibrium* (remembering that ν_n^{*} are *equilibrium frequencies*), a feature pointed out in [51] as also characterizing amorphous condensed matter behavior.

Based on the above predictions, one may reasonably conjecture the possibility of this theory providing a sound basis to model temperature-dependent amorphous condensed matter behavior for a wide range of timescales, much beyond those accessible to both direct experiments and molecular dynamics simulations, as documented by the ordinate values of Figs. 8.3 and 8.4 and the extreme ease and speed of all

---

[26] [Or even of γ or δ relaxations, as characterized by their respective activation energies and entropies, which determine the slopes and the intercepts of the plots, respectively].

[27] Just 10^{-10} Hz already corresponds to 1 cycle in just over 317 years!

computations. As pointed out later in Sects. 9.8 and 10.2, contrary to MCT, CTMD predicts *no divergence of characteristic timescales* at some $T > 0$ K (and even less at T_c or $T_g < T_c$) as temperature is reduced. As may be checked by analysis of (8.3), none of the cooperative frequencies $v_n^*(T)$ become zero, and so no average characteristic time will strictly diverge, except at 0 K, a relevant physical prediction that may question the existence of a *thermodynamic glass transition*. Of course, the accuracy of any theoretical predictions for the virtually unlimited timescales[28] accessible to CTMD cannot possibly ever be directly checked (and may, in any event, be irrelevant), but one may expect that the predictions (1) might approximately describe well documented experimental behavior at much shorter, experimentally accessible, timescales, (2) at least might make physical sense at much extended timescales, which (3) have not been (might they ever be?!) replicated nor disproved by direct molecular dynamics simulations.

As an additional and extreme illustration of the possible universality and applicability of this theoretical framework to a wide range of systems, Chap. 13 (in Part III) tentatively deals with the dynamics of a simple and highly flexible chemical structure, namely cyclohexane.

First CTMD Analysis of the Response to Dynamic Temperature Scans [17]

The above characterization of the experiment's timescale, through a horizontal line in Figs. 8.3 and 8.4 at a given frequency, is not applicable to *heating/cooling experiments*, where no other dynamic stimuli are imposed on the material. This situation, however, may also be analysed and depicted in an Arrhenius diagram. The condition that applies in heating/cooling experiments at the activation/de-activation of any elementary mode of characteristic time $\tau_n^* = 1/(2\pi v_n^*)$ may be expressed by $\frac{d\tau_n^*}{dt} = \mp 1$, respectively. The physical meaning of the above equality is that, at the time and temperature at which the mode is activated in heating, the process's characteristic time may be assumed to decrease at exactly the same rate at which actual time increases (-1 at the right-hand-side of the above equality); likewise, in cooling experiments, the de-activation may be reasonably assumed to occur at a time and temperature at which the process's characteristic time increases at the same rate as actual time. These conditions correspond to maximum/zero response (relaxation or susceptibility) rates of change for each given mode n, in heating/cooling experiments, respectively. The above conditions are equivalent to $\frac{d\tau_n^*}{dT} = -|r|^{-1}$, where r is the temperature scanning rate, dT/dt, and so the question is to determine what will be the *equivalent frequency path* followed by the material in its Arrhenius diagram,

---

[28] Within CTMD, the only limits to the timescales are those linked to the numbers' representation in the numerical processor's memory, not any specific short or long physical timescale limit. All computations leading to the data in Figs. 8.3 and 8.4, and all others in Part II of this book, were performed with conventional 2.2×10^{-16} precision mantissas and exponents between -308 and $+308$ such as to ensure 15–16 significant digits. Further, the actual *computation times*, of the order of *seconds* on any PC, *do not depend on the physical timescale* considered! Will this be possible by molecular dynamics simulations?

during a heating/cooling experiment at a given rate. If the last of (8.3) is differentiated with respect to temperature, one obtains,[29] for the scanning rate that activates/deactivates each individual mode, n, at temperature T, for $z_{1,r}^{\#}/z_1 = 1$,

$$|r|_n = \frac{2\pi v_n^*(T)}{n} k_B T^2 \left\{ E_{a,1} + \frac{h v_1^{\#}}{2} \left[\frac{1}{\tanh\left(\frac{h v_1^{\#}}{2k_B T}\right)} \right] \right\}^{-1}. \tag{8.8}$$

As physically expected, (8.8) yields fast increasing $|r|_n$ values with temperature, which may also be plotted in the same cluster frequencies diagram (dotted lines immediately above/to the left of the corresponding $v_n^*(T)$ lines, in Fig. 8.5). So, if we want to determine at which temperature, T_n, a given mode is activated/de-activated when the material is heated/cooled at rate r, we just have to make $|r|_n = r$ and solve (8.8) for $T = T_n$, the corresponding equivalent frequency being then obtained as $v_{eq} = v_n^*(T_n)$. Of course, $T_1(|r|)$ will be the ergodicity-making temperature, $T_{EM}(|r|)$, at the specified heating rate. If these calculations are accurately carried out for the same numerical example, and the results plotted in Fig. 8.5, which only differs from Fig. 8.3 in the different, zoomed-in, linear temperature and log-frequency scales for improved readability, one may obtain the curved equivalent frequency paths ($\bigcirc$, $\triangle$ and $\square$ symbols) followed by the material during thermal scans at 0.01, 0.1 and $1\,\mathrm{K\,s^{-1}}$, respectively, which cover the range of commonly used thermal scanning rates in differential scanning calorimetry. It may be seen that those equivalent frequency paths rise very steeply near T_c.

The iterative, Newton–Raphson, numerical procedure to accurately calculate the locus of $\left(T_n, v_n^*\right)$ with $\frac{z_{1,r}^{\#}}{z_1} = 1$, at each of the scanning rates—the mentioned *equivalent frequency path*, may be established as follows:

1. Assume $T_n = T_i$ (initial iteration on temperature);
2. Calculate $v_i = v_n^*(T_i)$ (8.3);

$$2.\quad \text{Calculate } X_i = k_B T_i \left\{ E_{a,1} + \frac{h v_1^{\#}}{2} \left[\frac{1}{\tanh\left(\frac{h v_1^{\#}}{2k_B T}\right)} - 1 \right] \right\}^{-1};$$

3. Calculate $r_i = |r|_n(T_i) = \frac{2\pi v_i^*}{n} X_i T_i$ (i.e. using 8.8);

$$4.\quad \text{Calculate } \frac{dr_i}{dT} = \left(\frac{d|r_n|}{dT}\right)_{T_i} = \frac{r_i}{T_i}\left\{ \frac{n}{X_i} + \left\{ 2 + \frac{X_i}{4}\left[\frac{\frac{h v_1^{\#}}{k_B T_i}}{\sinh\left(\frac{h v_1^{\#}}{2k_B T_i}\right)} \right]^2 \right\} \right\};$$

5. Calculate $T_{i+1} = T_i - \frac{r_i - |r|}{dr_i/dT}$, for a chosen $|r|$;
6. Follow steps 2–6, until convergence is achieved to the desired precision (16 significant digits, in this case, without any difficulty).

---

[29] [Again, excluding the factor 2 in the denominator of the last of (8.3) and in all derived results (just for consistency with all example calculations and diagrams)]. In any application of the theory to describe actual response data of materials, this will have no influence, as the parameter $z_{1,r}^{\#}/z_1$ will then be adjusted to known or estimated properties, like T_c, $E_{a,1}$ and v_c, as shown in Sect. 9.2.4 (9.17).

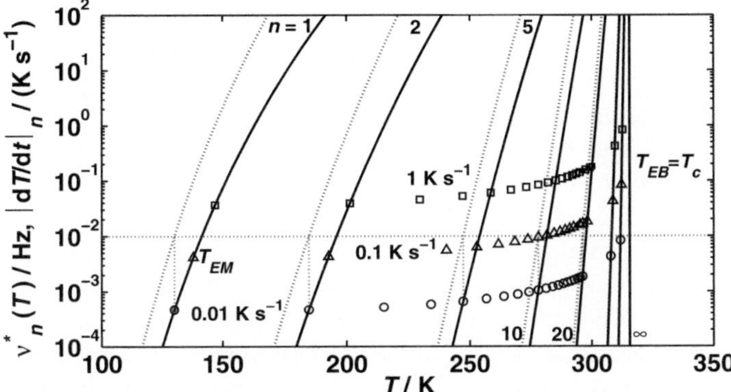

Fig. 8.5 Zoomed-in and re-scaled view of the results of Fig. 8.3 on a linear temperature scale, and the equivalent frequency paths at the temperature scanning rates of 0.01 (○), 0.1 (△) and 1.0 (□) K s^{-1}

The dotted horizontal and vertical lines in Fig. 8.5 provide a graphical illustration of the calculation for 0.01 K s^{-1}. The figure also shows all individual (T_n, v_n^*) calculated values for $1 \leq n \leq 20, 50$ and 100, for the three scanning rates considered.

It is important to note that the above equivalent frequency paths apply to both heating and cooling scans at the specified rates. This may seem strange, apparently ruling out the possibility of accurately predicting any hysteresis. That is not so, however. What happens is that the calculated equivalent paths are not even supposed to represent non-equilibrium behaviour; one must remember that the $v_n^*(T_n)$ are, as all $v_n^*(T)$, thermal equilibrium frequencies (in the same sense of Maxwell's molecular velocities in a gas), despite their association to some specific temperature scanning rate. The frequencies $v_n^*(T_n)$ represent just the last/first cluster frequency just being activated/de-activated during heating/cooling scans, respectively, at any of the rates and each specified T_n. The subtlety may be understood when one views those paths as those that would result if the heating/cooling were instantaneously stopped exactly at each of the calculated $T_n(|r|)$, the material allowed to reach full equilibrium (at least with respect to mode n), and then the scan re-initiated at the same rate. Obviously, this is in no way related to any actual non-equilibrium path. Such paths, and the associated hysteresis, might arguably one day be obtained by applying the methodology and procedure roughly outlined in the next section and in Chap. 14. In an uninterrupted controlled scan, what is physically expected and known to occur is that the material will evolve along a given thermal path, $T(t)$, always trying to reach local (and, obviously, continuously changing) equilibrium, characterized by a given full set of stabilized frequencies, $v_n^*[T(t)]$, with $1 \leq n \leq N_A$, for 1 mol of primitive relaxors.

Non-equilibrium will have to be quantitatively described by an incomplete set of active (and partly active) cluster transitions, as the temperature changes, where such sets are themselves expected to vary, at any one temperature, with the actual scanning rate. The detailed identification of those sets should be possible by taking into consideration that, to varying scanning rates, correspond different timescales, or time windows, within which one should detect the activity of some clusters (the fastest), while missing others (the slowest).

It is easily understood that, in this specific example, the crossover region cannot generally be probed by calorimetric techniques and will only be directly accessible using frequencies above 1 MHz. However, for studying the behavior at moderately low frequencies, down to 10^{-4} Hz, quantitative dynamic calorimetry (through accurate heat capacity measurements) could be one good choice for future tests of this theory. Still lower frequencies would require alternative experimental techniques. These considerations are important, as the most relevant feature of this theory is that, in principle, as already pointed out, it has no low or high frequency bounds, except any high frequency or temperature (and, of course, also forced excitation amplitude) structural durability limits.

The application of this theory to the actual variation of the internal energy and heat capacity (assuming, as above, constant volume conditions) will require tackling the challenge of its use in a meaningful and workable equilibrium (*) and non-equilibrium ($^{\pm}$) thermodynamic formulation of the contribution of the various cluster modes (for $n \geq 1$) to the total cooperative component of the molar internal energy, $\overline{U}^{*}_{coop}(T)$ and $\overline{U}^{\pm}_{coop}(T, r)$, respectively, as functions of temperature and, in the latter case, of the experimental timescale, for heating ($^{+}$) and cooling scans ($^{-}$) at rate $r = dT/dt$. The complete development is work for the future, but one possible strategy might be as tentatively outlined in the following section and in Chap. 14.

8.3.3 Clustering, CTMD and Thermodynamics—Glimpsing at a Future Equilibrium (ETG) and Non-equilibrium (NETG) Theory of Glasses

The reader may at this point genuinely question how the above TST-based treatment of dynamic clustering, where (as assumed) the individual molecular partition functions of *non-activated* clusters of size n are simply taken as the product of those of the n corresponding primitive relaxors, can possibly result in anything akin to an overall *loss of independence of the individual primitive relaxors* making up the whole system, which is what must be implied, and actually occur, in any realistic formulation of clustering. Will clustering, as modelled in Sects. 8.3.1 and 8.3.2, be just illusory and finally have no real effect whatsoever?

When the actual number of activated clusters of any size, n, is explicitly considered (implying that activated complexes, not only do indeed exist [27], but could in principle even be *counted*), one may write $N_n^{\#} = N_n e^{-\Delta a_n^{\#}/(k_B T)} = N_n v_n^{*}(T)/v_1^{\#}$ (cf. 8.2 and the two paragraphs that immediately follow it) and, when the total free energy of all activated clusters of that same size, $A^{\#}_{coop,n}$ (taken as representative of the total *cooperative* component of those clusters' Helmholtz free energy), is calculated relative to that of the non-activated component primitive relaxors, one obtains for the *average cooperative free energy per cluster of size n* (activated or non-activated alike, whose total number is N_n),

$$a^{*}_{coop,n} = \frac{A^{\#}_{coop,n}}{N_n} = \frac{N^{\#}_n}{N_n} \Delta a^{\#}_n = - k_B T \frac{v^{*}_n(T)}{v^{\#}_1} \ln\left[\frac{v^{*}_n(T)}{v^{\#}_1}\right], \qquad (8.9)$$

because $\frac{N^{\#}_n}{N_n} = \frac{v^{*}_n(T)}{v^{\#}_1}$ and $\Delta a^{\#}_n = - k_B T \ln\left[\frac{v^{*}_n(T)}{v^{\#}_1}\right]$.

This very subtle, unsuspected, but simple result enables us to define the *equilibrium logarithmic or geometric average cooperative* component factor of the *molecular partition function* of each cluster of size n (irrespective of its activated or non-activated state), *at equilibrium*—from $a^{*}_{coop,n} = - k_B T \ln z^{*}_{coop,n}(T)$, as[30,31]

$$z^{*}_{coop,n}(T) = \left[\frac{v^{*}_n(T)}{v^{\#}_1}\right]^{\frac{v^{*}_n(T)}{v^{\#}_1}}, \qquad (8.10)$$

which yields a value equal to unity at the crossover, T_c—suggesting a "correspondence" to a unique *dynamic* configuration[32] (all clusters synchronized, with $v^{*}_n(T_c) = v^{\#}_1 = v_c$ and $\Delta a^{\#}_n = 0$ at the compensation temperature, by its definition),

---

[30] This is the result when, as assumed, there will be only one transition path for each primitive relaxor, which may imply that no cooperative (just primitive relaxors') transitions will occur above T_c. These functions have their conventional functionality (as expressed by 8.9) and are indeed equivalent to their usual definition, as may be concluded from the following paragraph in the main text.

[31] This, we recall, is the result of assuming constant temperature, T, and volume, V (as in the canonical statistical ensemble of states). At constant T and pressure, P, one should instead consider the *isothermal-isobaric ensemble* and may expect that $z^{*}_{coop,n}(T)$ would be replaced by $z'^{*}_{coop,n}(T) = \int z^{*}_{coop,n}(T)e^{-\frac{Pv_n}{k_B T}} g(v_n)dv_n$, where $v_n = v_{o,n} + v_{f,n} = n(v_{o,1} + v_{f,1})$ (subscripts "o" and "f" denoting the occupied and free volumes—cf. Sect. 6.10 and Chap. 15) and $g(v_n)$ the volume distributions of the various clusters of n primitive relaxors, which may be taken approximetely uniform such as to simplify to $z'^{*}_{coop,n}(T) \neq z^{*}_{coop,n}(T)e^{-nP(v_{o,1}+v_{f,1})/(k_B T)}$, with $v_{f,1}$ identical to the equilibrium average value defined in Chap. 15 of the free volume of each primitive relaxor, $v_{f,\infty} = v_{f,1,0} - (2f_{u,0} - 1)v^{\#}_1$ (for the case where $E_{u,r} = E_{r,u}$), $v_{f,1,0}$ and $v^{\#}_1$ being the average initial free volume and the average activation volume per primitive relaxor, respectively, and $f_{u,0}$ the initial fraction of unrelaxed primitive relaxors. In $v_{o,1}$, we would need to account for the thermal expansion of the occupied volume. This correction of the $z^{*}_{coop,n}(T)$ would also require being carried throughout the treatment of non-equilibrium of Chap. 14, thus greatly increasing its computational complexity. The foregoing correction(s) will finally result in the replacement of the Helmholtz free energies $\overline{A}^{*}_{coop}(T)$ and $\overline{A}^{\pm}_{coop}(T)$ in the scheme of Fig. 14.1 by the corresponding Gibbs free energies, $\overline{G}^{*}_{coop}(T)$ and $\overline{G}^{\pm}_{coop,i}(T, r)$

[32] It also means that, at T_c, there is no decrease nor increase of the number of significantly populated molecular states relative to the non-cooperative ones (vibrational, rotational, etc.), here designated by the subscript "coop", if one considers that $z^{*}_{total,n}(T) = z^{*}_{coop,n}(T)z^{*}_{coop,n}(T)$.

and also at 0 K (because $\lim_{x=0} x^x = 1$)—now definitely meaning a completely *frozen* configuration, with no motions of any kind, as all ν_n^* are exactly 0 at, *and only at*, 0 K—cf. (8.3). "*Equilibrium*" means here, we remind, an *ideal infinitely long timescale*, so that even the lowest cluster frequencies might be accurately sampled, no matter how low the temperature might be.

The strange form of the cooperative partition functions just defined requires justification as to their equivalence to their usual definition.[33] If we go back to the second equality of (8.3), we see that $\frac{\nu_n^*(T)}{\nu_1^\#} = \frac{z_n^\#}{z_1^n} e^{-E_{a,n}/(k_B T)} = z_n^{\#\prime}(T)$ may be identified as the average partition function of each *activated cluster* of size n, *relative to the non-activated level* (if one notes the addition of the activation energy $E_{a,n}$ at 0 K—i.e. the difference between the zero-point energies of the activated and non-activated clusters—to all lower energy values, and the division by $z_1^n \sim z_n$). So, $\left(z_n^{\#\prime}\right)^{N_n^\#}$ is the partition function of the whole set of activated clusters of size n,[34] and $\left(z_n^{\#\prime}\right)^{\left(N_n^\#/N_n\right)}$ no other than the *logarithmic or geometric average* (at equilibrium) *of the cooperative component of the partition function per cluster* of the same size n, activated *and* non-activated, relative to the same level (i.e. the level that includes all lower, mainly vibrational, energy states), such that $z_{coop,n}^*(T) = \frac{z_{total,n}^*(T)}{z_{coop,n}^*(T)} = \left(z_n^{\#\prime}\right)^{\left(N_n^\#/N_n\right)}$—cf. also Footnote 33, where $z_{coop,n}^* \equiv z_{n,vib+etc.}'$, used in Sect. 8.2. Now, recalling that, as both $z_n^{\#\prime}(T)$ and $\frac{N_n^\#}{N_n}(T)$ are equal to $\frac{\nu_n^*(T)}{\nu_1^\#}$, we then obtain the same partition functions of (8.10).

Using the cluster frequencies of (8.3) (plotted in the examples of Figs. 8.3 and 8.4), these partition functions may themselves be plotted against temperature for various cluster sizes, n, as shown in Fig. 8.6 (for the data of Fig. 8.3), and all of them obviously differ from $\left(z_{coop,1}^*\right)^n$, except at $T = T_c$ and $T = 0$ K, given that

$$\left(z_{coop,1}^*\right)^n = \left[\frac{\nu_1^*(T)}{\nu_1^\#}\right]^{\frac{n\nu_1^*(T)}{\nu_1^\#}} = \left[\frac{\nu_n^*(T)}{\nu_1^\#}\right]^{\frac{\nu_1^*(T)}{\nu_1^\#}}$$

$$\neq z_{coop,n}^* = \left[\frac{\nu_n^*(T)}{\nu_1^\#}\right]^{\frac{\nu_n^*(T)}{\nu_1^\#}}, \tag{8.11}$$

---

[33] As $z_n^\#$ has all its energies relative to the ground or zero-point level of an *activated cluster* [#] of size n, they need the addition of the activation energy at 0 K [1] to obtain the partition function of the same *activated cluster*, $z_n^{\#\prime}$, relative to the zero-point state of its non-activated counterpart. And, by dividing $z_n^{\#\prime}$ by the partition function of the mentioned *non-activated cluster* relative to the same zero-point energy, $z_{coop,n}'$, one then obtains the (partial) *cooperative partition function* (or factor) *of each activated cluster*, i.e. $z_{coop,n}^* = z_n^{\#\prime}/z_{coop,n}'$, where in $z_{coop,n}' = z_n$ only the non-activated clusters' vibrational (and other non-cooperative) energies are included.

[1] This requires, and is equivalent to, multiplying $z_n^\#$ by the negative exponential of the activation energy at 0 K divided by $(k_B T)$.

[34] [Because distinguishable, *effective* (i.e. operating), clusters of any size are indeed independent from all other clusters (including those of the same size, otherwise the effective size of the cluster would be greater than n), thus justifying the factorization just made].

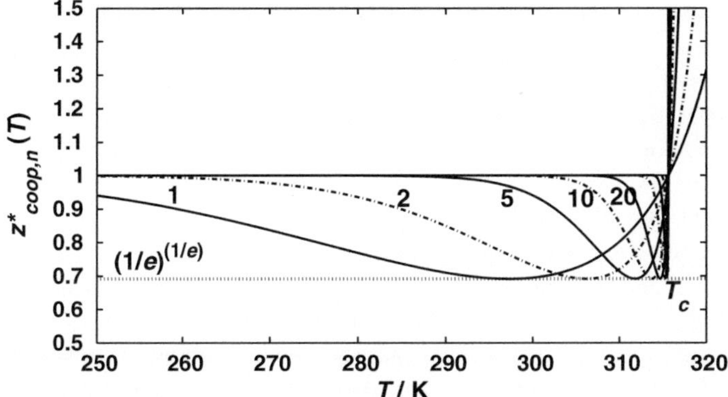

Fig. 8.6 Equilibrium cooperative partition functions, $z^*_{coop,n}(T)$ of the numerical example, for the indicated cluster sizes or n values, as functions of temperature

because $v^*_n(T) < v^*_1(T)$ for all $n > 1$ at $T < T_c$ (and because $v^*_n(T) > v^*_1(T)$ when $T > T_c$, but in these conditions we may have to discard all cluster frequencies except those of the primitive relaxors and, of course, the corresponding extended curves in Fig. 8.6). Therefore, the Inequality (8.11) proves beyond doubt that *clustering* corresponds, and effectively leads to, *loss of independence* of the whole set of primitive relaxors[35] that make up the material (for which their residual, non-cooperative—vibrational and other—degrees of freedom included in $z^*_{coop,n}(T)$ may be assumed to keep their original independence), as soon as random or assisted synchronization of their individual transitions occurs.

One should note that (in line with the development leading to 8.3), if we consider the partition function of the activated clusters only (cf. the expression for $\Delta a^{\#}_n$), it will reduce to the exact product of the functions of their constitutive primitive relaxors. However, the activated clusters of each size do not represent the whole set of (activated and non-activated) clusters of that same size, as these are distributed over the whole range of energies below those of the activated states down to the zero-point energy, and that is why the denominator in (8.9) is N_n (the total number of activated and non-activated clusters of each size), not $N^{\#}_n$ (just the number of the activated ones). Also, the frequencies $v^*_n(T)$ are *averages* over *all* clusters of each size, irrespective of their state.

Given the meaning of any *molecular partition function*—whose numerical value for a canonical ensemble is often taken as a rough measure of the number of molecular states that are significantly populated, one should recognize that, according to CTMD (cf. Fig. 8.6), the effect of clustering below the crossover temperature always appears one of reducing the average complete molecular partition function of any cluster (or distinguishable combination) of primitive relaxors relative to its complete non-cooperative partition function, and so clustering actually "steals" degrees of

---

[35] [As any of the primitive relaxors may locally participate in clusters of any size].

freedom and energy from the system,[36] thus being unfavorable to (i.e. freezing) materials' responses at any temperature below T_c ($z^*_{coop,n} < 1$)—what else? Above the crossover the opposite might arguably happen, in case CTMD's conjectural hypothesis of cooperative frequencies $v^*_{n>1}(T > T_c) > v^*_1(T > T_c) > v^{\#}_1 = v_c$ turn out correct, corresponding to an acceleration of all physical responses relative to those if primitive relaxors were the only contributing structures. This highlights the significant different dynamics below and above the crossover, making T_c a dynamically *critical* temperature, even at equilibrium.[37]

From (8.10) and Fig. 8.6, we see that, for temperatures well below T_c, one finds $z^*_{coop,n>1}(T \ll T_c) > z^*_{coop,n=1}(T \ll T_c)$, with the difference increasing with n and all $z^*_{coop,n>1}$ approaching 1, meaning that the negative effect of clustering is being reduced as the temperature is lowered or cluster size increased, i.e. clustering is being statistically (though not kinetically) favored, but a reversing of the above relationship is obtained as T_c is approached, until all cluster partition functions become exactly equal to 1 at T_c, where clustering becomes statistically indifferent with all $\Delta a^{\#}_n = 0$, except for the effect of the reduced numbers of large clusters within the system relative to that of uncoupled primitive relaxors, as we will see from the treatment of Sect. 9. 1. Above T_c, if and where "supercritical" frequencies might occur, we recover the above inequality for $T \ll T_c$ (corresponding now to a kinetic facilitation by increased activation entropies, $\Delta s^{\#}_n$, with $\Delta a^{\#}_n < 0$), but its reality and physical meaning, as previously pointed out, will depend on the validity or not of the cluster frequencies above the crossover. If valid, the corresponding cluster partition functions mean that clustering continues to be favored, but the overwhelming numbers of the smallest clusters and primitive relaxors take over the material's average response, as we will also see in Chap. 9. This would be consistent with the much-reduced slope ($n \sim 1$ or 2) of the effective (average) relaxation maps illustrated in Figs. 8.3 and 8.4 above T_c for the estimated α transitions.

Another curious observation from the above formulation of this numerical example, that only became evident after obtaining the plots of Fig. 8.6, is that all the $z^*_{coop,n}(T)$ show the same minimum value for a specific temperature that increases with cluster size. The reader should be able to locate that minimum[38] at such conditions that make $\frac{v^*_n(T_{z_{min},n})}{v^{\#}_1} = e^{-1}$, where all $z^*_{coop,n}(T_{z_{min},n})$ will in fact become identical and equal to $(1/e)^{1/e} = 0.6922006\ldots$, at a temperature that may be obtained for each cluster size, either graphically from the relaxation maps illustrated in Figs. 8.3 and 8.4, or (as with 8.5 and 8.7) by direct numerical solution of

---

[36] That could also explain the loss peaks that are obtained in any forced (out-of-equilibrium) dynamic (mechanical, dielectric, etc.) excitations.

[37] *If and where* "supercritical" frequencies (at $T > T_c$) might occur, we have met the theoretical possibility of the above freezing effect starting even above T_c! Under forced excitations (the subject of Chap. 9), however, one may expect a decreased freezing effect on the response.

[38] In working out the minimum, the reader will easily conclude that the derivative of $z^{\#}_{coop,n}(T)$ is also zero at 0 K, where all $\frac{dv^*_n(T)}{dT} = 0$ *and* also $v^*_n(T) = 0$. This means that all motions freeze very slowly indeed, as 0 K is approached—very much slower than at higher temperatures, e.g. below T_c or even T_g. Figure 9.13 in Chap. 9 illustrates that behavior.

$$T_{z_{min},n} = f\left(T_{z_{min},n}\right) = -\frac{E_{a,1} - h\nu_c/2}{k_B\left\{-\frac{1}{n} + \ln\left[2\sinh\left(\frac{h\nu_c}{2k_B T_{z_{min},n}}\right)/\left(z_{1,r}^{\#}/z_1\right)\right]\right\}}. \tag{8.12}$$

$T_{z_{min},n}$ is the temperature at which the greatest negative effect of clustering should be observed for each n value, and it is logical that such temperature should increase with cluster size. However, the effect is then reduced at lower temperatures and this suggests, if the formulation is right, that such relative *statistical facilitation* by clustering may now result from a *negative dynamic facilitation* (i.e. clustering by slowness)—a logical effect, explainable by increasingly slower transition rates, i.e. more significant progressive *freezing* on cooling (what else?) of the corresponding clusters, at temperatures $T < T_{z_{min},n}$[39] increasing with cluster size.

We may thus have in these systems both *positive dynamic facilitation* of clustering above $T_{z_{min},n}$, below (and arguably above) T_c by the *compensation effect* of larger activation entropies, and *negative dynamic facilitation* below $T_{z_{min},n}$ by overriding *motion freezing*. This may seem paradoxical, but both types of *facilitation* are relative to the vibrational behaviour of uncorrelated primitive relaxors. This will be true of all such large-scale motions, not only of clusters, but also of individual (uncorrelated) primitive relaxors, because, as already suggested, one expects that their vibrational energy might be partly "stolen" to help feed the activated transitions of all participating structures—clusters *and* primitive relaxors alike. That may be why *all* $z_{coop,n\geq1}^{*}(T)$ turn out less than 1 for all temperatures below the crossover.

It should be noted that, although $z_{coop,n}^{*} = 1$ at T_c may suggest as meaning *one single dynamic configuration* (and it does!), what happens is that, at T_c, the negative effect of clustering on the number of purely vibrational accessible states (or degrees of freedom) is exactly compensated by the positive effect on the number of new cooperative modes brought about by the clustering processes. This further highlights the critical importance and meaning of the crossover temperature, T_c. It is just below T_c that, as already suggested, the cluster modes start to "steal" more vibrational degrees of freedom and energy than those brought about by the newly generated cooperative modes themselves, because the latter become progressively slower, involving larger and larger clusters that require higher activation energies. Below T_c, the loss of independence of the primitive relaxors (cf. Inequality 8.11) entails partial loss of purely vibrational degrees of freedom and energy. However, as we have seen, these effects, after growing within a range of temperatures below T_c and above a cluster size-dependent $T_{z_{min},n}$, then attenuate below each $T_{z_{min},n}$, by increased "freezing" of the corresponding clusters and loss of their own cooperative degrees of freedom, thus leading to growing "savings" of the vibrational ones.

---

[39] After all, the equilibrium cluster partition functions directly depend on the dynamics of the system, through the $\nu_n^{*}(T)$, and remember that, as 0 K is approached, the material should behave as a set of reducing numbers of larger and larger slow clusters, until the whole system will end up as a unique, completely frozen one, at 0 K.

The mentioned minima of $z^*_{coop,n}$, where $v^*_n(T_{z_{min},n}) = v^{\#}_1/e$ correspond to having $\Delta a^{\#}_n = k_B T$, i.e. free energies of activation just matched by the available thermal energy. So, the increases of $z^*_{coop,n}$ above and below $T_{z_{min},n}$ mean, in the first case, that the negative effect of clustering is strongly attenuated and even reversed due to the more significant acceleration of the cooperative transitions than those of the isolated primitive relaxors at the same temperatures (given that $\Delta a^{\#}_n = n\Delta a^{\#}_1$), with sufficient thermal energy being available to feed those transitions without significant loss of vibrational degrees of freedom, while in the second case that negative effect might be attenuated at decreasing temperatures by the lengthening of the periods of time—proportional to $\left(v^*_{n>1}\right)^{-1}$—between successful synchronization of (mainly adjacent[40]) primitive relaxors and their participation in less and less frequent large scale motions relative to the vibrations and other motions of smaller clusters and single primitive relaxors. This may explain the larger negative slopes and higher ordinates of the $z^*_{coop,n}(T)$ curves for larger cluster sizes, below the respective $T_{z_{min},n}$ temperatures.[41]

The above partition functions values less than 1 were initially unexpected but were justified in the foregoing discussion. Their emergence within this theory warrants the following summarizing comments:

(1)   Their values do not invalidate the basic assumption of the possibility of factorizing the overall partition function into the non-cooperative and cooperative components.[42] We have seen that the cooperative components emerge as simple attenuation factors of the non-cooperative ones, meaning that they "steal" some of the non-cooperative degrees of freedom and energies to feed the new cooperative motions within the material, and the extent to which this happens will depend on the temperature. That these cooperative motions must be related to, and be partly fed by, the underlying vibrational spectrum is no surprise, as $v^{\#}_1$ (and possibly other different frequencies within a narrow range) are part of that same spectrum.[43]

(2)   At equilibrium and very low temperatures, the effect is negligible (except the one due to primitive relaxors, as shown in Fig. 8.6), but it increases with temperature, reaching a maximum—corresponding to minimum partition function(s)—at a given temperature, at which the activation free energies are matched by the available thermal energy and is then gradually reduced as those cooperative motions are fed to greater extent by the growing available thermal energy. The behavior just described, however, assumes that the system will be allowed sufficient time to reach full thermal equilibrium.

(3)   Out of thermal equilibrium and/or under some dynamic forced excitation (i.e. under finite timescales), that transfer of degrees of freedom and energy in favor of the emerging cooperative motions will not be as easy, as these motions need

---

[40] [As assumed in Chap. 9].

[41] We nevertheless acknowledge the still conjectural nature of the foregoing discussion.

[42] [Also discussed in Chap. 14].

[43] The reader may here revisit Footnote 18.

long times to develop and stabilize. However, the forced (mechanical or electrical) and/or thermal excitation (the heat flux input) might be able to feed those specific motions at some temperature below the ones defining the maximum effect at equilibrium conditions (i.e. the various $T_{z_{min},n}$), whose values should of course increase with the rate of the excitation (temperature scanning rate and/or forcing frequency), to secure that the frequency of the cooperative motions matches the timescale of the observation.[44] This effect will then show up in whatever dynamic experiment or measurement being made of heat capacities or mechanical/dielectric responses, by the detection of the onset and development of those new degrees of freedom and cooperative motions via heat capacity changes or specific absorption peaks.

The possible future usefulness of the above individual cooperative or cluster partition functions is of course not guaranteed, but could be suggested by the following conjectural formulation of an equilibrium overall cooperative partition function, $Z^*_{coop}(T)$, for 1 mol of primitive relaxors, as[45]

$$\ln\left[Z^*_{coop}(T)\right] = N_A \ln\left[z^*_{coop,1,avg}(T)\right]$$

$$= N_A \sum_{i=1}^{i_{max}} \left[\frac{F_i'^*(T)}{i}\right] \ln\left[z^*_{coop,i}(T)\right], \qquad (8.13)$$

where $z^*_{coop,1,avg}(T)$ plays the role of an effective logarithmic average equilibrium cooperative partition function of each primitive relaxor,[46] and i_{max} such that $F'_{i>i_{max}}(T) < \textit{machine precision}$. Except when $F_1'^* = 1$ and $F_{j>1}'^* = 0$, i.e. in the absence of any clustering, $Z^*_{coop}(T)$ differs from $\left[z^*_{coop,1}(T)\right]^{N_A}$ (because $z^*_{coop,i}(T) \neq \left[z^*_{coop,1}(T)\right]^i$, as already shown), where N_A is Avogadro's number and the $F_i'^*$ might be the weighing factors formulated in Sect. 9.1, or future better alternatives. In the above equality, the $F_i'^*(T)^{47}$ will be formulated as the fractions of the maximum number of distinguishable participations of primitive relaxors in clusters of all sizes that do so in clusters of size i in

---

[44] And that is why, for example, T_g is always below T_c and even most of the $T_{z_{min},n}$ of Fig. 8.6 and, as well established by experiment, T_g and other transition temperatures always increase as the timescale is shortened (or testing speed increased).

[45] The previously mentioned *statistical facilitation* (of whatever kind) through the $z^*_{coop,i}$ values at some temperatures may of course be changed and eventually even reversed through the values of the statistical weights $F_i'(T)$, to be considered in Chap. 9.

[46] The ratio $z^*_{coop,1,avg}/z^*_{coop,1}$ might arguably measure an overall *"loss of independence"* of the primitive relaxors, due to their partial clustering.

[47] The problem of the formulation of the statistical weights F_i' is considered in Chap. 9 (for situations of mechanical or other forced response), and it will be shown that the weighing must be proportionately to the total times each cluster's "activated state" or efective transition process is "visited" by the system. In the above equilibrium case at any given temperature, those weights, $F_i'^*(T)$, should accordingly be proportional to $\left[\left[v_n^*(T)\right]\right]^{-1}$, where of course $v_n^*(T)$ are the equilibrium cluster frequencies formulated at the beginning of this chapter.

1 mol of primitive relaxors of the material at equilibrium, i.e. when the observation time scale is much longer than any of the response times, and all system states and configurations are readily accessible and sampled. Their division by i is here required to convert $F_i'^*(T)N_A$ to the corresponding number of clusters of each size i. Once more, we stress that the $Z_{coop}^*(T) \neq \left[z_{coop,1}^*(T)\right]^{N_A}$ inequality ensures that the system will definitely not be equivalent to a set of N_A independent or uncorrelated primitive relaxors, not even at equilibrium at whatever temperature (cf. also Section "Physical Interpretation of the Detailed Dynamics of Amorphous Materials and First Rough Estimates of Glass Transition Temperatures, Tg [14]" and [51] on the emergence of dynamical heterogeneity also at equilibrium).

From the above $Z_{coop}^*(T)$, if correct, one might hope to calculate the *cooperative part* of the free and internal energies, entropies, heat capacities, etc., over and above (or below) the non-cooperative, e.g. vibrational, ones—for which something equivalent to, and including, Debye's heat capacity formulation is expected to apply, using conventional statistical physics, in what might eventually stand as a workable and realistic, even if only approximate, *equilibrium theory of glasses* (ETG).

One additional important problem to be solved will be, for each specific case, the chemical identification of the effective primitive relaxor, as the above suggested analysis (if successful) can only yield values of the cooperative components of relevant thermodynamic functions *per mol of primitive relaxors*, whatever their nature, which may nevertheless illustrate (or relate to) the intended universality of CTMD.

The above cooperative components, adequately modified for situations out of equilibrium, will be the ones that should determine the response dynamics, glass transition, etc. of the material, i.e. all aspects of its physical behaviour that depend on large scale motions at the molecular scale. A full *non-equilibrium* theory of glasses (likely corresponding to a reduced number of accessible states/configurations) is of course still several very long steps farther from CTMD's present stage of development, but a possible strategy is suggested in Chap. 14. That chapter also presents some more specific notes on the *quantum mechanics and statistical physics of amorphous condensed matter*, aiming at providing additional basis for the content of this section, and for the proposed strategy of tackling the future development of an *equilibrium* (ETG) *and non-equilibrium* (NETG) *theory of glasses.*

One is thus justified to state that *clustering*, as modeled here by CTMD, effectively leads to *loss of independence* by the participating primitive relaxors and also relative to any other distinguishable clusters (to an extent that depends on temperature), because clustering of primitive relaxors into any specific cluster size occurs *intermittently* in competition with, and at the expense of, their uncorrelated response, as well as of their association in clusters of any other different sizes.

To summarize in a convenient itemized way:

(1) The primitive relaxors are *independent prior to cooperative* ($n > 1$) *activation,* and after single ($n = 1$) activation. N' of these independent activated *and* non-activated primitive relaxors will have an overall (or joint) *average* partition function factor, relative to $\left[z^*_{coop,1}(T)\right]^{N'}$, given by $Z^*_{coop,1}(T) = \left[z^*_{coop,1}(T)\right]^{N'}$, with $z^*_{coop,1}(T) = \left[\frac{v_1^*(T)}{v_1^\#}\right]^{\frac{v_1^*(T)}{v_1^\#}}$, from (8.10) with $n = 1$,[48] but each of the $N_1^\# < N'$ activated single primitive relaxors will "acquire" a partition function factor $z_1^{\#\prime}(T) = \frac{v_1^*(T)}{v_1^\#}$ (from the expression of $\Delta a_1^\#$—in the second of 8.9) relative to the same $z^*_{coop,1}(T)$, while each of the remaining $\left(N' - N_1^\#\right)$ non-activated ones will keep $z^*_{coop,1}(T)$ as its complete partition function.

(2) *Upon joint activation,* by definition, the primitive relaxors that jointly activate *gain independence relative to all other remaining primitive relaxors.*

(3) As a result, *all primitive relaxors loose their independence overall,* each time any cluster of any number $n > 1$ of primitive relaxors (a) forms and (b) activates. Their overall cooperative partition function factor (for 1 mol of them) *becomes $Z^*_{coop}(T) \neq \left[z^*_{coop,1}(T)\right]^{N_A}$*, given by (8.13), which takes into account all cluster sizes and their objective distinguishability (through their simple factorization), and also their intermittency (arguably through the $F_i^{\prime*}$ formulated at the beginning of Chap. 9).

(4) The primitive relaxors that make up each activated cluster may also be described as *independent,* not only from all other clusters and primitive relaxors, but (*surprise!*) now also among themselves, for the same peculiar physical and mathematical reason why the frequencies in any Einstein, Debye or real crystal are independent, despite it being subject to standing waves propagating through the whole crystal. In the clusters case, such (assumed) waves will only extend to the boundaries of each cluster and will assist the clustering process itself, in competition with other waves coming from neighboring clusters or from the independent transitions of neighboring primitive relaxors. This may stand as the "materialization" of the recent concept of "clustrons" [34] as opposed to that of "phonons" (which extend to and propagate through the whole material). The logarithmic average cooperative partition function factor of each of the these activated clusters of size n is $z^{*\prime}_{coop,n}(T) = \left[z^*_{coop,1}(T)\right]^n = \left[\frac{v_1^*(T)}{v_1^\#}\right]^n = \frac{v_n^*(T)}{v_1^\#} = z_n^{\#\prime}(T)$, as we found in the paragraph following that of (8.10).

(5) Except for $z^*_{coop,1}$, all the above partition functions are relative to their non-cooperative components, i.e. prior to the activation of all clusters or primitive relaxors involved, and so they all emerge as attenuation factors of their non-cooperative components below T_c (or, arguably, expansion factors above T_c, when multiple activation paths might be possible). All the above would apply

---

[48] Activation of just *single* primitive relaxors does of course *not* destroy their independence; only clustered activations will do it.

to the situation where *clustering might be purely statistical*, as assumed so far in the development of CTMD. Should other factors influence clustering (such as specific physical entanglements of any nature), the loss of independence will be even clearer but will warrant separate future analysis.

An important note must be made: all statistical physics formulations in this chapter assumed *constant volume*. This will have a strong bearing when, in Sects. 9.8 and 9.11 and Chap. 11, we will discuss the relative importance of temperature and molecular packing on the materials' response dynamics.

References

1. M. Wolfgardt, J. Baschnagel, W. Paul, K. Binder, Phys. Rev. E **54**, 1535 (1996)
2. J. Baschnagel, M. Wolfgardt, W. Paul, K. Binder, J. Res. Natl. Inst. Stand. Technol. **102**, 159 (1997)
3. M. Pyda, B. Wunderlich, Macromolecules **32**, 2044 (1999)
4. G.P. Johari, J. Chem. Phys. **265**, 217 (2000)
5. L. Sante, W. Krauth, Nature (London) **405**, 550 (2000)
6. G.P. Johari, J. Non-Cryst. Solids **288**, 148 (2001)
7. F.H. Stillinger, P.G. Debenedetti, T.M. Truskett, J. Phys. Chem. B **105**, 11809 (2001)
8. M. Pyda, B. Wunderlich, J. Polym. Sci. Phys. Ed. **40**, 1245 (2002)
9. S.L. Simon, G.B. Mckenna, J. Non-Cryst. Solids **355**, 672 (2009)
10. E.-J. Donth, *The Glass Transition: Relaxation Dynamics in Liquids and Disordered Materials* (Springer-Verlag, Berlin Heidelberg New York, 2001)
11. K.L. Ngai, *Relaxation and Diffusion in Complex Systems* (Springer, New York Dordrecht Heidelberg London, 2011)
12. W. Götze, *Complex Dynamics of Glass-Forming Liquids—A Mode Coupling Theory* (Oxford Science Publications, Oxford, 2009)
13. J.J.C. Cruz Pinto, J.R.S. André, Polym. Eng. Sci. **56**, 348 (2016)
14. J.J.C. Cruz Pinto, A cooperative theory of amorphous materials dynamics, in *Online Supporting Material for Reference* 10 (2016)
15. K.L. Ngai, J. Non-Cryst. Solids **353**, 709–718 (2007)
16. T.F. Schatzki, J. Polym. Sci. **57**, 496 (1962)
17. J.J.C. Cruz Pinto, A.I.P. Conf. Proc. **982**, 452–457 (2008)
18. J.J.C. Cruz Pinto, J.R.S. André, Cooperative molecular and macroscopic dynamics of amorphous polymers—a new analytical view and its potential, in *Processing and Characterization of Multicomponent Polymer Systems—New Insights*, ed. by J. James, S. Thomas, N. Kalarikkal, Y. Weimin, K. Pal, Chap. 13 (Apple Academic Press, Taylor & Francis, 2019), pp. 237–260
19. S. Blonski, W. Brostow, J. Kubát, Phys. Ver. B **49**, 6494 (1994)
20. M. Warren, J. Rottler, Phys. Rev. E **76**, 031802 (2007); Phys. Rev. Lett. **104**, 205501 (2010)
21. M.L. Ferrer, C. Lawrence, B.G. Demirijan, D. Kivelson, D.C. Alba-Siminionesco, G. Tarjus, J. Chem. Phys. **109**, 8010 (1998)
22. D. Walton, *Abductive Reasoning* (The University of Alabama Press, Tuscaloosa, 2014)
23. H. Eyring, J. Chem. Phys. **3**, 107 (1935); K.J. Laidler, *Reaction Kinetics: Homogeneous Gas Reactions* (Pergamon Press, 1963)
24. S. Glasstone, K.J. Laidler, H. Eyring, *The Theory of Rate Processes* (McGraw-Hill, New York, 1941)
25. I.N. Levine, *Physical Chemistry*, 5th edn. (McGraw-Hill, New York, 2003)
26. A. Nitzan, *Chemical Dynamics in Condensed Phases—Relaxation, Transfer, and Reactions in Condensed Molecular Systems* (Oxford University Press, Oxford, 2006)

27. J.C. Polanyi, A.H. Zewail, Acc. Chem. Res. **28**, 119 (1995)
28. A.H. Zewail, Sci. Am. Dec. **76** (1990); J. Phys. Chem. A **104**, 5660 (2000)
29. D. Jiulin, Phys. A **391**, 1718, 2930 (2012)
30. Y. Cangtao, D. Jiulin, Phys. A **395**, 416 (2014)
31. A. Keys, A. Abate, S. Glotzer, D. Durian, Nat. Phys. **3**, 260 (2007)
32. L. Berthier, G. Biroli, J.-P. Bouchaud, L. Cipelletti, W. van Saarloos, *Dynamical Heteregeneities in Glasses, Colloids and Granular Materials* (Oxford University Press, Oxford, 2011)
33. G. Biroli, J.P. Garrahan, J. Chem. Phys. **138**, 12A301 (2013)
34. I. Pócsik, Phys. A **201**(1–3), 34 (1993)
35. M. Beiner, H. Huth, K. Schröter, J. Non-Cryst. Solids **279**, 126 (2001)
36. U. Bengtzelius, W. Götze, A. Sjölander, J. Phys. C Solid State Phys. **17**, 5915 (1984)
37. E. Leutheusser, Phys. Rev. A **29**, 2765 (1984)
38. W. Götze, L. Sjögren, Transp. Theor. Stat. Phys. **24**, 801 (1995)
39. W. Götze, L. Sjören, Rep. Progr. Phys. **55**, 24 (1992)
40. H.J. Sillescu, J. Non-Cryst. Solids **243**, 81 (1999)
41. A. Dufresne, C. Lavergne, C. Lacabanne, Solid State Commun. **88**, 753 (1993)
42. C. Lacabanne, A. Lamure, G. Theyssèdre, A. Bernes, M. Mourgues, J. Non-Cryst. Solids **172–174**, 884 (1994)
43. E. Marchal, J. Non-Cryst. Solids **172–174**, 902 (1994)
44. E.B. Starikov, B. Nordén, Chem. Phys. Lett. **538**, 118 (2012)
45. E.B. Starikov, Chem. Phys. Lett. **564**, 88 (2013)
46. C.A. Angell, Curr. Opin. Solid State Mater. Sci. **1**, 578 (1996)
47. M. Goldstein, J. Chem. Phys. **51**(9), 3728 (1969)
48. G.P. Johari, M. Goldstein, J. Chem. Phys. **53**(6), 2372 (1970)
49. K.L. Ngai, J. Chem. Phys. **109**(16), 6982 (1998)
50. K.L. Ngai, J. Chem. Phys. **120**, 857 (2004)
51. P. Harrowell, Phys. Rev. E **48**, 4359 (1993)

Chapter 9
CTMD's Calculation of the Materials' Responses to Forced Mechanical Stimuli

9.1 The Problem of the Distribution of Response (Relaxation and Retardation) Times

The specific objectives of this section have so far been the most difficult ones to pursue in the development of CTMD. The paragraphs that follow summarize the main steps followed in the line of thought, which, despite their provisional character, are leading to reasonable and encouraging results.[1] Paths to be followed in required future updates will be considered in Chap. 12.

The procedure started with conventional and conceptually simple combinatorial arguments because clustering was first assumed to be a random association of primitive relaxors, as defined and explained in Chap. 8, but physically expected to involve mostly or even exclusively *adjacent* and perhaps in some degree "entangled" primitive relaxors (whatever the entanglement mechanism that might be operating). In fact, any sporadic synchronization of the transitions of non-adjacent primitive relaxors would probably occur by mere coincidence—not as the result of (difficult to explain) action at distance—and would not be physically and statistically distinguishable from independent and uncorrelated transitions by single primitive relaxors at their specific average frequency ν_1, to be formulated below in Sects. 9.2 and 9.3 (or ν_1^* if at equilibrium, as already calculated in Chap. 8).

---

[1] We refer to (1) the CTMD's predictions of response time spectra with approximate log-normal contours (when there is just one single type of primitive relaxors), where (2) both the average and breadth increase as the temperature is lowered, (3) such as to give the VTF and WLF types of temperature dependence that are experimentally found—cf. Sect. 9.8—, and (4) which yield values of various properties (glass transition temperatures, crossover temperatures and frequencies, fragility indexes, etc., as well as stress relaxation moduli from the above properties and approximate spectra (cf. Chap. 11), in encouraging approximate agreement with experimental values for amorphous polymers.

J. J. Cruz Pinto and J. R. dos Santos André, *Analytical Molecular Dynamics of Amorphous Condensed Matter*, Springer Series in Materials Science 342, https://doi.org/10.1007/978-3-031-56517-5_9

9.1.1 First Combinatorial Analysis of Clustering

In a first stage of the argument, the clusters' relative weights *per primitive relaxor*, F_k—in the whole set of the primitive relaxors (at equilibrium), crossing the resisting area (in stress relaxation), or along a given reference length (in creep)—were taken proportional to the total number of distinguishable combinations of k individual *adjacent* primitive relaxors containing any given one of them, $f(k)\dbinom{N_A}{k}/N_A$, multiplied by k (assuming that each individual primitive relaxor will, in any situation, have the same well defined maximum contribution to the overall physical response), where N_A is Avogadro's number (assuming 1 mol of primitive relaxors) and $f(k)$ is the fraction of those combinations or clusters that are exclusively made of adjacent primitive relaxors. Each of the k primitive relaxors of any given cluster will, at equilibrium, be "vibrating" at the same synchronized frequency $v_k^*(T)$ given by (8.3), to which may also correspond a yet to be determined specific individual contribution to the mean internal energy and other thermodynamic properties,[2] as well as to the maximum, final, physical response of the material (in stress relaxation, creep, etc.) close to, or more or less far from, equilibrium or linear viscoelastic conditions, in which case the above frequency will be modified depending on the excitation. It will therefore be useful to refer all calculations to each one of the material's primitive relaxors (in the bulk, crossing the resisting area, or along a given reference length), as set out at the beginning of this paragraph.

In probabilistic terms, $\mathcal{P}$ standing for *"probability"*, what we want to estimate is $\mathcal{P}$ (*participation in clusters of k <u>adjacent AND somehow physically entangled</u> primitive relaxors per primitive relaxor in the system*), taking as reference the total number of possible clusters of whatever size (including $k = 1$), in which each individual primitive relaxor may participate, "imagining" for now that all clusters of different sizes would have identical statistical weights,[3] apart from the further multiplication by k, already mentioned in the preceding paragraph. This probability may therefore be expressed as the product.

$\mathcal{P}$ (*combination of k <u>adjacent</u> relaxors per primitive relaxor in the system among all possible combinations of any number of primitive relaxors*) $\times$ $\mathcal{P}$ (*actual physical <u>entanglement</u> of k adjacent primitive relaxors*), the second of these factors being a conditional probability (i.e. of effective *entanglement* or *clustering given the adjacency*).

Should the adjacency of any k primitive relaxors be a sufficient condition for clustering at the specific frequencies given by 8.3 or should the above conditional entanglement probability be constant and uniform for all clusters of adjacent primitive relaxors of any size, the formulation would be reduced to that of the first factor of the above product. That is what is assumed at this initial stage, leaving a discussion

---

[2] [Important in a possible future development of a thermodynamic equilibrium theory of glasses (ETG)].

[3] [Which obviously they have *not*, because clusters of different sizes will not become active with equal frequencies, as discussed and taken into account further below].

on the possible need to allow for some specific entanglement-assisted clustering to Chap. 12 and possible future work.

G. Tarjus, in Chap. 2 of *"Dynamic Heterogeneities of Glasses, Colloids and Granular Media"* (OUP, Oxford, 2011), makes the suggestion that a possible fruitful strategy to fully understand glassy behavior could (and perhaps should) involve devising models that might exaggerate collective behavior, i.e. cooperativity, to enable proving or disproving their asymptotic predictions. We entirely agree and hold the view that, as needed, such collective features might afterwards be scaled down and calibrated until acceptable agreement is reached with real, experimental, behavior. This is the strategy adopted here, to be evaluated and given its first, very conjectural, adjustments later, in Sect. 11.9.3 and Chap. 12.

$\binom{N_A}{k}$ being the number of distinguishable combinations of any kind of k (adjacent *and* non-adjacent) primitive relaxors, $k.\binom{N_A}{k}/N_A$ is the number of distinct participations by primitive relaxors in those combinations, per primitive relaxor in the bulk, crossing the resisting area (in stress relaxation), or along a given length (in creep), i.e. $(N_A - 1)!/[(k - 1)!(N_A - k)!]$.[4] One would then need to estimate $f(k)$. For that, let us consider any *one* of the primitive relaxors to which this calculation is referred (taken as, say, an "anchor" in the present calculation) and any *one* particular of its own assigned dense clusters with other $k - 1$ *adjacent* relaxors. Next, let us try to calculate the total number of possible different ways in which those $k - 1$ adjacent relaxors might ideally be substituted by the $N_A - 1$ relaxors other than the "anchor" (by direct exchange, one by one, by any order, with any of those $N_A - 1$ other relaxors), to generate (and be replaced by) all other possible clusters with the same number of adjacent or non-adjacent primitive relaxors that will include the "anchor" one. Formally, the above procedure may appear to at most yield a crude approximation to the objective of determining the reciprocal of $f(k)$, because one would be apparently neglecting all non-dense clusters where only a fraction of those $k - 1$ primitive relaxors were replaced by others external to the "anchor" cluster considered. However, it is shown in Appendix A that, for large systems, the above apparent approximation may be upheld.[5]

Under the validity of the above approximation, the reciprocal of $f(k)$ would then be $(k - 1).(N_A - 1).(k - 2).(N_A - 2)\ldots 1.\frac{[N_A-(k-1)]}{(k-1)!} = (N_A - 1)!/(N_A - k)!$, only valid for $k \geq 2$,[6] where the division by $(k - 1)!$ in the left-hand-side of the above equality ensures that all possible permutations of the $k - 1$ primitive relaxors

---

[4] This considers that the primitive relaxors may be taken as uniformly distributed among all combinations of the same size, which may raise questions for small systems, where such uniform distribution is not expected to hold near the systems' boundaries. Therefore, a safe assumption here will be of an extended material sample (very large N_A, for example, 1 mol).

[5] To avoid breaking this discussion with further lateral and lengthy combinatorial arguments, this matter is left to Appendix A to check and confirm the final result.

[6] Primitive relaxors ($k = 1$) are, of course, compact "clusters" of just one primitive relaxor, for which $f(k = 1) = 1$.

initially adjacent to the "anchor" will in the end count as belonging to one and the same "shuffled" cluster. From the above result, one would obtain $F_k \propto kf$ $(k)\binom{N_A}{k}/N_A \propto 1/(k-1)!$[7] which, making $\sum_{k=1}^{N_A} F_k = 1$, would lead (recalling that N_A is a very large number) to[8]

$$F_k = e^{-1}/(k-1)!. \tag{9.1}$$

The simplicity of the above result may emerge as a surprise, but it might be helpful to note that (1) it predicts lower weights or probabilities for larger clusters of adjacent relaxors and that (2) such probabilities may be expected to vary proportionately to the reciprocal of the number of different ways of "dispersing" or "de-clustering" the cluster's k relaxors, $k!$, e.g. by random uncorrelated vibrations or local motions of varied nature, thus yielding $k/k! = 1/(k-1)!$ for the F_k, as previously obtained, after accounting for the additional proportionality of the physical responses to the cluster sizes, k.

The above scheme to estimate $f(k)$ just considers the fraction of clusters of adjacent relaxors relative to the total number of clusters with the same number of primitive relaxors (adjacent or not), without giving any additional, physically based,[9] statistical weight to adjacency. Towards the end of Chap. 11 and in Chap. 12, this result will warrant a discussion as to whether it may be representative or not of effective (not just potential) clusters exclusively made of adjacent and somehow entangled, primitive relaxors, in cases where clustering (instead of being random, as so far assumed) might be driven or assisted by some unspecified *structural entanglements*.

Apart from any later refinements that may (and most certainly will) prove necessary, the above F_k could be taken as very first (trial) estimates of the fraction of the maximum possible number of distinguishable *potential* participations or transitions of individual primitive relaxors that are able to do so within clusters of k adjacent relaxors. They would not count any of the primitive relaxors more than once for each k-relaxor cluster but would still count each of them multiple times in relation to their participation, at *distinct times*, in *clusters* of any *other* sizes. All these physically distinct clusters or combinations of adjacent primitive relaxors do provide separate positive or negative contributions to the system's response, by means of transitions that occur at randomly *distinct times*, involving *materially distinct fractions* of the system, leading to the so-called *dynamic heterogeneity* that characterizes condensed

---

[7] Check it as an exercise, remembering that $f(k = 1)$ must be equal to 1, and so also F_1 prior to the normalization of the F_k, which will then yield (9.1).

[8] This entire procedure, including the one used to estimate $f(k)$, ensures that no cluster is counted more than once, because (1) all combinations included in $\binom{N_A}{k}$ and $\binom{N_A}{k}/N_A$ are, by definition, distinguishable, and (2) if the above calculation procedure were repeated taking as reference any other primitive (and "anchor") relaxor, the resulting "shuffled" clusters would differ from any other similarly obtained by at least their reference and anchor primitive relaxor.

[9] [E.g. by taking into account physical entanglements of whatever nature].

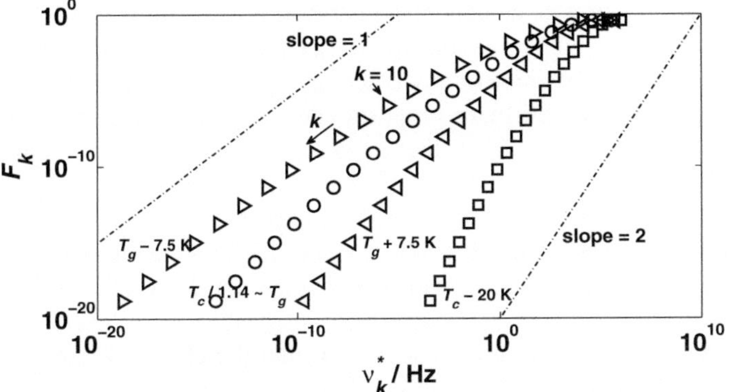

Fig. 9.1 First (trial) estimate of the cooperative frequency distribution below the crossover and its variation with cluster size, k, and temperature, taking $T_g \sim T_c/1.14$ (as noted in the next subsection, Fig. 9.2, and Footnote 16)

matter behavior, whose investigation has been considered the forefront of the theoretical and experimental effort in this field [1]. How to account for this dynamic heterogeneity will be considered in the next subsection.

It may be observed that the F_k just obtained decrease rapidly with k and, as the ν_k^* also decrease with k below the crossover temperature (cf. Figs. 8.3 and 8.4), they would yield a frequency or state density distribution, $F(\nu)$, that increases with ν, as expected at low frequencies.[10] At the crossover temperature, $T = T_c$, as all ν_k^* equal the crossover frequency, ν_c, the above distribution becomes infinitely sharp. Figure 9.1 shows plots of the F_k values obtained with (9.1) with the ν_k^* calculated by (8.3), for the same numerical example leading to Fig. 8.3, for k between 1 and 21, as a function of frequency, which illustrates the behavior just described for four temperatures below the crossover. The variation of the F_k with the ν_k^* is nearly exponential, with a temperature-dependent exponent. The significant temperature dependence of this partial, cooperative, frequency spectrum is itself a very first logical and encouraging result.

9.1.2 Accounting for Dynamic Heterogeneity

The above cluster weighing scheme cannot be final. The F_k values just obtained, if correct, would quantify the *total potential instantaneous "composition"* of the system (or the integrated real one over all possible times), irrespective of its role via the clusters' effective participation or not in actual transitions at random times. The aspect that still needs quantitative consideration relates to the discussion two paragraphs above, and is the fact that, once a given individual cluster yields its positive

---

[10] We will nevertheless see that these $F(\nu)$ are not yet final.

or negative contribution to the final response, its primitive relaxors may, in principle, yield further negative or positive contributions, respectively, as part of other (distinct) individual clusters of the same or any other size. Thus, we propose that it should be necessary to further weigh them proportionately to their effective *persistence or active time*, or to the reciprocal of the frequency at which each individual cluster is "functionally dismantled" or disappears by actual participation in a transition, i.e. (as we suggest) proportionately to their response times, $\theta_k(T)$[11,12] in stress relaxation (to yield the relaxation spectrum) or $\tau_k(T)$ in creep (to yield the retardation spectrum), in order to obtain more reasonable estimates of their correct weights in the material's final response.

One may therefore suggest (or conjecture) that the cluster weights sought may in principle be approximately given by $F'_k(T) = F_k\theta_k(T)/\theta'_{avg}(T)$, with $\theta'_{avg}(T)$ as defined just below, i.e.

$$F'_k(T) = \frac{1}{e(k-1)!}\frac{\theta_k(T)}{\theta'_{avg}(T)}, \qquad (9.2)$$

with

$$\theta'_{avg}(T) = \sum_{k=1}^{N_A} \frac{1}{e(k-1)!}\theta_k(T). \qquad (9.2a)$$

The latter is akin to a (but *not* necessarily the) working average characteristic response time of the material, as explained later in this chapter. The above series summation may be easily carried out up to any adjustable temperature-dependent k_{max} (physically, a temperature-dependent *maximum "detectable" cooperativity level*[13]), such as to ensure a specified relative precision (for example, up to 1×10^{-2} of the full machine precision—2.2×10^{-16}—, as used in all our calculations) in the normalization condition $\sum_{k=1}^{k_{max}} F'_k(T) = 1$, as a perfectly acceptable approximation to $\sum_{k=1}^{N_A} F'_k(T) = 1$. The above proportionality to the θ_k turns out formally or at least functionally analogous to the textbook-classical response time spectra formulation change from a θ-dependent, $f(\theta)$, to the preferred $(\ln\theta)$-dependent spectral function, $H(\ln\theta)$, or $L(\ln\tau)$ –, such that $f(\theta)d\theta = H(\ln\theta)d\ln\theta$, with $H = \theta f(\theta)$, or $L = \tau g(\tau)$, in stress relaxation and creep, respectively. Bearing in mind that, as shown later in this chapter, the CTMD's θ_k turn out equally spaced on a log-scale (a feature that will later determine the definition of the system's effective average

---

[11] In view of the strict scale similarity of the (1st-order, Debye) response functions of every set of individual clusters (cf. Sects. 9.2–9.4), that "active time" may be taken proportional to those clusters' specific characteristic response times, $\theta_k(T)$, used everywhere in this section and in stress relaxation, or $\tau_k(T)$ in creep.

[12] [Or $\theta^*_k \propto 1/\nu^*_k$ at equilibrium].

[13] This theory presumes that any super-molecular scale may be arbitrarily large compared with the molecular scale, and thus follows a recent call (cf. Ref. [1]) for "exaggerating the relevant collective properties, so that asymptotically precise statements can be verified or falsified".

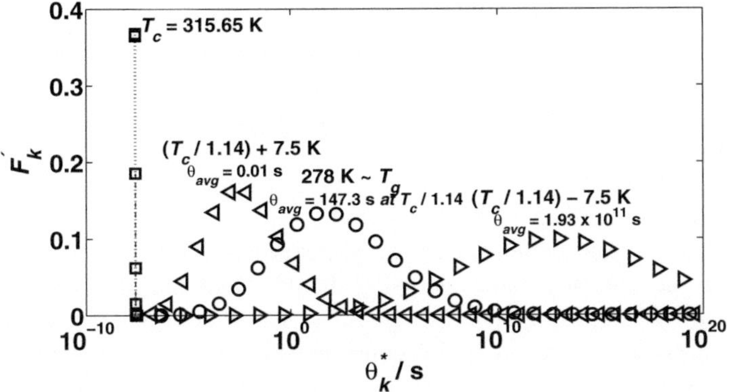

Fig. 9.2 Predicted discontinuous[15] relaxation spectra at equilibrium, or within the linear viscoelastic domain, for the Numerical Example specified in Chap. 8 (cluster sizes, k, are counted from the leftmost symbol at each temperature)

response time in Sects. 9.2–9.10 and 10.1), the F_k of (9.1) are analogous to $f(\theta)$ and the F'_k of (9.2) to $H(\ln \theta)$ or $L(\ln \tau)$, divided by conventional scaling factors related to the maximum relaxation modulus or creep compliance variations, respectively.

It should be stressed that the above significant $F_k\prime$ dependence on temperature (cf. (9.2)) indeed is a physical pre-requisite of any realistic attempt to describe the temperature-dependent cooperative nature of the behavior, and the above analysis does meet such requirement. Bearing in mind the form of the relevant first-order response functions,[14] this proportionality to the individual response times means that the fastest and most probable processes should turn out to be the least/most significant in determining the dynamic behavior for long/short timescales. As previously pointed out, the same primitive relaxors that may have already taken part (positively or negatively) in the response of the smallest and fastest clusters will still retain their ability to (now negatively or positively, respectively) participate, at any later times, in larger and slower clustered motions. The above physical arguments are entirely consistent with the recognized intermittency of activity of the primitive relaxors as parts of widely varying clusters, as previously pointed out. The actual mathematical relationships developed so far will most probably be only approximate and must of course be evaluated by comparing the resulting model predictions with actual experimental response measurements, which will be dealt with in Chap. 11.

Figure 9.2 shows the distributions calculated by (9.2) for the same Numerical Example used in Chap. 8 to illustrate the concepts and first predictions of CTMD.

---

[14] $[e^{-\frac{t}{\theta_k}}$ in relaxation or $1 - e^{-\frac{t}{\tau_k}}$ in susceptibility or compliance situations].

[15] It is the ordinates of the points that add up to 1, not the area under each curve.

The plots clearly illustrate the ability of this first model to predict a distribution with an approximate log-normal contour, spreading and increasing in cooperativity[16] as the temperature is decreased, as is theoretically expected and experimentally found. One may immediately note that these distributions, if approximately correct, do not lend any support whatsoever to past claims of rheological simplicity as the main reason behind time–temperature superposition. This specific matter will be dealt with and clarified in greater detail in Chap. 11.

These or any future better cluster weight expressions should of course also be applicable to non-linear conditions, using the adequate strain-dependent response times, $\theta_k(\varepsilon_0, T)$ in stress relaxation, where ε_0 is the constant applied strain, and $\tau_k(\sigma_0, T)$ in creep, where σ_0 is the constant applied stress. This will hopefully confirm the known dependence of the response time spectra on the level of excitation. Another note is that, within limits, and subject to specific validation, it might prove possible to apply this same or similar formalism to obtain the cluster weights for systems with $k_{max} \ll N_A$, for which $\sum_{k=1}^{k_{max}} \theta_k/(k-1)!$ or any corresponding summations may come out truncated at a relative value of $\theta_{k_{max}}/(k_{max}-1)!$ larger than the processor's precision, requiring an adjusted normalization condition for the F_k', because (despite the strongly decreasing F_k) the θ_k strongly increase with k. However, by contrast, the validity of the statistical mechanical formulation that led to the equilibrium cluster transition frequencies, v_k^*, may then legitimately be questioned for such small systems. At this stage, awaiting extensive validation of all consequences of this formulation, we should limit our expectations to an approximate description of such complex behavior for *large systems*. The approach and the abundant and specific predictions of CTMD in the remainder of this chapter and Chap. 11, despite their apparent reasonableness and encouraging results so far, do involve some physical conjectures,[17] and therefore risks, but are susceptible of and open to falsification, as required in any physical theory.

Highlight 9.1

In the expression of $F_k = \left[kf(k) \binom{N_A}{k} \right] / \sum_{j \geq 1} jf(j) \binom{N_A}{j}$, which led to (9.1), the numerator is the total number of distinguishable potential participations by primitive relaxors of the system[18] in compact clusters of k adjacent primitive relaxors, and the denominator is the total number of such potential participations in compact clusters of all possible sizes $(k, j \geq 1)$, and so F_k is the corresponding probability.

---

[16] Counting the number of primitive relaxors in each cluster as indicated in the Figure's caption, note that the average effective cooperativity changes from 6–7 to 9–10 between 278 ($\sim T_g$) and 270.5 K. Also note that the T_g value corresponds here to a timescale of ~ 150 s.

[17] [As the adopted definition of clustering, the corresponding relative probabilities and, less critically, also the reasonable but stringent assumption that active clusters may only (or predominantly) be made of adjacent primitive relaxors].

[18] The system taken as reference may be (*a*) e.g. 1 mol of primitive relaxors in the bulk, when describing thermodynamic properties, (*b*) the primitive relaxors crossing a given resisting area in tensile or shear stress relaxation, or (*c*) the primitive relaxors along a given reference length in the direction of the applied stress in tensile or shear creep.

Multiplying the numerator and all terms in the denominator's sum by the corresponding average times during which each complete set of clusters of any given size will be "active", it is assumed that one has then obtained F'_k—(9.2)—as a measure of the effective relative weight or "activity" of the clusters of the size specified in the numerator, in whatever response behavior we may be considering. Both F_k and F'_k result normalized, as required.

The three critical assumptions in this formulation are:

1. The $f(k)$, as calculated by $(N_A - k)!/(N_A - 1)!$, are accurate (or, at least, reasonable) measures of the fraction of compact clusters in the whole set of distinguishable combinations of k primitive relaxors.
2. The "activity times" of each set of clusters of any given size are directly proportional to their characteristic response times in the (equilibrium or non-equilibrium) situation being considered.
3. No other topological, entanglement or other factors are so far assumed to determine or interfere with the clustering process.

9.1.3 Further Remarks on the Cluster Size Distributions

Two additional notes should be made about this distribution of cooperative frequencies or response times:

(1) By its definition, $F(\nu)$ or $F'(\nu, T)$, where $\nu = 1/(2\pi\theta)$ or $1/(2\pi\tau)$, include only those frequencies that contribute to the glass transition and other associated phenomena, i.e. that directly relate to the observed changes in the structure's degrees of freedom at a molecular scale (responsible, for example, for the measured heat capacity changes when approaching the glass transition), through the onset or freezing of structure-specific cluster transitions, and so it is of course expected to differ from the overall frequency spectrum, which must include all types of motions and vibrations within the structure. In other words, as already pointed out in Sect. 8.3.3 and later in Chap. 14, the above frequency spectrum may be assumed to add itself to a vibrational background (object of Debye's or other more exact heat capacity formulation), and determines the cooperative component of the internal energies, heat capacities, entropies, etc., in principle obtainable from the corresponding overall cooperative equilibrium or out-of-equilibrium partial partition functions—$Z^*_{coop}(T)$ or $Z^{\pm}_{coop}(T, |r|, \ldots)$, respectively, if and when they might become accessible.[19]

(2) The spatial variability of the structural parameters E_{01}, $\nu^{\#}_1$ and $z_{\#,r}/z_1$ within any real material, especially E_{a1}, will result in spatial variations of the basic cluster frequencies and, therefore, one expects a smearing of the discontinuous spectrum defined by (9.2), together with the relevant frequencies. A continuous and somewhat wider spectrum will thus be the practical result, which might also be

---

[19] [Following, for example, the strategy tentatively outlined towards the end of Sect. 8.3.3 and in Chap. 14].

described in an improved theory. These local fluctuations will not only be spatial, but also temporal, and will assume increased relevance (higher probability) for smaller cluster sizes, where the individual theoretical ("pure") frequencies are wider apart (cf. Figs. 8.3 and 8.4), as the averaging effect of cluster size will tend to attenuate such fluctuations. The resulting continuous distributions should still be approximately consistent with a log-normal or truncated—somewhere near/below an average ν_1—log-normal frequency or characteristic time spectrum, as often assumed in relaxation or susceptibility studies (cf. Chap. 4).

We recall that it should be our long-term goal to be able to describe any aspect of the physical behavior of an amorphous material (at least at constant volume[20]), including the dynamics of the glass transition in controlled frequency experiments and, in the future, in simple heating and cooling scans, by formulating the material's cooperative component of the internal energy and other thermodynamic functions in terms of the frequencies $\nu_k^*(T)$. In the case of forced excitations—the subject of the next six main sections of this chapter—, what we will need to quantify in detail is the changes imparted on the above frequencies by the applied strain or stress, considering that such changes may and will generally be different for the two possible directions of the underlying transitions.[21]

As already pointed out, in obtaining the above statistical weights F_k and F_k', it is so far assumed that all primitive relaxors within any given cluster will have identical specific contributions (either all positive *or* all negative[22]) to the response, which was already explicitly taken into account in the imposed (additional) proportionality of the above weights to the clusters' size, k. Therefore, the maximum contribution to the stress relaxation modulus decrease by the complete set of k-primitive relaxor clusters (at $t = \infty$) will be taken as $F_k'(E_0 - E_\infty)$, where E_0 and E_∞ are the instantaneous and fully relaxed moduli, respectively, while the corresponding maximum contribution to the creep compliance increase will be $F_k'(D_\infty - D_0)$, where D_0 and D_∞ are the instantaneous and final creep compliances.[23]

Closing Notes

At the end of this section, it is important to stress that, given the difficulty of the subject, the novelty of the present proposals, and the assumptions made, the treatment may well in the future require significant update, despite the promising results illustrated in Chap. 11. Likewise, in the following presentation of stress relaxation

---

[20] The statistical treatment of clustering in Chap. 8 is valid at constant volume and temperature of the system.

[21] [Specifically, those leading to local stress reduction or increase, in stress relaxation, and strain increase or reduction, in creep].

[22] This is a very stringent and most certainly inaccurate condition, but in Appendix B the possibility of a (binomial) distribution of positive and negative contributions by the primitive relaxors within the same cluster will also be analyzed.

[23] This neglects viscous flow or assumes a very lightly crosslinked material. Sect. 9.5 will consider specific effects of crosslinking and partial crystallization.

and creep, the complexity of the microscopic behavior requires the adoption of a step-wise approach in this text and in Appendixes B and C, where the latter are intended to provide, in a complementary reading, a more accurate and complete picture.

In some physical research, *phenomenology* has somewhat unfavorable but arguably undeserved connotations. However, *good and sound phenomenology*— defined by the most accurate possible description and formulation of *real phenomena*, with the minimum number of physically significant parameters, all of them identifiable and (where possible) measurable *physical properties* (not "fitting constants"), capable of providing detailed "microscopic" insight – is of the utmost importance in all areas of physics and engineering. This is the view taken in the present text.[24]

9.2 Uniaxial Tensile Stress Relaxation [4]

In stress relaxation, the ideally instantaneous applied strain, ε_0, is immediately supported by the materials' molecular skeleton (which also instantaneously[25] becomes subject to corresponding local stresses, here assumed uniform[26] and equal to $E_0\varepsilon_0$, where E_0 is the instantaneous elastic modulus), and these stresses relax as local delayed elastic strains increase driven by those same local stresses, under the same overall strain (kept constant in time), because those increasing delayed elastic strains result in decreased real elastic strains in the molecular skeleton.[27] At all times, all local stresses (resulting from the corresponding local elastic strains) are of course also related to the latter through the instantaneous elastic modulus, E_0, characterizing the molecular skeleton. So, although in stress relaxation there is no external work done on the system from its surroundings, there are local positive/negative exchanges of stored elastic energies, which will decrease/increase the activation energy barriers that must be locally overcome by the system, to allow the local delayed elastic strain increases necessary to relax the local stress.

The line of thought of the next paragraphs and sections is also quite elaborate, and so a gradual (piecemeal) approach will be easier to follow. The initial strategy will therefore be to consider separately each type of contributors (primitive relaxors, first, and then any specific set of clusters of k primitive relaxors), as if there were no other simultaneous contributors[28] (or competitors) to the overall response of the material. The adoption of this strategy implies that, to obtain the final overall response, the

---

[24] Section 9.2 greatly expands on sections of Ref. [4] by permission of John Wiley and Sons (Copyright © 2016).

[25] [Not at all exact, as discussed in detail in Chap. 5].

[26] This assumes that no initial residual stresses are present.

[27] The reader should relate this behavior with that of the standard linear solid (modified Maxwell or modified Voigt-Kelvin) model, discussed in Sects. 3.1.1–3.1.4.

[28] The reader should also relate this to the discussion in various sections (for example, Sects. 8.3.3 and (9.1.2)) on the *independence* of each set or "family" of clusters of any given size relative to any other set of clusters of different sizes or primitive relaxors and on their *intermittency*, leading to *dynamic heterogeneity*.

individual contributions to the total stress by each set of clusters of size k, σ_k, taking the role of "cluster-private local stresses", must first be weighed by their respective F'_k (tentatively calculated in Sect. 9.1.2) and then added, to make $\sigma = \sum_{k \geq 1} F'_k \sigma_k$, as if the clusters that were first assumed to cover or cross (though many times over, really[29]) the entire resisting area effectively covered or crossed (also many times over) only a fraction F'_k of the total.

Let us then first consider the behavior in stress relaxation of the whole set of primitive relaxors crossing the unit resisting cross-sectional or tangential area (in tension or in shear, respectively) of a test sample of a given material, assuming for now that they are the only relaxors contributing to the response, with no clustering, and try to quantitatively formulate how they collectively relax the stress from an initial value of $\sigma_{1,0} = \sigma_0$, upon applying a given constant strain, ε_0, before generalizing for any set of clusters of k primitive relaxors, and then combining their responses weighed by the F'_k obtained in Sect. 9.1. We will be considering the simple case where there is only recoverable (on an infinite time scale, if necessary) delayed elastic or viscoelastic stress relaxation, i.e. without any contribution from irreversible translation or pure viscous flow, as in very lightly cross-linked amorphous polymers.[30] We will also be excluding cases where some other specific stress relaxation mechanism might be operating, such as covalent or other bond scission.

9.2.1 The Basis of the Formulation

Considering that, by definition, after each "relaxing transition" the molecular skeleton of each primitive relaxor will be in a state of a slightly reduced local stress (to be quantified below), one needs to consider:

(a) the exact initial numbers of *un-relaxed* and *relaxed* primitive relaxors, depending on the material's initial state, all of them initially subject to the same local stress, $\sigma_{1,0} = \sigma_0 = E_0 \varepsilon_0$ (if we neglect any prevailing residual strains/stresses—a critical but inevitable assumption at this stage), where E_0 is the material's instantaneous modulus, and ε_0 the applied constant strain;

(b) the specific activation energies of the (forward and backward) transitions between the two possible states of each relaxor, in both the unstrained and strained material (i.e. before and after ε_0 is applied, in the latter case as functions of ε_0);

(c) the total frequencies of forward and backward transitions of all relevant types as functions of time, t, together with the actual local stress decrease or increase after each of the above transitions.

E_0 characterizes what we may call the material's skeleton (considered Hookean at very short times), the strain ε_0 being initially supported mainly by increased

---

[29] [Due to the multiple distinguishable combinations of k primitive relaxors within the same resisting area].

[30] As already suggested, significant crosslinking or crystallization will require a modified analysis, as outlined further below, in Sect. 9.5.

intermolecular distances and (in polymers) possibly also by some contribution of increased covalent bond angles, particularly in cases of significant macromolecular orientation. During the stress relaxation process, ε_0 will then gradually transfer to an increasing number of relaxed (at constant overall strain), and a decreasing number of remaining un-relaxed, primitive relaxors crossing the resisting area (cross-sectional, in tension, tangential, in shear), as the stress reduces in proportion to the effective number of forward *minus* backward transitions.

The above cumbersome "un-relaxed and relaxed relaxor" phraseology may initially confuse the reader, but one way of avoiding it (as in recent stress relaxation [2] and creep [3] work on polymers) is by considering an analogous but purely ideal problem of the evolution of an assumed set of interchanging *"gauche"* and *"trans"* polymer chain conformers—just crossing in parallel the resisting area, in this case of stress relaxation. Such idealized single, uncorrelated, conformational transitions are of course totally excluded in bulk polymer materials, as they would require simultaneous motions involving the entire or large sections of the same and neighboring chains (a clear call for clustering!), but they may provide a much simpler analogous picture of how the set of primitive relaxors might behave, while not changing the essence of the physical argument, as we have done in Sect. 4.4.2. In fact, both the gradual reduction of the stress supported by the material's skeleton in stress relaxation at constant strain [2, 4] and the gradual chain unfolding in creep [3] are more easily visualized, described, and formulated by means of such (would be) *"gauche"* to/from *"trans"* localized transitions.

In the present specific case of stress relaxation, we have to consider by how much the local, non-affine, skeleton strain, ε', and stress, σ' (for whatever single type of relaxor sets—primitive relaxors, as first considered here, or clusters of any specific size), are reduced from the initial ε_0 and $\sigma_0 = E_0\varepsilon_0$ values, as a result of each forward transition (and by how much they are increased after each reverse transition), and this was first analyzed in some detail [2] for the mentioned simple set of *"gauche"*/ *"trans"* conformers. As soon as the final response of such idealized set was obtained (Eq. 8 in Ref. [2]), the conformers were completely discarded and forgotten, and the equation taken to represent the response of the set of actual primitive relaxors of whatever nature within the material. In reformulating that development [4], now assigning and referring it to the actual primitive relaxors (not to any *"gauche"* or *"trans"* conformers) or to any set of clusters of any given size, the scheme of Fig. 9.3 depicts the four possible states (of structure, stress, and strain) of those primitive relaxors or clusters at any given time t, and also the four types of transitions that may occur [4].

It should be clear that, locally within the material, the skeleton's local strains and stresses will decrease from u to r (from left to right along each solid arrowed line of the scheme, representing all possible direct, stress-reducing, and reverse, stress-increasing, transitions), and increase from subscript 1 to 2. It now remains to

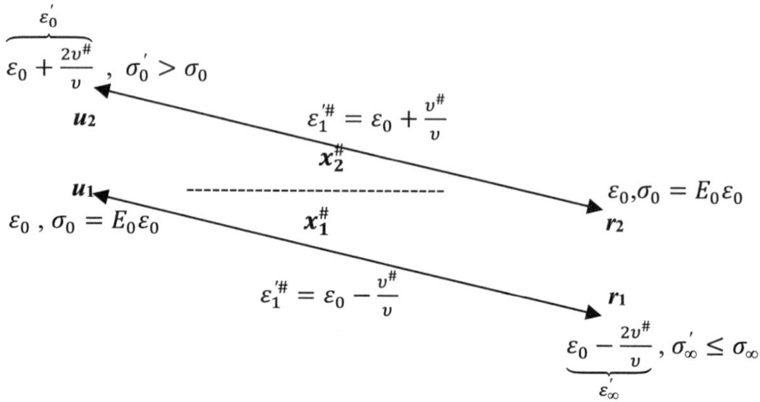

Fig. 9.3 Scheme of the four possible structural/stress/strain local states of the material's primitive relaxors or any set of clusters of the same size in a uniaxial tensile stress relaxation situation: u_1 and r_2 represent the initial un-relaxed and relaxed states, respectively (characterized by identical local strains and stresses—cf. connecting broken line), immediately after the strain ε_0 is applied, and r_1 and u_2 represent the ensuing relaxed and un-relaxed states of the initial u_1 and r_2 relaxors (characterized by decreased/increased local strains and stresses, respectively), after the occurrence of a stress decreasing (direct) or increasing (reverse) transition, respectively. This scheme applies to primitive relaxors or clusters of any number of primitive relaxors, and so the stresses indicated in the diagram are meant to denote the partial values pertaining to anyone of the single sets of clusters of any number ($k \geq 1$) of primitive relaxors. Adapted from Fig. 1 of Ref. [4], by permission of John Wiley and Sons (Copyright © 2016)

justify the actual local strain and stress variations specified in the scheme, including half-way[31] through each transition, i.e. at each corresponding transition state.

With reference to the $u_1 \leftrightarrow r_1$ transitions, we may represent the local skeleton strain and stress decreases after each direct (stress-reducing, $u_1 \rightarrow r_1$) transition by $(\varepsilon_0 - \varepsilon_\infty')$ and $(\sigma_0 - \sigma_\infty')$, respectively, where $\sigma_0 = E_0\varepsilon_0$ and $\sigma_\infty' = E_0\varepsilon_\infty' \leq \sigma_\infty = E_\infty\varepsilon_0$, , given that all local stresses must at all times be supported by the material's skeleton (whose stiffness is characterized by E_0), $\sigma_\infty \geq \sigma_\infty'$[32] being the final, fully relaxed, overall or average plateau-stress at $t = \infty$. Similar strain and stress decrease assignments may be approximately made for the $u_2 \rightarrow r_2$ transitions, and identical stress increases may also approximately apply to the $r_1 \rightarrow u_1$ and $r_2 \rightarrow u_2$ transitions. It is in this way recognized that, even assuming ideal perfectly uniform local initial strains, ε_0, and stresses, σ_0 (a not at all accurate assumption, depending on the material's initial state—but just acceptable, to avoid a much more complex treatment), a range of local strains (from $\varepsilon_\infty' < \sigma_\infty/E_0$ to $\varepsilon_0' > \varepsilon_0$) and stresses (from $\sigma_\infty' < \sigma_\infty$ to $\sigma_0' > E_0\varepsilon_0$) may develop during the relaxation process. This range was [2, 4] and will be here only approximately estimated, but the strategy

---

[31] We will soon be assuming that the initial and final states of each transition will be approximately symmetrical relative to its activated state, to avoid excessive complexity at this stage.

[32] [Because we are expected to end up at $t = \infty$ with a mixture of unrelaxed ($\sigma' = \sigma_0$) and relaxed ($\sigma' = \sigma_\infty\prime$) relaxors of any kind, contributing to the final average stress, $\sigma_\infty > \sigma_\infty'$].

for its more accurate treatment will be described in Appendix C. It should be noted that σ'_∞ and ε'_∞ may therefore become negative, meaning that the molecular skeleton could then be subject locally to actual compression after each $u_1 \to r_1$ transition.

Assuming that *transition (activated) states* are always identifiable between u_1 and r_1, and r_2 and u_2, and that the initial and final states of each transition may be taken as symmetrical relative to the corresponding transition states from the standpoint of the local skeleton strains, one may write as an approximation that $\left(\varepsilon_0 - \varepsilon'_\infty\right) = \left(\sigma_0 - \sigma'_\infty\right)/E_0 \cong 2\upsilon^\#/\upsilon$, where $\upsilon = \upsilon_o + \upsilon_f$ may be taken as the average total (occupied, υ_o, *and* free, υ_f) volume of each primitive relaxor *or* of anyone of their clusters of specific size, and $\upsilon^\#$ as the adopted definition of some sort of corresponding *activation volume*—arbitrarily assumed identical for all transitions, as a measure of the volume swept (either in contraction, $u \to r$, or in expansion, $r \to u$) by the relaxor's skeleton, i.e. by the intermolecular packing and bond angles, up to half-way through each transition—hereby defined as $\upsilon^\# = \upsilon\left(\varepsilon_0 - \varepsilon_1'^\#\right) = \upsilon\left(\varepsilon_1'^\# - \varepsilon'_\infty\right) = \upsilon\left(\varepsilon_2'^\# - \varepsilon_0\right) = \upsilon\left(\varepsilon'_0 - \varepsilon_2'^\#\right)$, $\varepsilon_1'^\#$ and $\varepsilon_2'^\#$ being the local skeleton strains at each specific type of transition or activated state. The above description defines all items in the scheme of Fig. 9.3. For clusters of k primitive relaxors, υ and $\upsilon^\#$ will obviously be proportional to k.

However, this scheme may be reduced to a simplified (and, of course, less accurate) form, assuming that all relaxors are initially at state u_1 (i.e. no r_2 relaxors exist and, therefore, no u_2 at any time either), with $\sigma_0 = E_0\varepsilon_0$. Let us for the moment also guess that they will all end at state r_1, with $\varepsilon'_\infty \approx \sigma_\infty/E_0$ (just an approximation, we stress, soon to be lifted), after stress relaxation is complete, i.e. at $t = \infty$, corresponding to considering, and simplifying as much as possible, just the lower part of the diagram. Assuming that no relaxed relaxors are assumed initially present within the material means that the material is considered microscopically uniform and perfectly isotropic, which of course may most often not be the case, as briefly analyzed further below, though more fully only in Appendix C. Later it will be confirmed that the introduction of the above volume variables, $\upsilon^\#$ and υ_o, reveals that the difference between the limiting plateau-moduli, $(E_0 - E_\infty)$, may decisively influence all relaxation times, and may (together with the activation energies quantified below) also contribute to accounting in some way for an important effect of intermolecular packing.

9.2.2   *The Rate of Stress Relaxation*

The formulation may then proceed to quantifying, first, the specific frequency of all relevant transitions of primitive relaxors[33] (only $u_1 \leftrightarrow r_1$ transitions in the present simplification) and the resulting overall rate of stress relaxation by[34]

$$-\frac{d\sigma_1}{dt} = \omega_1\left(\sigma_{1,0} - \sigma_{1,\infty}\right)\left[f_{u_1} e^{-E'_{ur}/(k_B T)} - f_{r_1} e^{-E'_{ru}/(k_B T)}\right]$$
$$= \omega_1\left(\sigma_{1,0} - \sigma_{1,\infty}\right)\left\{f_{u_{1,0}}\left[1 - c_1\left(\sigma_{1,0} - \sigma_1\right)\right]e^{-E'_{ur}/(k_B T)}\right.$$
$$\left. - f_{u_{1,0}} c_1\left(\sigma_{1,0} - \sigma_1\right)e^{-E'_{ru}/(k_B T)}\right\}, \qquad (9.3)$$

where $\sigma_{1,0} = \sigma_0$, σ_1 (and later σ_k being such that $\sigma = \sum_{k\geq 1} F'_k \sigma_k$, where σ is the total stress), and all F'_k are assumed to be given by (9.2) and (9.2a), k_B is Boltzmann's constant, T is the temperature, f_{u_1} and f_{r_1} are the fractions of unrelaxed and relaxed primitive relaxors (respectively) in all those crossing a unit resisting area (here assumed initially equal to 1 and 0, respectively), and c_1 is a constant calculated below from the final, no-relaxation, condition at ($t = \infty$, $\sigma_1 = \sigma_{1,\infty}$)—because, as justified above, the stress relaxes in direct exact proportion to the number of direct minus reverse transitions. As to E'_{ur} and E'_{ru}, they are the strain-dependent activation energies that must be overcome in each $u_1 \to r_1$ and $r_1 \to u_1$ transition, and $\omega_1(T, v_c)$ is an angular frequency factor (in rad s^{-1}) characterizing the material's structure and dynamics, whose value and dependence on temperature and other properties will be obtained by applying CTMD to the motions of the primitive relaxors, as well as, later, of any of its clusters of size k (integer).

The values of E'_{ur} and E'_{ru} must of course take into account that the applied strain, ε_0, may often make $E'_{ur} < E_{ur}$ and $E'_{ru} > E_{ru}$ by the difference in the *stored elastic energies* of the un-relaxed and the activated, and the activated and the relaxed states, respectively, as formulated in (9.4) and (9.5), under the same (overall, not local) applied strain, with the unprimed energies being those in the unstrained material. This is consistent with, and physically justified by, the fact that positive (tensile) stresses and strains are indisputably responsible for measurable gap reductions between the characteristic vibrational energy levels characterizing most or all material structures, as experimentally observed by various quantitative infra-red spectroscopic methods [5, 6]. Therefore, similar behavior should also apply to the much lower vibrational and/or rotational energies involved in materials deformation

---

[33] The subscript 1 in the stress values in the rate equations that follow denote that the equations refer to primitive relaxors ($k = 1$). Afterwards, the equations will be generalized to any cluster size, k.

[34] In case there will be more than one unrelaxed state (u_1) and only one r_1 state, f_{u_1} will be a sum of identical terms and f_{r_1} will have to be multiplied by that number. This and other possibilities are not considered in the present simplified formulation but (for example), if stress relaxation in a polymer were to be only driven by "*gauche*"-to- "*trans*" conformational transitions, f_{u_1} and f_{r_1} would both be multiplied by 2 (the number of direct and reverse transition paths).

and flow, near and even below the glass transition temperature. As to E_{ur} and E_{ru}, to avoid unessential complexity at this stage,[35] they will be assumed approximately equal in the present idealization. In a more accurate formulation, one would need to consider $E_{ru} \geq E_{ur}$ and their difference could introduce an extra material parameter, in addition to a possible corresponding variation in the relevant activation volumes (cf. (9.4) and (9.5)).

So, it is considered that an imposed unidirectional extensional or shear strain disturbs the dynamics, by increasing the $u_1 \rightarrow r_1$ relative to the $r_1 \rightarrow u_1$ transitions, through changes in the corresponding activation energies, as Eyring and many other authors have proposed and successfully used since long ago in some of the few existing and very simplified stress relaxation models [7–10], as well as in models of creep of varying complexity [3, 11–13]. However, by contrast with creep, now the process is not at constant stress, but at constant overall strain, and so the above activation energy changes should *not* be crudely formulated as stress multiplied by an activation volume, and could also become asymmetric, i.e.—$(E'_{ur} - E_{ur}) >$ $(E'_{ru} - E_{ru})$, thus potentially contributing to faster relaxation than creep, as known from conventional, phenomenological, linear viscoelastic theory (cf. comparison of relaxation and retardation times of standard linear solids in Sect. 3.1.3), and also from relevant related work on electrical stimuli [14].

Recalling that the local material's skeleton is purely elastic (Hookean), and using the above definition(s) of the activation volume, $v^{\#}$, one has

$$E'_{ur} - E_{ur} = -\frac{v}{2}\left(\sigma_0 \varepsilon_0 - \sigma'^{\#}_1 \varepsilon'^{\#}_1\right)$$

$$= -\frac{v}{2} E_0 \left(\varepsilon_0 + \varepsilon'^{\#}_1\right)\left(\varepsilon_0 - \varepsilon'^{\#}_1\right) = -E_0 v^{\#}\left(\varepsilon_0 - \frac{v^{\#}}{2v}\right) \qquad (9.4)$$

and

$$E'_{ru} - E_{ur} = \frac{v}{2}\left(\sigma'^{\#}_1 \varepsilon'^{\#}_1 - \sigma'_\infty \varepsilon'_\infty\right)$$

$$= \frac{v}{2} E_0 \left(\varepsilon'^{\#}_1 + \varepsilon'_\infty\right)\left(\varepsilon'^{\#}_1 - \varepsilon'_\infty\right) = E_0 v^{\#}\left(\varepsilon_0 - \frac{3v^{\#}}{2v}\right), \qquad (9.5)$$

where we must note that, being v the total volume of any of the primitive relaxors or clusters being considered, it must be multiplied by half the product of the relevant

---

[35] [Like the one that would be brought about by the different energies of true "*gauche*" and "*trans*" conformers in the stress relaxation or extension of a polymer chain (cf. Sect. 4.4.2] It should be borne in mind that this simplification, though very convenient at the present initial stage of the theory (avoiding additional mathematical complexity), may contrast with the identity and behavior of many real systems. Lifting this assumption (beyond the very specific application outlined in Chap. 13) may nevertheless be mandatory in future updates of CTMD, bearing in mind that, at least for the specific case where polymer creep was described by simple transitions between "*gauche*" and "*trans*" chain conformations (in Sect. 4.4.2 and Ref. [3]), no additional parameter proved necessary, and so it willt be worthwhile a future revision of the foregoing treatment without the mentioned assumption.

elastic strains and stresses, to obtain the values of the local stored elastic energies. The rate of stress relaxation then becomes

$$-\frac{d\sigma_1}{dt} = \omega_1\big(\sigma_{1,0} - \sigma_{1,\infty}\big)e^{-E_{ur}/(k_B T)}\Big\{\big[1 - c_1\big(\sigma_{1,0} - \sigma_1\big)\big]e^{\alpha_1}$$
$$-c_1\big(\sigma_{1,0} - \sigma_1\big)e^{-\alpha_2}\Big\}, \tag{9.6}$$

recalling that $f_{u_{1,0}}$ is here still assumed equal to 1, with $\alpha_1 = -\big(E'_{ur} - E_{ur}\big)/(k_B T)$ and $\alpha_2 = \big(E'_{ru} - E_{ur}\big)/(k_B T)$ or, after careful expansion of (9.4) and (9.5) with the previous definitions of $v^{\#}$ and the obvious relationships $\sigma_{1,0} = \sigma_0 = E_0\varepsilon_0$ (the initial condition in the absence of residual stresses), and also assuming a final uniform equilibrium state, $\sigma_{1,\infty} = \sigma_\infty = E_0\varepsilon_\infty$,[36] the reader will be able to obtain for α_1 and α_2

$$\alpha_1 = \frac{v}{8k_B T}\frac{(E_0 - E_\infty)(3E_0 + E_\infty)}{E_0}\varepsilon_0^2 \quad \text{and} \quad \alpha_2 = \frac{v}{8k_B T}\frac{(E_0 - E_\infty)(E_0 + 3E_\infty)}{E_0}\varepsilon_0^2. \tag{9.6a}$$

Figure 9.4 eases the identification of the stored elastic energies within the material's skeleton (in any set of relaxors of any size) that are exchanged in any of the four types of possible transitions and determine the above activation energy changes (9.4 and 9.5), already considering the general case where $r_2 \leftrightarrow u_2$ transitions would also have to be included.

As to the constant c_1, it is calculated from (9.6) (taking $-d\sigma_1/dt = 0$, at $t = \infty$) as

$$c_1 = \frac{e^{\alpha_1}}{\big(\sigma_{1,0} - \sigma_{1,\infty}\big)(e^{\alpha_1} + e^{-\alpha_2})}, \tag{9.6b}$$

yielding

$$-\frac{d\sigma_1}{dt} = \omega_1\big(\sigma_1 - \sigma_{1,\infty}\big)e^{-E_{ur}/(k_B T)}e^{\alpha_1} \quad or \quad \frac{d\sigma_1}{\sigma_1 - \sigma_{1,\infty}} = -\omega_1 e^{-E_{ur}/(k_B T)}e^{\alpha_1}dt$$

---

[36] At the final equilibrium (with equality of all direct to the corresponding reverse transitions' frequencies), the average stress should be constant in time and reasonably uniform, because all primitive relaxors had time to participate in clusters of all possible sizes, and so both the cumulative final average result of the whole set of transitions ($k \geq 1$) by clusters of any specific size (k), may be expected to be one and the same. Of course, given the intermittency of the transitions of clusters of the different sizes, this does not exclude fluctuating stresses in time at any location, namely before (but also after) the equilibrium is reached. The individual transitions of primitive relaxors or of any of their clusters in materials' responses are Poisson processes, like individual atomic disintegrations in radioactive materials. They occur sporadically at a characteristic average frequency, and collectively follow a first order process. What is quantified here when following the corresponding kinetics is the system's overall rate of change by the specific processes being considered—i.e. the number of transitions by primitive relaxors or whole sets of clusters of any specific number of primitive relaxors –, not the sequence of the individual transitions, which are completely random.

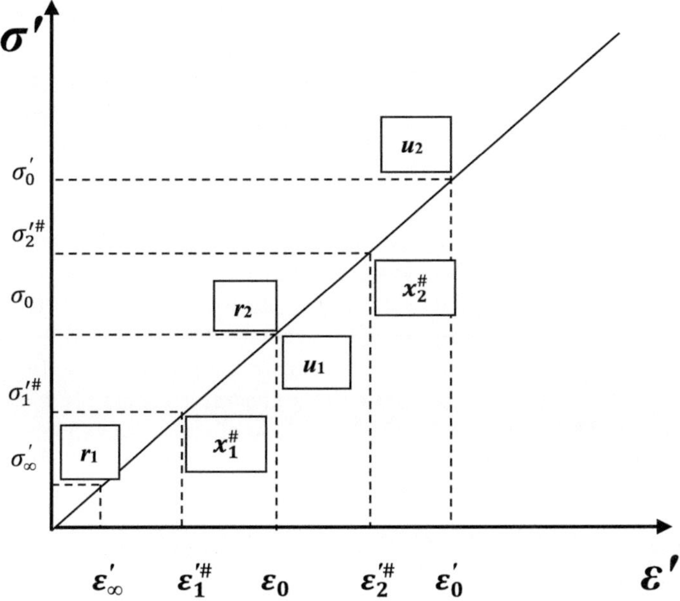

Fig. 9.4 Sketch of the local uniaxial tensile elastic skeleton stresses (σ') and strains (ε') at the four initial (u_1, r_2) or final (r_1, u_2) states and two transition states ($x_1^{\#}, x_2^{\#}$) of Fig. 9.3, under a constant overall strain, ε_0, for each set of clusters of any size (including for primitive relaxors). Adapted from Fig. 2 of Ref. [4], by permission of John Wiley and Sons (Copyright ©2016)

and, by integration, for the primitive relaxors's contribution to the stress relaxation modulus, $E_{r,1}(t) = \sigma_1(t)/\varepsilon_0$,

$$E_{r,1}(t) = E_{1,\infty} + \left(E_{1,0} - E_{1,\infty}\right)e^{-t/\theta_1},\tag{9.7}$$

with

$$\theta_1 = \frac{1}{\omega_1}e^{E_{ur}/(k_B T)}e^{-\frac{\upsilon}{8k_B T}\frac{(E_0-E_\infty)(3E_0+E_\infty)}{E_0}\varepsilon_0^2},\tag{9.8}$$

for the characteristic relaxation time of the entire set of primitive relaxors of the material, whatever their exact nature. It is this inverse proportionality of ω_1 to the relaxation time θ_1, apart from the temperature- and strain-dependent factors, that should assign it the nature of an angular (transition) frequency, as well known from general systems' dynamics. Equations (9.7) and (9.8) give the stress relaxation modulus of what may be called a *standard non-linear solid*, whose behavior may thus be characterized at any temperature, T, and applied strain, ε_0, by four physically meaningful parameters—E_0 and E_∞ (both overall, or total, moduli), E_{ur} and υ, as previously defined, plus the value of ω_1 to be quantified in Sect. 9.2.4. We thus see that the formulation already predicts that, through υ, molecular packing also influences the relaxation time of all individual relaxors—a relevant outcome of the present analysis.

We recall that, so far, this model was meant to describe stress relaxation of a uniform set of primitive relaxors within the material, considering that no other types and sizes of relaxors contribute to the behavior. However, this limitation may be lifted as developed and discussed three sections below, by weighing the contribution of the whole range of clusters of primitive relaxors that may form and thereby influence the overall relaxation behavior, by means of the corresponding relaxation spectrum obtained in Sect. 9.1 (or future better alternative).

Before proceeding to formulate the frequency factor ω_1, we will sketch the procedure and result of applying the full, more accurate, four-level process scheme of Fig. 9.3. Its complete treatment, however, is extremely involved and its presentation must be left to Appendix C, leading to a general and possibly more accurate *analytical* (or, if necessary, *numerical*) description of stress relaxation.

9.2.3　Accounting for the Full Range of Local Stresses and Strains

We recall that, in (9.3) and (9.6)–(9.8), one has hastily assumed, together with an all-un-relaxed initial structure, $f_{u_1,0} = 1$, that, at the final equilibrium ($t = \infty$), $f_{r_1,\infty}$ also had to be 1. This may seem reasonable, but turns out unnecessary and in fact inaccurate, because what determines the final structural composition is the stress relaxation dynamics itself, specifically the dynamic equilibrium that is then reached between the frequencies of direct and reverse transitions. This condition must be applied to the equivalent of (9.6) after the true stress decay per transition is included in the formulation, i.e. $\left(\sigma_{1,0} - \sigma'_{1,\infty}\right)$, where $\sigma'_{1,\infty}$ is now *not* necessarily made equal to $\sigma_{1,\infty}$, but may be lower (and even negative). In fact, a stress balance at any given time t over all primitive relaxors crossing the resisting area will give, for the complete and more accurate scheme of Fig. 9.3,[37]

$$\begin{aligned}
\sigma_1(t) &= \sigma'_{1,\infty} f_{r_1} + \sigma_{1,0}\left(f_{u_1} + f_{r_2}\right) + \sigma'_{1,0} f_{u_2} \\
&= \varepsilon_0 E_{r,1}(t) = \sigma'_{1,\infty} f_{r_1} + \sigma_{1,0} \\
&\quad \times \left[1 - \left(f_{r_1} + f_{u_2}\right)\right] + \sigma'_{1,0} f_{u_2} = \sigma_{1,0} - f_{r_1}\left(\sigma_{1,0} - \sigma'_{1,\infty}\right) + f_{u_2}\left(\sigma'_{1,0} - \sigma_{1,0}\right),
\end{aligned}$$

where $E_{r,1}(t)$ is the *relaxation modulus* if all stress relaxation were due to the primitive relaxors only, without any clustering, or

---

[37] The present stress balance only neglects the activated relaxors, as the small stress reductions brought about by the extremely small numbers of activated relaxors half-way through the *positive* transitions may be nearly compensated by the similarly small stress increases brought about by the extremely small numbers of activated relaxors half-way through their *negative* transitions.

$$\sigma_1(t) = \sigma_{1,0} - 2E_0 \frac{v^{\#}}{v}\left[f_{r_1}(t) - f_{u_2}(t)\right] \tag{9.9}$$

with, of course, $\sigma_{1,\infty} = \sigma_{1,0} - 2E_0 \frac{v^{\#}}{v}\left(f_{r_1,\infty} - f_{u_2,\infty}\right)$, given that the elementary stress reduction or increase per transition is now (from the previous section, including Fig. 9.3, and 9.9)

$$\sigma_{1,0} - \sigma'_{1,\infty} = \sigma'_{1,0} - \sigma_{1,0} = 2E_0 \frac{v^{\#}}{v} = \frac{\sigma_{1,0} - \sigma_{1,\infty}}{f_{r_1,\infty} - f_{u_2,\infty}}, \tag{9.10}$$

not $\left(\sigma_{1,0} - \sigma_{1,\infty}\right)$, as inaccurately assumed in writing the right-hand-side of (9.3) and (9.6), not even when $f_{u_1,0}$ is equal to 1 and $f_{r_2,0} = 0$ (and when, as a result, f_{u_2} would also be zero at all times). The symmetry of the stress variations in (9.10) result from the simplifying assumption that $v^{\#}$ is approximately the same for the $u_1 \leftrightarrow r_1$ and the $r_2 \leftrightarrow u_2$ transitions.

The stresses $\sigma'_0 > \sigma_0 > \sigma_\infty > \sigma'_\infty$ and strains $\epsilon'_0 > \epsilon_0 > \epsilon_\infty > \epsilon'_\infty$ illustrate that there will generally be a wide range of *non-affine stresses and strains* within the material, which need to be accounted for. In the simplest possible situation, however, where $f_{u_1,0} = 1$, there will be no primitive relaxors at stress $\sigma'_{1,0}$, and $f_{r_1,\infty}$ may be < 1, given by (cf. 9.6b)

$$f_{r_1,\infty} = c_1\left(\sigma_{1,0} - \sigma_{1,\infty}\right) = e^{\alpha_1}/\left(e^{\alpha_1} + e^{-\alpha_2}\right). \tag{9.11}$$

The working variable c_1 must of course satisfy the same previous equilibrium no relaxation condition, as may now be checked in the more accurate version of (9.6),

$$-\frac{d\sigma_1}{dt} = \omega_1 \frac{\sigma_{1,0} - \sigma_{1,\infty}}{f_{r_1,\infty}} e^{-E_{ur}/(k_BT)}\left\{\left[1 - c_1\left(\sigma_{1,0} - \sigma_1\right)\right]e^{\alpha_1} - c_1\left(\sigma_{1,0} - \sigma_1\right)e^{-\alpha_2}\right\}$$

$$= \frac{\omega_1}{c_1} e^{-E_{ur}/(k_BT)}\left\{\left[1 - c_1\left(\sigma_{1,0} - \sigma_1\right)\right]e^{\alpha_1} - c_1\left(\sigma_{1,0} - \sigma_1\right)e^{-\alpha_2}\right\}, \tag{9.12}$$

confirming for c_1 the same result as (9.6b).[38,39]

Therefore, a more accurate overall stress relaxation rate now becomes

$$-\frac{d\sigma_1}{dt} = \omega_1\left(\sigma_1 - \sigma_{1,\infty}\right)e^{-E_{ur}/(k_BT)}\left(e^{\alpha_1} + e^{-\alpha_2}\right), \tag{9.13}$$

---

[38] The curious reader may additionally verify that, in these same conditions ($f_{u_1,0} = 1$), $c_1 = \left(2E_0 v^{\#}/v\right)^{-1}$, which therefore turns out independent of cluster size, i.e. $c_k = c_1 = c$, and also $\frac{\left(\sigma_{k,0}-\sigma_{k,\infty}\right)}{f_{r_k,\infty}} = constant$, for any cluster size, $k \geq 1$.

[39] Strictly speaking, the expression for $f_{r_1,\infty}$ (9.11) implies that, though $\alpha_1 > \alpha_2$ (later seen to be $\alpha_1 \gg 3|\alpha_2|$), $f_{r_k,\infty}$ would not be the same for all cluster sizes, and neither would $\sigma_{k,\infty}$ be the same, for the same invariant value of c_1. So, Footnote 36 would also appear to be an approximation. This might be clarified only by considering that not all primitive relaxors within any given cluster are in the same state (relaxed or unrelaxed), as formulated in Appendix B.

which, as $e^{\alpha_1} + e^{-\alpha_2} = 2e^{(\alpha_1-\alpha_2)/2}\cosh\left(\frac{\alpha_1+\alpha_2}{2}\right)$,

may be recognized to result in the following more accurate relaxation time for each set of primitive relaxors,[40]

$$\begin{aligned}
\theta_1 &= \frac{1}{\omega_1}e^{E_{ur}/(k_BT)}\left(e^{\alpha_1} + e^{-\alpha_2}\right)^{-1} \\
&= \frac{1}{2\omega_1}e^{E_{ur}/(k_BT)}e^{-(\alpha_1-\alpha_2)/2}\mathrm{sech}[(\alpha_1 + \alpha_2)/2],
\end{aligned} \tag{9.14}$$

to be used in the same *standard non-linear solid* relaxation modulus of (9.7), with adequately modified α_1 and α_2, according to

$$\begin{aligned}
\alpha_1 &= \frac{\upsilon}{8k_BT}\frac{(E_0 - E_\infty)\left[\left(4f_{r_1,\infty} - 1\right)E_0 + E_\infty\right]}{E_0 f_{r_1,\infty}^2}\varepsilon_0^2 \\
\alpha_2 &= \frac{\upsilon}{8k_BT}\frac{(E_0 - E_\infty)\left[\left(4f_{r_1,\infty} - 3\right)E_0 + 3E_\infty\right]}{E_0 f_{r_1,\infty}^2}\varepsilon_0^2,
\end{aligned} \tag{9.15}$$

as may be obtained from (9.4), (9.5) and (9.10) (with $f_{u_2,\infty} = 0$). Equations (9.15) reduce to (9.6a) whenever $f_{r_1,\infty} = 1$. When performing these calculations, $f_{r_1,\infty}$ may initially be taken ~ 1 and proceed with the iterations, by calculating α_1 and α_2, recalculating $f_{r_1,\infty} = e^{\alpha_1}/\left(e^{\alpha_1} + e^{-\alpha_2}\right)$, etc., until convergence is reached to within the desired precision.

Two sections below, this analysis will be extended to a complex real material, characterized by the participation of a wide range of relaxor clusters, not just uncorrelated primitive relaxors. However, such extension will initially assume that the above formulation does apply to all sizes of clusters of primitive relaxors, which implies (cf. (9.10) and the discussion preceding it) that all individual relaxors of a given size are assigned the same quantitative contribution, in magnitude and sign, to the stress relaxation process. Given that, in fact, the various primitive relaxors within any relaxing cluster may contribute differently (depending on their own un-relaxed or relaxed states prior to any given transition), a more exact formulation may still be needed, as considered in Appendix B. This might prove the best or the only way to resolve the slight inconsistency pointed out towards the end of Footnote 39.

It is seen from the above treatment that the behavior is now (as it should) influenced by both α_1 and α_2. To our knowledge, this is the first attempt at looking in more detail into the calculation of the local non-affine strains and stresses within the material, as expressed by the relationship preceding equation (9.9). A similar analysis will of course apply to an initial condition of the material for which $f_{u_1,0} < 1$ and $f_{r_2,0} = 1 - f_{u_1,0}$, requiring the use of the full scheme of Fig. 9.1 and the calculation of the frequencies of all four (not just two) types of transitions, to obtain $-d\sigma_1/dt$. The calculations become even more complex (as shown in Appendix C), requiring

---

[40] Relaxing the $E_{ur} = E_{ru}$ assumption may, in future revised formulations, lead to an adjusted dependence of θ_1 on the applied strain via α_1 and α_2.

the calculation of, not just the two activation energy changes of (9.4) and (9.5), but also the two additional ones associated with the $u_2 \leftrightarrow r_2$ transitions, with $v^{\#}/v$ still assumed to be given by (9.10). As to $\left(f_{r_1,\infty} - f_{u_2,\infty}\right)$, it may now be obtained together with all the above four activation energy changes (themselves dependent on the applied strain, ε_0) by solving a set of five non-linear equations as a function of the parameter $f_{u_1,0}$ $(0 < f_{u_1,0} \leq 1)$ that characterizes the initial state of the material. This set of five simultaneous equations will always have a solution and reduce to (9.15) when $f_{u_1,0} = 1$ (cf. Appendix C).

We may here just point out that this new θ_1, proportional to a "sech" of an argument proportional to ε_0^2, ensures a perhaps better defined linear viscoelastic *domain* (rather than a sharply defined limit) as the applied strain is reduced, than the approximation that led to (9.8).[41] Equations (9.7) and (9.14), complemented by the formulation of ω_1 in the next subsection, stand so far as our best definition of a *standard non-linear solid* under stress relaxation.

9.2.4 Evaluation of the Angular Frequency Factors, ω_1 and $\omega_{k>1}$, by CTMD

If to E_{ur}—that may now be replaced by $E_{pr} = E_{a,1}$, the minimum activation energy, referring to each primitive relaxor (pr)—we assign the nature of an activation internal energy at constant volume (or, alternatively, enthalpy at constant pressure), it is expected from well-established kinetics theory that ω_1 in (9.3), (9.6), (9.8) and (9.12)– (9.14) will approximately be the product of some other, more fundamental, frequency factor characterizing the system[42] by a positive exponential whose argument should have an entropic (and thus *extensive*) nature. Therefore, the ω_1-factor in the above equations—from now onwards generalized to $\omega_k(T)$, if referred to any cluster of k primitive relaxors—may be expected to vary, not only with temperature, but also exponentially with the cluster size, k.

It may then now be anticipated (as argued in the next subsection) that *each set of relaxors of size k will collectively also behave as a *standard non-linear solid* with its specific relaxation time, θ_k (replacing θ_1 in (9.8) and (9.14)), given by

---

[41] This is because both the first and second partial derivatives of the "sech" with respect to ε_0 are zero at the limit $\varepsilon_0 \to 0$—by contrast to the negative limit of the second derivative of the strain-dependent exponential of (9.8), and so the updated formulation of θ_1 may succeed in predicting (when $\varepsilon_0 \to 0$) a limiting (though narrow) *region* or *domain*, where the stress relaxation curves may appear coincident, i.e. strain-independent, within experimental error, as observed. However, depending on the value of $(\alpha_1 - \alpha_2)/2$, the corresponding negative exponential in (9.14) may limit the mentioned advantage of the sech function. In any of the formulations, the partial derivatives of $E_r(t, \varepsilon_0)$ with respect to $\varepsilon_0 > 0$ are of course always negative, because $\alpha_1 \gg 3|\alpha_2|$ (as discussed in Sect. 9.4), except when $\varepsilon_0 = 0$.

[42] The curious reader may here start "guessing" which frequency it should be.

$$\theta_k = \frac{1}{\omega_k(T)} e^{kE_{a,1}/(k_BT)} \times \left[e^{-k\alpha_1} \, or \, 2\left(e^{k\alpha_1} + e^{-k\alpha_2}\right)^{-1} \right], \tag{9.16}$$

depending on adopting the approximate *or* more exact formulation,[43] respectively, given that both the activation energy, $E_{a,k}$, and the volume, υ_k, associated to each cluster size should of course be proportional to k. The two factors that multiply the expression(s) within the square brackets of (9.16) make up the linear viscoelastic limit of θ_k. We stress that the factors $\upsilon_k = k\upsilon_1$, with υ_1 the volume of each primitive relaxor, in the argument of the exponentials within [] in the above equation entail, by themselves, a non-negligible influence of molecular packing on the response behavior of all clusters of primitive relaxors, in addition to also expected increases in $E_{a,1}$ and $E_{a,k} = kE_{a,1}$ (corresponding to a more "rugged" potential energy landscape[44]), as molecular packing is tightened, where the effect on the activation energies might in the end prove the most significant. These effects of the molecular packing, directly through υ_k and $E_{a,k}$, will warrant further discussion in Sect. 9.11.

One now needs to formulate the dependence of $\omega_k(T)$ on temperature, T, and number of primitive relaxors, k, within a given cluster, using CTMD by means of (8.3) and the definition of the crossover temperature, T_c, in the preceding chapter, from which

$$\left(z_{1,r}^{\#}/z_1\right) = 2\sinh\left(\frac{h\nu_c}{2k_BT_c}\right) e^{\left[(E_{a,1}-h\nu_c/2)/(k_BT_c)\right]}, \tag{9.17}$$

yielding, by substitution into

$$\omega_k(T) = 2\pi\,\nu_c \left[\left(\frac{z_{1,r}^{\#}}{z_1}\right) \frac{e^{h\nu_c/(2k_BT)}}{2\sinh\left(\frac{h\nu_c}{2k_BT}\right)} \right]^k \tag{9.18}$$

(the same as (8.3) without the $E_{a,1}$ exponential that is already separated in (9.16)),

$$\omega_1(T) = 2\pi\,\nu_c \frac{\sinh\left(\frac{h\nu_c}{2k_BT_c}\right)}{\sinh\left(\frac{h\nu_c}{2k_BT}\right)} \exp\left[\frac{h\nu_c}{2k_B}\left(\frac{1}{T} - \frac{1}{T_c}\right)\right] \exp\left(\frac{E_{a,1}}{k_BT_c}\right), \tag{9.18a}$$

and

$$\omega_k(T) = 2\pi\,\nu_c \left[\frac{\omega_1(T)}{2\pi\,\nu_c}\right]^k, \tag{9.18b}$$

---

[43] In the latter case, we will see that $\omega_k(T)$ should be half of those calculated in (9.18) below.

[44] [Somehow the opposite effect of extensional stresses and deformations on the gaps between vibrational energy levels].

where T_c and v_c are, respectively, the crossover temperature and frequency (in Hz). So, $\omega_c = 2\pi v_c$ would turn out the frequency factor that the reader would rightly "guess" in Footnote 41.

However, we should note that the above $\omega_k(T)$ include vibrations that, at equilibrium, yield transitions in two possible "directions" (corresponding to both positive and negative contributions to the materials' forced responses, as for example in stress relaxation and creep). So, assuming as final, or as accurate as possible, the second form of (9.16) (equivalent to (9.14)), and noting that one always finds, in stress relaxation and in creep (cf. (9.24a) further below), as well as in all first-order physical or chemical processes, that

$$\mp \frac{d(Observable\ Quantity,\ Q)}{dt} = \frac{Shift\ of\ Q\ from\ Equilibrium}{(Sum\ of\ Specific\ Rates\ of\ Direct\ and\ Reverse\ Processes)^{-1}}$$

(as may be proved by working through the *Study Question 9.1* proposed at the end of Sect. 9.3.2), one should consider replacing $\omega_k(T)$ in the most accurate form of (9.16) by half the value given by (9.18), equivalent to halving ω_c, as suggested in Footnote 42. This way, we rightly include in the specific rates of the direct and reverse processes in the above expression only the vibrations that give rise to each of those specific processes—half of the total in equilibrium conditions (i.e. under negligible excitation—strain, stress, etc.) when, as assumed, the activation energies of the direct and reverse processes are identical ($E_{ur} = E_{ru}$). When this condition will not apply, the process becomes more complex and the above expressions may require modification.

One interesting and arguably relevant advantage of the above proposition to halve all $\omega_k(T)$ in the most accurate response expressions is that, in the linear viscoelastic limit, the system will then be made to obey the *fluctuation–dissipation relationship* (FDR)[45] in stress relaxation. In the non-linear viscoelastic domain, however, FDR violations will be met, as expected, depending on the intensity of the excitation.[46]

9.2.5   The Final Stress Relaxation Modulus

The basic stress relaxation model for each complete set of clusters of k relaxors—behaving like a *standard non-linear solid*—may therefore be expressed by (9.7) and

---

[45] [First mentioned and defined in this book in Sect. 7.2].

[46] For calculations in the non-linear viscoelastic domain, however, there will not be any simple and sure way of deciding whether or not the $\omega_k/2$ pre-factors in all response rates will accurately apply to both direct and reverse transitions, as they may slightly depend on the actual state of strain or stress of the relaxors involved. This is also linked to the possibility of not one single $v_1^{\#}$ (as assumed), but a more or less narrow range of values, determining the local cluster transition frequencies [The reader my relate this with Footnote 18 of Chap. 8].

(9.8) or, more accurately, (9.14), with ω_1 replaced by half the $\omega_k(T)$ of (9.18) and θ_1 by θ_k, and has 6 physical parameters (the same for *all* cluster sizes)—E_0, E_∞, T_c, ν_c, $E_{a,1}$ and υ_1—all of which, at least for some materials, may be approximately estimated (directly or indirectly) from available experimental data. It should also be noted that this model, with the above $\omega_k/2$, predicts for the relaxation times of each set of clustered relaxors results that are compatible with various other earlier cooperative formulations, like Adam and Gibbs's [15] and Ngai's coupling model [16–21], as shown for the latter case in Sect. 10.1, at least in the linear viscoelastic domain, the only one theoretically and experimentally explored in comparable earlier works, i.e. when, at very low strains, α_1 and α_2 in (9.14) and (9.16) approach zero.

The full CTMD may therefore be applied to quantify (and, if necessary, smooth out the complete relaxation spectra) and calculate overall stress relaxation moduli. In fact, the same equation (9.3) may be written for any cluster size, k, as

$$-\frac{d\sigma_k}{dt} = \frac{\omega_k}{2}\left(\sigma_{k,0} - \sigma_{k,\infty}\right)\left[f_{u_1,k}e^{-E'_{ur,k}/(k_BT)} - f_{r_1,k}e^{-E'_{ru,k}/(k_BT)}\right], \qquad (9.3a)$$

which will, after adequate expansion and integration, yield an analogue to (9.7) with θ_k given by (9.16). Combining all these cluster- specific results, with the discontinuous spectra of Sect. 9.1, and considering that the stresses σ_k have been formulated as if only k-clusters covered or crossed the resisting area, the overall stress relaxation moduli would then result as

$$E_r(t) = E_\infty + (E_0 - E_\infty)\sum_{k\geq1} F'_k(T)e^{-t/\theta_k(T)}, \qquad (9.19)$$

with of course $E_\infty = \sum_{k\geq1} F'_k E_{k,\infty}$ and $E_0 = \sum_{k\geq1} F'_k E_{k,0}$, where $E_{k,0} = E_0$ and $\sum_{k\geq1} F'_k = 1$, and additionally assuming that $E_{k,\infty} = E_\infty$ (notwithstanding the doubts raised in Footnote 39 [47]), as may be obtained from a weighed sum (by means of the F'_k) of the given solutions of (9.13) and of its counterparts for all cluster sizes (9.3a above). And the reason why it is not unreasonable to first integrate the above equations and then obtain the weighed sum of their solutions is because, in addition to the obvious fact that we must always have $\sigma(t) = \sum_{k\geq1} F'_k\sigma_k(t)$ (cf. the meaning of the F'_k, functionally equivalent to fractions of the resisting area covered/crossed by clusters of size k), effective clustering requires by definition (and therefore implies) *independence*, i.e. *uncorrelated behavior*, relative to any other set of clusters of different size or single primitive relaxors. The system is thus assumed to behave as a "fluctuating" parallel association of standard non-linear solids, where the $\sigma_k(t)$ may be different for the various cluster sizes, k, at any given finite time.

It is the wide spectra of response times, themselves the result of the prevailing significant dynamic heterogeneity of the materials behavior, that yield the well-known experimental KWW (Kohlrausch–Williams–Watts) extended exponential

---

^{47} [Arguably clarified in Appendix B].

behavior [22, 23]. KWW-type expressions, however, are semi-empirical two-parameter equations of mainly historical relevance with somewhat limited underlying physics, except that one of the parameters locates the timescale at which the response shows its maximum rate of change, or inflection point, on a log(time) scale (defining some sort of average response time), while the other (an exponent < 1) determines the actual value of such maximum rate of change. Within Sect. 10.1, a detailed analysis of KWW behavior in the light of CTMD will enable to finally specify that *average response time*, and physically identify the nature of KWW's exponent—Ngai's CM "quintessential" parameter [21]—as the reciprocal of the *average cooperativity level*, or *average cluster size*.

Highlight 9.2

The equivalent to (9.3) for any cluster size, k, is

$$-\frac{d\sigma_k(t)}{dt} = \left(\sigma_{k,0} - \sigma_{k,\infty}\right)\frac{\omega_k}{2}\left[f_{u_1,k}(t)e^{-E_{ur,k'}/(k_BT)} - f_{r_1,k}(t)e^{-E_{ru,k'}/(k_BT)}\right]$$

with $\sigma_{k,0} = \sigma_0$, being also assumed that $\sigma_{k,\infty} = \sigma_\infty$, where $f_{u_1,k}(t) = f_{u_1,k,0}\left[1 - c\left(\sigma_{k,0} - \sigma_k(t)\right)\right]$, $f_{r_1,k}(t) = f_{u_1,k,0}c\left(\sigma_{k,0} - \sigma_k(t)\right)$ and $c = c_{k\geq1} = \left(2E_0 v^{\#}/v\right)^{-1}$. From the result of its integration (analogous to that of (9.12) and (9.13)) and the definition of the corresponding θ_k (9.16), the above differential equation may in the end[48] be seen to be equivalent to $-\frac{d\sigma_k(t)}{dt} = (\sigma_0 - \sigma_\infty)\frac{1}{\theta_k}e^{-t/\theta_k}$.

As clusters, during their individual transitions respond independently of clusters of any other sizes,[49] their contributions may be simply weighed and added, yielding $-\frac{d\sigma}{dt} = -\sum_{k\geq1} F'_k\frac{d\sigma_k}{dt} = (\sigma_0 - \sigma_\infty)\sum_{k\geq1}\frac{F'_k}{\theta_k}e^{-t/\theta_k}$, with the F'_k of (9.2) (if correct). Its integration between $(0, t)$ and (σ_0, σ) gives $\sigma = \sigma_\infty + (\sigma_0 - \sigma_\infty)\sum_{k\geq1} F'_k e^{-t/\theta_k}$ and (9.19) for the stress relaxation modulus.

9.3 Uniaxial Tensile Creep

As for stress relaxation, we will first consider the creep behavior of the whole set of primitive relaxors (assuming no clustering) in a unit length along the direction of the applied stress in tensile straining or, in the case of shear, in the whole length in the direction of the applied stress at a given position within the thickness of a test sample of a given material.

---

[48] [After (1) separate integration, (2) weighing by the F'_k, and (3) summing].

[49] Section 8.3.3 discusses this at length in the framework of statistical physics. All cluster size-dependent contributions, $d\sigma_k/dt$, $(\sigma_0 - \sigma_k)$ and σ_k to stress relaxation and the overall stress are independent, because all primitive relaxor participations in different clusters are intermittent and, by definition, independent of the participations of the same primitive relaxors in other clusters at different times.

9.3.1 The Basis of the Formulation

In creep, an instantaneous uniform stress, σ_0, is applied and kept constant in time, and so the molecular skeleton remains everywhere and always under that constant uniform stress (assuming again that no residual stresses were initially present), but local delayed elastic strains gradually increase. So, in this case, work is being done *on* the system by the applied constant stress wherever and whenever the local delayed elastic strain increases, but work is done *by* the system against the stress when and where that local strain decreases. As a result, in those instances and local processes leading to decreased local delayed elastic strains (i.e. in the reverse transitions), their activation energy barriers will undoubtedly be *increased* by the work done *by* the system against the applied stress, and in those processes leading to increased local delayed elastic strains (the direct transitions), it has always in the past been argued [3, 11–13] that all the work done *on* the system by the applied stress should be *subtracted* from the corresponding activation energy barriers. We argue in Sect. 9.4 that it may *not* be the case, but first we will adopt the same classical view to extract and discuss one relevant consequence.

As the stress supported by the molecular skeleton is everywhere and at all times equal to σ_0, and only the total strain changes with time along a given direction, the system turns out more easily described, as illustrated in Fig. 9.5 by the horizontal arrowed line representing the constant stress σ_0, where the two possible undeformed (u_1 and u_2) and two possible deformed (d_1 and d_2) states of a hypothetical primitive relaxor are represented and assigned the corresponding total (instantaneous + delayed elastic) strains, as well as the transition or activated states $x_1^\#$ and $x_2^\#$ of the transitions $u_1 \leftrightarrow d_1$ and $d_2 \leftrightarrow u_2$. As for stress relaxation, we simplify and assume that those activated states have delayed elastic strains halfway between those corresponding to the initial and final states, namely $v^\#/v$ and $-v^\#/v$, respectively, if we adopt an analogous definition for the activation volume to that of Sect. 9.2.1 for stress relaxation.

The situation therefore "appears" physically much simpler in creep than in stress relaxation,[50] not just as to the local stress and strain distributions (assuming no residual initial values), but also as to the activation energy changes imparted by the applied stress, which (according to the first paragraph of this section, and considering the simplified two-state model,[51] corresponding to the right-most half of Fig. 9.5) become $E'_{ud} - E_{ud} = -\sigma_0 v^\#$ (according to the classical assumption), for the direct transitions leading to increases in the delayed elastic strain, and $E'_{du} - E_{du} = +\sigma_0 v^\#$, for the reverse, strain-decreasing, transitions, where one again assumes that the activation energies for the unstressed material are such that $E_{ud} \cong E_{du} = E_{a,1}$.[52]

---

[50] Later in this chapter (Sect. 9.4) we will argue that it might not be so.

[51] An analysis of the complete four-state model would follow along similar lines (but maybe simpler, due to the constant local stress) of the brief outline in Appendix C for stress relaxation but has not yet been carried out.

[52] As in stress relaxation (cf. Footnote 35), this simplifying assumption may of course influence the results of this analysis of creep and may in the end prove unnecessary (not requiring an extra

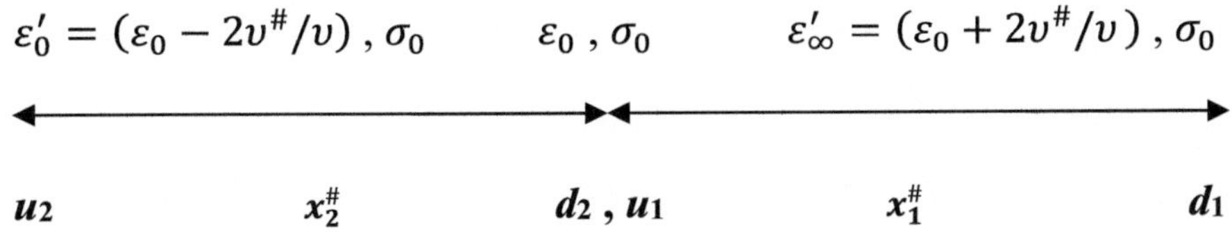

$$\varepsilon_0' = (\varepsilon_0 - 2v^{\#}/v)\,,\sigma_0 \qquad \varepsilon_0\,,\sigma_0 \qquad \varepsilon_\infty' = (\varepsilon_0 + 2v^{\#}/v)\,,\sigma_0$$

$$\longleftarrow \qquad\qquad\qquad\qquad \longrightarrow\!\!\longleftarrow \qquad\qquad\qquad\qquad \longrightarrow$$

$$\boldsymbol{u_2} \qquad\qquad \boldsymbol{x_2^{\#}} \qquad\qquad \boldsymbol{d_2\,,u_1} \qquad\qquad \boldsymbol{x_1^{\#}} \qquad\qquad \boldsymbol{d_1}$$

Fig. 9.5 Sketch of the local uniaxial tensile elastic skeleton stress, σ_0, and total (instantaneous + delayed elastic) strains, ε', at the four initial (u_1, d_2) and final (d_1, u_2) states and two transition states ($x_1^{\#}$, $x_2^{\#}$) of any primitive relaxor or cluster of any size, in a creep situation, immediately after the stress σ_0 is applied, for the complete four-state model. The arrows represent the four possible (two strain- increasing and two strain-decreasing) transitions contributing to creep

We may guess that this $E_{a,1}$ may be the same or very similar to the corresponding value for stress relaxation, implying that the primitive relaxors will not be significantly different (structurally and energetically) in the two situations, but we cannot be absolutely sure that differences might not need to be considered.

A strain balance at any given time t over all primitive relaxors (denoted by subscript 1) specified in the first lines of Sect. 9.3 may be written as

$$\begin{aligned}
\varepsilon_1(t) &= f_{u_1}\varepsilon_{1,0} + f_{d_1}\varepsilon_{1,\infty}' + f_{d_2}\varepsilon_{1,0} + f_{u_2}\varepsilon_{1,0}' = \sigma_0 D_1(t) \\
&= f_{d_1}\varepsilon_{1,\infty}' + \varepsilon_{1,0}\big[1 - (f_{d_1} + f_{u_2})\big] + f_{u_2}\varepsilon_{1,0}' \\
&= \varepsilon_{1,0} + f_{d_1}(\varepsilon_{1,\infty} - \varepsilon_{1,0}) - f_{u_2}(\varepsilon_{1,0} - \varepsilon_{1,0}'), \quad \text{or}
\end{aligned}$$

$$\varepsilon_1(t) = \varepsilon_{1,0} + (2v^{\#}/v)\big[f_{d_1}(t) - f_{u_2}(t)\big], \tag{9.20}$$

where the f's are the fractions of a given length along the direction of the stress, in tensile or shear creep, occupied by primitive relaxors in each of the undeformed or deformed states, and $D_1(t)$ is the creep compliance contribution by the primitive relaxors if no other relaxor clusters could be formed. The resulting strain increase per direct transition $u_1 \rightarrow d_1$ within the two-state model (for which $f_{u_{1,0}} = 1$ and $f_{d_{2,0}} = 0$, i.e. assuming initially no primitive relaxors fully strained at $\varepsilon_{1,\infty}'$) will therefore be $\varepsilon_{1,\infty}' - \varepsilon_{1,0} = \left(\frac{2v^{\#}}{v}\right)$, and the final overall strain increase $\varepsilon_{1,\infty} - \varepsilon_{1,0} = \left(\frac{2v^{\#}}{v}\right) f_{d_1,\infty}$ (from (9.20)), where $f_{d_1,\infty} = f_{u_1,0}c_1'(\varepsilon_{1,\infty} - \varepsilon_{1,0}) = c_1'(\varepsilon_{1,\infty} - \varepsilon_{1,0})$,[53] as we have to consider that the delayed elastic strain increases proportionally to the number of direct transitions in the two-state model. Of course, we will also have $f_{u_1}(t) = f_{u_1,0}\big[1 - c_1'(\varepsilon_1 - \varepsilon_{1,0})\big] = 1 - c_1'(\varepsilon_1 - \varepsilon_{1,0})$. In these conditions, this parameter c_1' (which may also be obtained from the final equilibrium, zero creep rate, condition) may be seen to obey the relationship $c_1' = (2v^{\#}/v)^{-1} = c'$, again the same for the primitive relaxors as for any of their clusters of whatever size. Above and

---

parameter), as in our early formulation of polymer creep by simple chain unfolding (cf. Ref. [3] and Sec. 4.4.2).

[53] We now do not need the initial step taken for stress relaxation of provisionally assuming, at $t = \infty$, 100% of completed direct transitions by the primitive relaxors.

below in this analysis, the instantaneous elastic strain (not the total strain) is assumed uniform and the same for every primitive relaxor (and cluster), i.e. $\varepsilon_{1,0} = \varepsilon_0 (= \varepsilon_{k,0})$, irrespective of its relaxed or un-relaxed state.

9.3.2   The Primitive Relaxors' and Clusters' Joint Rate of Creep

The overall rate of creep resulting from the $u_1 \rightleftharpoons d_1$ transitions of the primitive relaxors may (analogously to the stress relaxation case and already considering the $\omega_1/2$ pre-factor) be expressed by[54]

$$\frac{d\varepsilon_1}{dt} = \frac{\omega_1}{2}\left(2v^{\#}/v\right)\left[f_{u_1}e^{-E'_{ud,1}/(k_BT)} - f_{d_1}e^{-E'_{du,1}/(k_BT)}\right]$$

$$= \frac{\omega_1/2}{c'}\left\{f_{u_1,0}\left[1 - c'(\varepsilon_1 - \varepsilon_{1,0})\right]e^{-E'_{ud,1}/(k_BT)} - f_{u_1,0}c'(\varepsilon_1 - \varepsilon_{1,0})e^{-E'_{du,1}/(k_BT)}\right\},$$

$$(9.21)$$

with $f_{u_1,0} = 1$. By substituting the activation energy changes specified in the previous section, one obtains

$$\frac{d\varepsilon_1}{dt} = \frac{\omega_1/2}{c'}e^{-E_{ud,1}/(k_BT)}\left\{e^{\alpha}\left[1 - c'(\varepsilon_1 - \varepsilon_{1,0})\right] - e^{-\alpha}c'(\varepsilon_1 - \varepsilon_{1,0})\right\}, \qquad (9.21a)$$

with $\alpha = \sigma_0 v^{\#}/(k_BT)$. As c' must also satisfy the condition of zero creep at $t = \infty$, it also results

$$c' = \frac{e^{\alpha}}{\left(\varepsilon_{1,\infty} - \varepsilon_{1,0}\right)(e^{\alpha} + e^{-\alpha})} \qquad (9.21b)$$

and (9.21a) will lead to $\dfrac{d\left(\frac{\varepsilon_1 - \varepsilon_{1,0}}{\varepsilon_{1,\infty} - \varepsilon_{0,1}}\right)}{1 - \left(\frac{\varepsilon_1 - \varepsilon_{1,0}}{\varepsilon_{1,\infty} - \varepsilon_{1,0}}\right)} = \omega_1\cosh(\alpha)e^{-E_{ud,1}/(k_BT)}dt$, whose integration between $(0, t)$ and $[0, \varepsilon_1(t) = \sigma_0 D_1(t)]$ yields the creep compliance[55]

$$D_1(t) = D_{1,0} + \left(D_{1,\infty} - D_{1,0}\right)\left[1 - e^{-t/\tau_1(T)}\right], \qquad (9.22)$$

with

---

[54] As in stress relaxation, one could multiply f_{d_1} by the number of inverse paths, n_d, and consider $f_{u_1,0}$ a sum of n_d identical terms.

[55] The result of an updated formulation with $E_{ud} \neq E_{du}$ may eventually yield $\tau_1(T, \sigma_0) \propto \frac{1}{\omega_1(T)}e^{\frac{E_{a,1}}{(k_BT)}}\frac{\sigma_0}{\sinh(\alpha\sigma_0)}$ with $\alpha = \frac{\sigma_0 v^{\#}}{k_BT}$, as in our early polymer creep studies (cf. Ref. [3] and Sect. 4.4.2), but this will have to be confirmed (or not) in the future.

$$\tau_1(T, \sigma_0) = \frac{2}{\omega_1(T)} e^{E_{a,1}/(k_B T)} \operatorname{sech}(\alpha(T, \sigma_0)), \tag{9.23}$$

where $\omega_1(T)$ is given by (9.18a) and $\alpha = \frac{\sigma_0 v^{\#}}{k_B T}$, with $v^{\#} = \frac{(\frac{v}{2})(\varepsilon_{1,\infty} - \varepsilon_{1,0})}{f_{d_1,\infty}} = \frac{v}{2c'} = \frac{v\sigma_0}{2}(D_{1,\infty} - D_{0,1})(1 + e^{-2\alpha})$, giving $\alpha \sim \sigma_0^2 v/(2k_B T E_{\infty})$, if we take the creep compliance of the primitive relaxors (or of any of its clusters) at $t = \infty$ to be the same that characterizes the whole system, and $D_{\infty} = 1/E_{\infty}$.

Not at all surprisingly, the set of primitive relaxors behave in creep also as a *standard non-linear solid*, with a characteristic *retardation time* given by (9.23). It should be noted the similarity as well as the differences between τ_1 and θ_1, the most obvious of which is the negative exponential in θ_1 that may be suspected to contribute to reduced relaxation times relative to the corresponding retardation ones. This will be given a more detailed discussion in the next section.

The final creep compliance of the whole range of clusters of relaxors making up the material, which behaves as a series association of standard non-linear solids, will result as

$$D(t) = D_0 + (D_0 - D_{\infty}) \sum_{k \geq 1} F_k'(T) \left[1 - e^{-t/\tau_k(T)}\right], \tag{9.24}$$

with

$$\tau_k(T, \sigma_0) = \frac{2}{\omega_k(T)} e^{k E_{a,1}/(k_B T)} \operatorname{sech}(k\alpha(T, \sigma_0)), \tag{9.24a}$$

$\omega_k(T)$ given by (9.18), $D_0 = \sum_{k \geq 1} F_k' D_{k,0}$ and $D_{\infty} = \sum_{k \geq 1} F_k' D_{k,\infty}$, where $D_{k,0} = D_0$, $D_{k,\infty} = D_{\infty}$ and $\sum_{k \geq 1} F_k' = 1$. $E_{ud,1}$ was just replaced by $E_{a,1}$, the minimum activation energy.

How the cluster *size* and *response time distributions* in (9.19) and (9.24) will be related in stress relaxation and creep is not an obvious and easy question but will be given in the next section the analysis that is possible at this early stage of the development of CTMD. An obvious note is that, in stress relaxation, the F_k' are functions of k and $\theta_k(T)$, while in creep they are functions of k and $\tau_k(T)$—where, as we may anticipate, $\tau_k(T) \neq \theta_k(T)$.

Study Question 9.1 In physics and engineering it is conceptually important and useful to relate and unify concepts across subject matters within related fields.

The reader is therefore invited to:

1. Recognize that, in modelling non-linear stress relaxation and creep, one ends up with general kinetics relationships (cf. (9.13) and those following (9.21) for stress relaxation and creep, respectively, as well as the relevant ones in Appendixes B and C) of the form

$$\mp \frac{d(Observable\ Quantity,\ Q)}{dt}$$

$$= \frac{Shift\ of\ Q\ from\ Equilibrium}{(Sum\ of\ Specific\ Rates\ of\ Direct\ and\ Reverse\ Processes)^{-1}}$$

where the denominator is the *characteristic time* of the composite, reversible, process. This applies to all *first-order reversible* (chemical *or* physical) *processes*.

2. To illustrate the above general applicability, show that, in chemical kinetics, for a *first-order reversible reaction*, $A \underset{k^-}{\overset{k^+}{\rightleftharpoons}} B$, the rate of the reaction may likewise be expressed as above, i.e. that $-\frac{d[A]}{dt} = \frac{d[B]}{dt} = (k^+ + k^-)([A] - [A]_\infty) = (k^+ + k^-)([B]_\infty - [B])$ {*Prove it!*}, where k^+ and k^- are the specific rates, or *rate constants*, of the direct and reverse reactions, and so the reciprocal of their sum is the *characteristic time*, θ or τ of the composite, reversible, process, i.e. $\theta\ or\ \tau = (k^+ + k^-)^{-1}$.

3. Recognize that the disappearance of A is akin to a *"relaxation"*, and the formation of B to a *"compliance"*, with $[A](t) = [A]_\infty + ([A]_0 - [A]_\infty)e^{-t/\theta}$ and $[B](t) = [B]_0 + ([B]_\infty - [B]_0)(1 - e^{-t/\tau})$, with $\theta = \tau$. The identity of τ and θ in this case is the result of the *"relaxation"* of A and the *"compliance"* of B being the *same single physical process*, which is not the case of stress relaxation and creep, as previously considered and further analyzed below.

9.4   The Problem of the Relative Rates of Stress Relaxation and Creep

In Sect. 3.1, we have shown that the relaxation time of a standard linear solid is always *shorter* than its retardation time (and it is known that the same relationship is valid for any other linear dynamic system of whatever kind [14]), so linear materials' stress relaxation is faster than creep, at similar strain and stress levels. Also experiment and engineering practice recognize the same general relationship in non-linear stress relaxation and creep. This, combined with the fact that, in general, excessive creep is more readily and early detected than stress relaxation without special instrumentation or specific checks, justifies the more critical character and potentially unanticipated effects of the latter.[56]

It may therefore be argued that, given that the same "relaxors" or "moving structures" within the material are expected to participate in both stress relaxation and creep, the same general relationship between relaxation and retardation times would be predicted by any realistic non-linear dynamic models of materials. Should that be so? That is what will be discussed next.

---

[56] Creep may often be visually suspected before failure, while stress relaxation in general requires periodical maintenance inspections and checks (or remote instrumented monitoring). Think on the example of the need for periodical inspections of pipe gaskets, to check and correct for leaks.

A first indication that the same relationship *might* be predicted by CTMD and the associated non-linear viscoelastic models is provided by (9.14) and (9.23), which show that the primitive relaxors, as any of its clusters, appear to behave in stress relaxation with characteristic times shorter than the corresponding retardation ones (assuming the same level of stresses and strains), by a factor given by a negative exponential whose argument is proportional to the square of the applied strain, *but only if* we could find that $\alpha \sim (\alpha_1 + \alpha_2)/2$, as $(\theta_1/\tau_1) = e^{-(\alpha_1-\alpha_2)}\cosh(\alpha)/\cosh[(\alpha_1 + \alpha_2)/2]$. Will that be the case? On the other hand, in the linear viscoelastic region ($\varepsilon_0 \to 0, \sigma_0 \to 0$), the two predicted characteristic times would in the limit merge, in apparent contrast with what would be expected.

If we compare the orders of magnitude of α, α_1 and α_2, for example for a typical amorphous polymer (with no crosslinks or crystallinity, and no significant orientation – $f_{u_1,0} = 1$), we obtain $\alpha > \sigma_0^2 \upsilon/(2k_B T E_\infty)$, while (cf. (9.15)) $\alpha_1 \sim 3\sigma_0^2 \upsilon/(8k_B T E_0)$ for $f_{r_1,\infty} \sim 1$, $\sim 4\sigma_0^2 \upsilon/(9k_B T E_0)$ for $f_{r_1,\infty} \sim 3/4$, and 0 for $f_{r_1,\infty} = \dfrac{\left(1-\frac{E_\infty}{E_0}\right)}{4} \sim 1/4$, while $\alpha_2 \sim \sigma_0^2 \upsilon/(8k_B T E_0)$ for $f_{r_1,\infty} \sim 1$, $\sim 3\alpha_1 E_\infty/(2E_0)$ for $f_{r_1,\infty} = 3/4$, and < 0 for $f_{r_1,\infty} < 3/4$, at the same level of initial stress, showing that $\alpha \gg \alpha_1 > 3|\alpha_2|$ (thus widely different from the above guess, equivalent to $\alpha \sim |\alpha_2|$ or $2|\alpha_2|$) and, as a result, $(\theta_1/\tau_1) = (e^\alpha + e^{-\alpha})/(e^{\alpha_1} + e^{-\alpha_2}) \sim e^{(\alpha-\alpha_1)} \gg 1$, which would contradict known behavior.

The classical formulation that was here adopted to obtain (9.23) therefore *might* need to be revised. What should perhaps be acknowledged is that the work $\sigma_0 \upsilon^{\#}$ supplied by the applied stress while an un-deformed relaxor is being raised to its activated state is not to be discounted (at least totally) from the thermal energy fluctuation necessary to drive the relaxor through the barrier, as a substantial barrier would still need to be overcome if the relaxor deformed "in the void", against no intermolecular friction, or if some pure rotations were necessary in the bulk, during creep.[57] The work $\sigma_0 \upsilon^{\#}$ might thus be substantially used to compensate for friction within the material (between each relaxor and neighboring ones in the orthogonal directions, in extension or in shear, relative to the direction of the stress), while keeping the same elastic deformation of the molecular skeleton imparted at $t = 0$. In these cases, the total or a significant fraction of the original activation potential energy could still be required to drive the relaxor through the transition, depending on the material. So, at least approximately, one may either need to keep the same activation energy that applies to the unstressed material, i.e. $E_{ud,1}$ for primitive relaxors or, at most, apply to it a much smaller but unknown correction to formulate the rate of the direct, strain-increasing, transitions.

By doing so, the exponential e^α in (9.21a) could, in the most extreme situation, be dropped (i.e. replaced by 1), keeping the $e^{-\alpha}$ one, to yield the following replacements to (9.21b) and (9.23), respectively for the parameter c' and the characteristic retardation time of the primitive relaxors, τ_1, of the corresponding standard non-linear solid:

---

[57] Consider, for example, in creep, the uncoiling of a single polymer chain, or some rotations that might be necessary in the bulk of a low molar mass material.

$$c' = 1/\left[\left(\varepsilon_{\infty,1} - \varepsilon_{0,1}\right)\left(1 + e^{-\alpha}\right)\right], \tag{9.21b'}$$

$$\tau_1 = \frac{2}{\omega_1} e^{E_{a,1}/(k_B T)} \left(1 + e^{-\alpha}\right)^{-1}. \tag{9.23'}$$

Comparing now the relaxation and retardation times, one obtains

$$\frac{\theta_1}{\tau_1} = \frac{1 + e^{-\alpha}}{e^{\alpha_1} + e^{-\alpha_2}} \sim e^{-\alpha_1} < 1, \tag{9.25}$$

maybe even $\ll 1$, which now could make better physical sense. We must however note that (1) $E_{ud,1}$ might instead still require some negative correction, though smaller than $\sigma_0 v^{\#}$, and that (2) in the above approximate estimates of the parameters, in addition to considering similar (initial) stress conditions, σ_0, we assumed similar definitions and values for the activation and total volumes of the relaxors in stress relaxation and creep. How the foregoing discussion on the relative values of relaxation and retardation times might need revision when $E_{ud} \neq E_{du}$ will be left for future analysis.

One apparent though minor problem still remains: the identity of both response times in the limit of very low strains and stresses (therefore well inside the linear viscoelastic domain[58]) which, although complying with the fluctuation–dissipation relation [24–26], i.e. $\theta_1 = \tau_1 \sim \frac{1}{\omega_1^*}$,[59] may not exactly agree with the general behavior of linear dynamic systems. Might there be some physical reason (additional to those implied in the above modified formulation(s) of creep) that will make $\theta_1 < \tau_1$ always, or at least $\theta_{k>1} < \tau_{k>1}$, including in the limit of very low strains or stresses (even if not strictly at zero stress or strain)?

What follows is very conjectural and difficult to formulate without introducing at least one additional empirical adjustable parameter. In stress relaxation, clustering does not only involve adjacent relaxors crossing the resisting area (as assumed in CTMD's treatment so far), but also neighboring ones in the orthogonal direction, not directly involved at the same time in other clusters. These additional relaxors will of course increase the actual sizes and response times of the clusters that are responsible for the stress relaxation in any given fraction of the resisting area because the individual activation energies of the primitive relaxors participating in any such cooperative transitions are all additive. Also, a similar effect occurs in creep, but in this case the additional clustering is in any of the orthogonal directions (two in both tensile and shear creep), and so an even larger effect is expected in creep.

So, while the statistical weights, F_k', in stress relaxation denote fractions of the resisting areas crossed by k relaxors, the actual response times of the clusters effectively involved will be that of tridimensional clusters of size βk, with $\beta > 1$, therefore

---

[58] It is true that, at $\varepsilon_0 \sim 0$ or $\sigma_0 \sim 0$, there is no stress relaxation or creep, but the fact is that $\theta_1 < \tau_1$ is universally valid for any standard linear solid down to any low (non-zero) values of strain or stress.

[59] $\omega_1^*(T) = 2\pi v_1^*(T)$ is not to be confused with $\omega_1(T)$ of (9.18a).

much longer (due to their exponential variation with cluster size), the effect being of course even more significant in creep than in stress relaxation. Further, to be accurate, the F'_k (or rather the $F'_{kk'}$) should in fact be calculated using k to quantify the F_k but $k' = \beta k$ to multiply the F_k by the $\theta_{k'}$, and thereby account for intermittency and dynamic heterogeneity (cf. Sect. 9.1.2). Given the very strong decrease of the F_k with k (cf. (9.1) and Fig. 9.1), this will lead to increased weights and longer individual and average response times in creep relative to stress relaxation, which would agree with actual behavior. The response time of un-clustered primitive relaxors would however remain unchanged.

For each given tridimensional cluster size k' (and given $\theta_{k'}$ and $\tau_{k'}$), being $\beta_{creep} = \frac{k'}{k_{creep}} > \beta_{relax} = \frac{k'}{k_{relax}}$, one would have $k_{relax} > k_{creep}$ (for the cluster's projections onto the resisting surface in stress relaxation than onto the stress direction in creep), $F_{k,relax} < F_{k,creep}$, and therefore $k_{relax,max} > k_{creep,max}$ (to ensure that $\sum_{k \geq 1} F_k = \sum_{k \geq 1} F'_k = 1$, which could result in narrower retardation than relaxation spectra (as already theoretically predicted [14]).

These new parameters β_{creep} and β_{relax} would however be very difficult to quantify (even as averages) without a very complex treatment of tridimensional clustering, coupled with even more complex evaluation of bi-dimensional and one-dimensional projections, and so would need to be dealt with (at least in an initial treatment) as adjustable parameters. Could we conjecture a relationship of the type $\beta_{creep} \sim \beta_{relax}^{3/2}$ for any given bidimensional or unidimensional projection of any size k_{relax}, or k_{creep}, onto the resisting area in stress relaxation, or onto the stress direction in creep, respectively? With or without any such conjectures, this would lead in CTMD to a somewhat undesirable mix of very specific physical properties $\left(\nu_c, E_{a,1}, T_c\right)$, with mere fitting constants.

The formulations and discussions in this entire chapter (plus those that most probably will in the future still be found missing in this first development of CTMD) will of course bear on the final and important relationship between the relaxation and retardation spectra of non-linear amorphous materials, $H(\theta_i)$ and $L(\tau_i)$, both represented here by either $F'_i(\theta_i)$ or $F'_i(\tau_i)$, arguably with $\theta_i < \tau_i$.

However, what one must recognize is that the considerations of the present section may not settle the issue of the relationship between relaxation and retardation times of amorphous condensed matter, not even in the linear viscoelastic domain. The reader will have registered that (1) in Chap. 3, idealized standard linear solids of polymeric nature were predicted to have retardation times 10–1000 times longer than their relaxation times and (2) in this chapter, we could not yet find a clear way of allowing for or obtaining a similar relationship in the mechanistic model of creep just formulated within CTMD. That may require considering that, in creep, as in stress relaxation, the local stresses that drive the actual cluster transitions and the overall process are not constant but decreasing (though more slowly in creep than in stress relaxation), as in any standard linear solid (cf. Sect. 3.1.3), which may then preclude reaching an analytical solution to the problem.[60] Therefore, contrary to past assumptions, linear

---

[60] Resorting to a simple correction of all relaxation times obtained in the preceding sections by multiplying them by the ratio E_0/E_∞ (as verified for linear standard solids of whatever nature)

and non-linear creep behavior turns out more difficult to formulate than the stress relaxation one. Another consequence of this contrast between creep and stress relaxation might be that agreement with the fluctuation–dissipation relationship (FDR) will not be met by both processes in the linear viscoelastic domain.

9.5 How to Account for Partial Crosslinking and/or Crystallization

Partial crosslinking and partial crystallization will have qualitatively similar effects in influencing how a material relaxes stress or creeps. Assuming, as from the beginning in this chapter, that bond scission and/or crystal disruption processes are excluded, stress relaxation and creep will only involve the materials' amorphous non-crosslinked fraction. So, all we have to take into account is that, at any given time, t (and considering the simple *two-state* part of the schemes of Figs. 9.3 and 9.5), while the sum of the number of un-relaxed (or un-deformed) and relaxed (or deformed) primitive relaxors in stress relaxation (or creep) is still equal to the total initial number of both types of primitive relaxors, there will always still be a significant number of crosslinked or crystallized primitive relaxors that remain subject to a constant stress σ_0 (in stress relaxation) or strain ε_0 (in creep), without contributing to the materials' response. The expected modifications to the previous analyses are therefore relatively minor, but the results will obviously be significantly different, depending on the extent of crosslinking or crystallization.

9.5.1 Uniaxial Tensile Stress Relaxation

Considering first, as before, the relaxation response limited to the uncorrelated primitive relaxors crossing the unit resisting area in tensile or shear stress relaxation, a stress balance may be written as.

$$\sigma_1(t) = \varepsilon_0 E_{r,1}(t) = \sigma'_{1,\infty} f_{r_1} + \sigma_{1,0}\left(f_{u_1} + f_X\right) = \sigma'_{1,\infty} f_{r_1} + \sigma_{1,0}\left(1 - f_{r_1}\right) = \sigma_{1,0} - 2E_0 \frac{\upsilon_{\#}}{\upsilon} f_{r_1},$$

where f_X is the fraction of the resisting area crossed by primitive relaxors involved in crosslinks or crystalline structures, yielding the same equation (9.9), with $f_{u_2} = 0$ (two-state model).

Therefore, the reader may see that the second equality of (9.12) will also remain valid because, though $f_{u_1,0} = 1 - f_X < 1$ and $f_{u_1} = f_{u_{1,0}}\left[1 - c_1\left(\sigma_{1,0} - \sigma_{1,\infty}\right)\right]$, the factor $f_{u_{1,0}}$ will also appear in its left-hand-side—which now becomes $\left(-f_{u,0} \frac{d\sigma_1}{dt}\right)$, because all σ in the equation are now local stresses in *relaxable* primitive relaxors,

---

could be tempting but would only shift the creep curves to longer log-time values, not leading to different widths of relaxation and retardation spectra as theoretically expected, or at least suggested in Ref. [14]. It would also appear as a somewhat artificial solution, instead of sound phenomenology, but we will provisionally resort to that questionable solution in some of the calculations in Chap. 11.

and appears again in the numerator and denominator of $\sigma_{1,0} - \sigma'_{1,\infty} = \frac{[\sigma_1(t) - \sigma_1(\infty)]}{f_{r_1,\infty}} = 1/c_1$, because the stresses in the numerator are average values (among relaxable *and* non-relaxable structures) and so now

$$f_{r_1,\infty} = \frac{f_{u_{1,0}} e^{\alpha_1}}{(e^{\alpha_1} + e^{-\alpha_2})}. \tag{9.26}$$

This will now have to be substituted in the expressions for α_1 and α_2 of (9.15).

As to E_∞, it must be replaced by the effective fully relaxed modulus of the crosslinked or crystalline material, $E^X_{1,\infty}$—which will be (often orders of magnitude) higher than the one characterizing a very lightly crosslinked similar material—, or by the incipient relaxed plateau modulus of its un-crosslinked variety before the onset of viscous flow, made apparent in a log(time) scale,[61] which must satisfy the relationship

$$E^X_{1,\infty} = E_{1,0}\left(1 - f_{r_1,\infty}\right) + f_{r_1,\infty} E'_{1,\infty}, \tag{9.27}$$

as results from the stress balance at the beginning of this section with $t = \infty$, where $E'_{1,\infty} = \sigma'_{1,\infty}\varepsilon_0 \leq E_{1,\infty} = \sigma_{1,\infty}\varepsilon_0$, because $\sigma'_{1,\infty} \leq \sigma_{1,\infty}$, as explained in Sect. 9.2.3.

To summarize, the dynamics of the whole set of primitive relaxors will be that of the same non-linear standard solid described by (9.13), (9.14), (9.15), (9.26) and (9.27), which of course require (as in Sect. 9.2.3 in the absence of crosslinks) that (9.15), (9.26) and (9.27) (four equations in four unknowns—α_1, α_2, $f_{r_1,\infty}$ and $E^X_{1,\infty}$) be simultaneously solved as functions of $f_{u_{1,0}}$, without much difficulty,[62] starting with an initial estimate of $f_{r_1,\infty} \cong f_{u_{1,0}} = 1 - f_X$, for any given set of values $E_{1,0}$, $E'_{1,\infty}$ and f_X (in addition to υ and all other CTMD parameters—cf. Chap. 8 and Sect. 9.12), at each specific temperature, to obtain

$$E_{r,1}(t) = E^X_{1,\infty} + \left(E_{1,0} - E^X_{1,\infty}\right)e^{-t/\theta_1}, \tag{9.28}$$

where $E^X_{1,\infty} \equiv E^X_\infty$ may now be much higher than the relaxed modulus, E_∞, of the corresponding un-crosslinked or very lightly crosslinked material.

We do not need to perform these calculations separately for each family of clusters, as their α_1 and α_2 are proportional to k, and $f_{r_1,\infty}$ is as before assumed to apply to any cluster size ($f_{r_k,\infty} = f_{r_1,\infty}$), and so the combined cooperative response of the whole material may be formulated as

$$E_r(t) = E^X_\infty + \left(E_0 - E^X_\infty\right) \sum_{k \geq 1} F'_k e^{-t/\theta_k}, \tag{9.29}$$

---

[61] Cf. Fig. 3.8 in Chap. 3.

[62] [As confirmed in preliminary calculations with experimental data for un-crosslinked PMMA and PC, when $f_{u_{1,0}} < 1$].

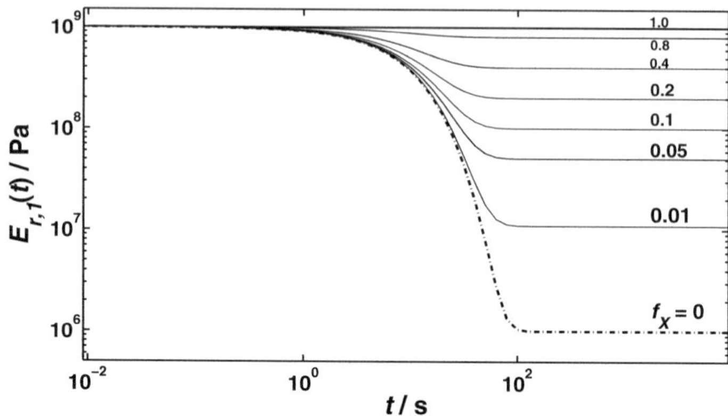

Fig. 9.6 Uniaxial tensile stress relaxation moduli for a hypothetical crosslinked homogeneous standard linear solid made of uncorrelatable relaxors with a relaxation time of 10 s, with variable crosslink densities,[63] where $f_X^{3/2}$ (cf. text) might be a more approximate estimate of the number fraction of relaxors involved in crosslinks and unable to move, in the whole mass of the solid

because, as previously explained, $E_\infty^X = \sum_{k\geq 1} F_k' E_{k,\infty}^X$, $E_{k,0} = E_0$, $\sum_{k\geq 1} F_k' = 1$ and $E_r(t) = \sum_{k\geq 1} F_k' E_{r,k}(t)$.

The reader may now, for example, see how crosslinks modify the stress relaxation behavior of a highly hypothetical standard linear solid made of totally "uncorrelatable" primitive relaxors (if such could be possible) of, for example, $E_0 = 1$ GPa and relaxed modulus (in the limit of negligible crosslinking) $E_\infty \sim E_\infty' = 1$ MPa, with a relaxation time of $\theta_1 = 10$ seconds, for a range of assumed values of $f_X = 1 - f_{u_{1,0}} \cong 1 - f_{r_1,\infty}$, by using (9.28), with $E_{1,\infty}^X$ given by (9.27), to calculate and plot the values of $E_{r,1}(t, f_X) = E_\infty + (E_0 - E_\infty)\left[f_X + (1 - f_X)e^{-t/\theta_1}\right]$, as in Fig. 9.6, where the case for $f_X = 0$ (no crosslinking), or $E_{r,1}(t) = E_\infty + (E_0 - E_\infty)e^{-t/\theta_1}$, is also shown for reference.

Within the framework of this formulation, f_X may stand for the fraction of relaxors involved in crosslinks and/or in crystalline regions, crossing the resisting area, and the corresponding fraction in the bulk (volume or mass) of the solid, assumed uniform and isotropic, may be estimated as $\sim f_X^{3/2}$. The qualitative similarity of the above curves with the real experimental behavior should however not lead the reader to overlook that, for a real *non-linear material* (or even a hypothetical non-linear solid, with one single relaxation time), the curves shown neglect the effect of f_X on the relaxation time itself, because increasing f_X values may not only entail significantly higher E_∞^X, also lower α_1 and α_2 (despite somewhat lower $f_{r_1,\infty}$—cf. (8.15)), and therefore longer relaxation times. So, in addition to the rising plateau for very long

---

[63] The reader should note that the significant separations of the long-time plateaus (to show the sensitivity of the analysis) for low crosslinkings is just the result of the logarithmic scale of the ordinates. On a linear scale, the diagram would look much more familiar, similar to experimental ones, showing an exact linear variation of the fully relaxed modulus with the density of crosslinks, f_X.

times, thus reducing the relaxation strength, $(E_0 - E_\infty^X)/E_0$, the downward transition in the curves will also be displaced to longer times, as physically expected and experimentally found, because crosslinks (and crystallization) do retard stress relaxation. The present formulation may rightly (even if not exactly) account for the effect of f_X at non-zero strains, ε_0, but does not predict any lengthening of the response times (by decreasing cluster transition frequencies) at equilibrium ($\varepsilon_0 = 0$), which will not be exact, requiring some future update of the effect of crosslinks and crystalline regions on the mentioned cluster frequencies, possibly by accounting for increased values of the minimum activation energy, $E_{a,1}$.

An equation analogous to that for $k = 1$ will be valid for any $k > 1$, i.e.

$$E_{r,k}(k, t, f_X) = E_\infty + (E_0 - E_\infty)\big[f_X + (1 - f_X)e^{-t/\theta_k}\big], \qquad (9.30)$$

yielding

$$E_r(t, f_X) = \sum_{k \geq 1} F_k' E_{r,k}(k, t, f_X), \qquad (9.31)$$

where f_X, if uniform within the material, will of course be the same for all clusters of primitive relaxors.

Study Question 9.2[64] For an also hypothetical *standard crosslinked non-linear solid*, use the values given in Chap. 11 for the relevant properties of a PMMA in (9.14) (with ω_1 replaced by $\omega_1/2$), (9.15) and (9.18a), and substitute θ_1 in the above expression of $E_{r,1}(t)$ by a recalculated $\theta_1\big(E_{a,1}, v_c, T_c, v, E_0, E_\infty^X, T, \varepsilon_0\big)$, to compute and plot the values of $E_{r,1}(t)$ for varying f_X, at one chosen strain and two different temperatures.

In any *real standard crosslinked non-linear solid*, with wide scale clustering possible, we will of course have a much more gradual transition from E_0 to E_∞^X, spanning several decades of timescale, as experimentally found, and this is what is also predicted by CTMD in Chap. 11 for two non-crosslinked amorphous polymers – PMMA and PC.

Study Question 9.3 After reading Chap. 11, try to repeat the calculations of the previous Study Question for PMMA with its whole predicted relaxation spectrum at $T = 380$ K, for example, or any other temperature. [You may use or base yourself on the Matlab® computer code listed in Sect. 11.8 just for the calculation of the $F_k'(T)$. The spectrum values are those of the variable F_i_p_E(i,j) in that listing, i being the cluster size (number of primitive relaxors) and j the index specifying the temperature value. In your case, you may ignore j, as you will be using just one temperature.]

---

[64] The reader may easily search and collect the relevant property values in Chap. 11 or revisit this study question after reading that chapter.

9.5.2 *Uniaxial Tensile Creep*

In this case, the reasoning is like the one for stress relaxation but applied to the strains. The strain balance along a longitudinal reference length (i.e. in the direction of the applied stress), assuming un-clustered primitive relaxors only, is $\epsilon_1(t) = \sigma_0 D_1(t) = \left(1 - f_{d_1}\right)\epsilon_{1,0} + f_{d_1}\varepsilon'_{1,\infty} = \varepsilon_{1,0} + f_{d_1}\left(2\upsilon^\#/\upsilon\right) = \varepsilon_{1,0} + \left(2\upsilon^\#/\upsilon\right)f_{u_1,0}c'_1\left(\varepsilon_1 - \varepsilon_{1,0}\right)$, and so $\frac{2\upsilon^\#}{\upsilon} = 1/\left(f_{u_1,0}c'_1\right)$. As a result, for the alternative formulation, we may again write (9.21a) without the first exponential, e^α, and we may also keep (9.21b′) for c'_1, and (9.23′) for the retardation time τ_1.

Like in the stress relaxation case, the solution procedure requires the simultaneous numerical solution of, not four, but three equations

$$f_{d_1,\infty} = f_{u_1,0}/\left(1 + e^{-\alpha}\right), \tag{9.32}$$

$$D^X_{1,\infty} = D_{1,0}\left(1 - f_{d_1,\infty}\right) + f_{d_1,\infty}D'_{1,\infty}, \tag{9.33}$$

and

$$\alpha = \frac{\sigma_0\upsilon^\#}{k_B T} = \frac{\sigma_0\upsilon}{2k_B T}\left(D^X_{\infty,1} - D_{1,0}\right)\left(1 + e^{-\alpha}\right), \tag{9.34}$$

to obtain what is necessary to compute the creep compliance $D_1(t)$ by (9.22), with $D^X_{1,\infty}$ replacing $D_{1,\infty}$. Once again, this does not need to be done separately for each family of clusters and primitive relaxors ($k \geq 1$), because $D_{0,k} = D_{1,0} = D_0$, $D'_\infty = \sum_{k\geq 1} F'_k D'_{k,\infty}$, $D^X_\infty = \sum_{k\geq 1} F'_k D^X_{k,\infty}$ and $\sum_{k\geq 1} F'_k = 1$. Then, with all retardation times obtained by

$$\tau_k = \frac{2}{\omega_k}e^{-kE_{a,1}/(k_B T)}\left(1 + e^{-k\alpha}\right)^{-1}, \tag{9.35}$$

and ω_k given by (9.18), the overall *creep compliance* may finally be calculated by

$$D(t) = \sum_{k\geq 1} F'_k D_k(t) = D_0 + \left(D^X_\infty - D_0\right)\sum_{k\geq 1} F'_k\left(1 - e^{-t/\tau_k}\right). \tag{9.36}$$

The reader will have no difficulty in illustrating the effect of crosslinks or crystallinity on the creep behavior of a hypothetical linear standard solid made of totally "un-correlatable" primitive relaxors ($k = 1$), of similar properties to that considered above in stress relaxation and obtain a creep compliance expression and plots homologous to those that led to Fig. 9.6 in stress relaxation.

9.6 Thermally Stimulated Stress Relaxation and Creep

Just as the stress relaxation modulus of a standard linear or non-linear solid, $E_r(t) = E_\infty + (E_0 - E_\infty)e^{-t/\theta}$ is the integral of $\frac{dE_r(t)}{dt} = -\frac{(E_0 - E_\infty)}{\theta}e^{-t/\theta}$, and the corresponding rate of variation with temperature in a scanning experiment, from $T = T_0$ at constant $r = \frac{dT}{dt}$, is $\frac{dE_r(T,r)}{dT} = -\frac{(E_0 - E_\infty)}{r\theta(T)}e^{-(T-T_0)/[r\theta(T)]}$, the predicted rate of variation of the relaxation modulus by CTMD in a *thermally stimulated stress relaxation* of a real non-linear material must be

$$\frac{dE_r(T,r)}{dT} = -\frac{(E_0 - E_\infty)}{r}\sum_{k\geq 1}\frac{F_k'(k,T)}{\theta_k(k,T)}e^{-(T-T_0)/[r\theta_k(k,T)]}. \tag{9.37}$$

Knowing or having determined (as approximately done in Chap. 11) the optimized values of all CTMD parameters, and E_0 and E_∞, (9.37) may be integrated numerically from T_0 to any temperature to obtain scanning rate-dependent curves of $E_r(T,r)$ (to include any transition region) and, if required, back to any lower temperature, to obtain the stress relaxation behavior under (pseudo-)*thermo-mechanical hysteresis.*[65] Of course, both variables—relaxation times, $\theta_k(k,T)$, and weights, $F_k'(k,T)$—must be recalculated at each step of the integration, continuously adjusting the number of terms in the above series to always ensure its convergence within full machine precision, and then carry out the integration with any other less stringent but adequate specified tolerance. At low temperatures, the very large values of $\theta_k(k,T)$ for large clusters in the denominator, combined with their low values of $F_k'(k,T)$ may strongly reduce the number of terms necessary in the above series. On the other hand, at high temperatures, the significant values of $\frac{F_k'(k,T)}{\theta_k(k,T)}$ may be strongly attenuated by the negative exponential (due to the high temperatures and shorter relaxation times), and so the computations should again, in principle, not run into serious numerical difficulties.

Similarly, we may calculate the creep compliance values in *thermally stimulated creep* situations, by means of

$$\frac{dD(T,r)}{dT} = \frac{(D_\infty - D_0)}{r}\sum_{k\geq 1}\frac{F_k'(k,T)}{\tau_k(k,T)}e^{-(T-T_0)/[r\tau_k(k,T)]}, \tag{9.38}$$

now using the retardation times, $\tau_k(k,T)$, and $F_k'(k,T)$ values for creep.

The modifications necessary to predict similar thermally stimulated stress relaxation or creep behavior for crosslinked or semi-crystalline materials will be analogous to those described in Sect. 9.5.

---

[65] Note, however, that in the above procedure we will also be assuming that the scanning rate, r, will be slow enough to always keep the system in thermal equilibrium. The only response lags that could thus be captured would the mechanical ones, resulting from the continuous changes of the individual cluster response times. No true *thermal hysteresis* would be obtainable by this means. Thermal hysteresis is given a very preliminary (and still very conjectural) discussion in Chap. 14.

9.7 Dynamic Mechanical Response

In Sects. 3.1.1.4 and 3.1.2.4, the dynamic mechanical behavior of linear materials was described and formulated in detail, in the absence of irreversible viscous flow, while the case where viscous flow is present was dealt with in Sect. 3.2.4. As may be easily understood, only *linear viscoelastic* dynamic mechanical behavior is worth being analyzed, otherwise we could even end up with fatigue (including thermal fatigue) situations, leading to rupture, and these are left outside the scope of this book. Further, in the present stage of development of CTMD, accounting for viscous flow was not yet considered, but how one may envisage such development in the future will be given a very brief comment in the closing chapter(s) of this book, relating it to the presentation of Sect. 3.2.4.

Now that the basis and present development of CTMD and of its application to stress relaxation and creep have been described, the reader will easily also apply them to the *linear dynamic mechanical behavior* of materials, with reference to (3.4), (3.5), (3.7) and (3.8). So, the *dynamic* (storage and loss) components of the *stress relaxation modulus* may be expressed by

$$E'(\omega, T) = E_\infty + (E_0 - E_\infty) \sum_{k \geq 1} F_k'[k, \theta_k(k, T)] \frac{[\omega\theta_k(k, T)]^2}{1 + [\omega\theta_k(k, T)]^2}, \qquad (9.39)$$

$$E''(\omega, T) = (E_0 - E_\infty) \sum_{k \geq 1} F_k'[k, \theta_k(k, T)] \frac{\omega\theta_k(k, T)}{1 + [\omega\theta_k(k, T)]^2}, \qquad (9.40)$$

respectively, and those of the *dynamic creep compliance* by

$$D'(\omega, T) = D_0 + (D_\infty - D_0) \sum_{k \geq 1} F_k'[k, \tau_k(k, T)] \frac{1}{1 + [\omega\tau_k(k, T)]^2}, \qquad (9.41)$$

and

$$D''(\omega, T) = (D_\infty - D_0) \sum_{k \geq 1} F_k'[k, \tau_k(kT)] \frac{\omega\tau_k(k, T)}{1 + [\omega\tau_k(i, T)]^2}, \qquad (9.42)$$

while

$$\tan[\delta_{E\,or\,D}(\omega, T)] = \frac{E''}{E'} \left(or \, \frac{D''}{D'} \right), \qquad (9.43)$$

but where the latter two results for $\tan[\delta(\omega, T)]$ will not be identical, as shown and discussed in Chaps. 2 and 11.

In all the above, of course, $\omega = 2\pi\nu$, where ν is the frequency of testing in Hz, and the influence of temperature and cluster sizes are taken into account in $\theta_k(k, T)$, $\tau_k(k, T)$, $F_k'[k, \theta_k(k, T)]$ and $F_k'[k, \tau_k(k, T)]$.

For a procedure whereby the material is subject to a *controlled* (very slow) positive or negative *temperature scan*, while being dynamically tested at some constant frequency (as could, if wanted, be implemented in advanced instrumentation), easily obtained differential forms of (9.39)–(9.42) (somewhat analogous to those of (9.37) and (9.38)) would enable to describe the behavior and interpret the corresponding data. But, again, such procedure would of course not give any thermal hysteresis information. The simplest and surest alternative analysis practice, as implemented in conventional instrumentation, involves (1) slowly shifting the temperature to an assigned value, (2) waiting for the fullest possible thermal and mechanical signal stabilization (to some instrument-dependent precisions, (3) taking the readings, and (4) repeating the procedure from (1) to (3) to cover the intended temperature range, for which (9.39)–(9.42) may be directly used. The same set of equations may of course also be directly used to describe the behavior in frequency sweeps at constant temperature, for which the current experimental procedure is like the one above, with the frequency now the shifting variable.

A still outstanding problem at the present stage of CTMD is, again (cf. Sect. 9.4), the fact that, in the absence of some future better alternative, these calculations for the dynamic mechanical behavior will have to use for the response times $\theta_k = \tau_k = \frac{2}{\omega_k} e^{k E_{a,1}/(k_B T)}$, instead of $\theta_k = \frac{2}{\omega_{\beta k}} e^{\beta k E_{a,1}/(k_B T)}$ and $\tau_k = \frac{2}{\omega_{\beta' k}} e^{\beta' k E_{a,1}/(k_B T)}$, requiring one or two extra parameters to be adjusted to experimental data. Some minor influence of temperature on E_∞ and D_∞ may also have to be considered, namely in the case of polymers, depending on molecular weight and the properties imparted by the crosslinks or the crystalline regions on the amorphous matrix.

In Sect. 11.4, the results of the above CTMD theoretical predictions for the dynamic mechanical behavior are presented and discussed, particularly as to the role of clustering in determining the shape of the response curves, in the absence of viscous flow effects, such that, for example in the case of the loss compliance, what is calculated is $D'' - 1/(2\pi \nu \eta)$, where η is the system's viscosity. This will first be done (in the linear viscoelastic domain for all cases, of course) for the same Numerical Example already used in Chap. 8 in the first application of the theory, and then also for the PMMA and PC materials studied in stress relaxation experiments that allowed to quantify their CTMD parameters $(E_{a,1}, T_c, \nu_c, \upsilon)$[66] and relevant mechanical properties (E_0, E_∞).

The reader will not face any conceptual difficulties in establishing how the dynamic mechanical responses of crosslinked or semi-crystalline materials should be formulated, by replacing E_∞ by E_∞^X and D_∞ by D_∞^X in all the above equations and applying the same given procedures of Sects. 9.5.1 and 9.5.2 to estimate E_∞^X, D_∞^X and the corresponding response times, though still neglecting the expected direct influence of f_X on the response times.

---

[66] [where the parameter υ will not be necessary in the linear viscoelastic limit].

9.8   The Effect of Temperature—Explaining Well Established Experimental Results

In Sects. 6.3 (and 6.10) we described and discussed all aspects of the sensitivity to temperature of the behavior of amorphous condensed matter, namely the reasonably widespread (textbook and much published research) view, whereby temperature appears or is claimed to influence the behavior mostly through the level of *molecular packing* or *free volume*. Temperature would thus not seem to result in any strong independent effect other than through volume changes. Many reports, however, have tried to dispute this simplistic view [27–31].

As at the present stage of CTMD every system is considered at constant volume and at some specified (but variable) temperature, let us see what may further result from the significant effect of temperature as formulated within this theory, already illustrated in Figs. 8.3, 8.4, 9.1 and 9.2. What will result if one presents CTMD-predicted data as conventional WLF (Williams-Landel-Ferry) and VTF (Vogel-Tammann-Fulcher) plots, i.e. $\dfrac{1}{\log_{10}\left[\dfrac{\theta^*_{avg}(T_{ref})}{\theta^*_{avg}(T)}\right]} = \dfrac{1}{\log_{10}\left[\dfrac{v^*_{avg}(T)}{v^*_{avg}(T_{ref})}\right]}$ versus $1/(T - T_{ref})$ and $\log_{10}\theta^*_{avg}(T)$ versus $1/(T - T_0)$, where $\theta^*_{avg}(T)$ is taken, for example, as the logarithmic (or generalized geometric) average of the whole range of cluster linear viscoelastic (or equilibrium) response times, $\theta^*_k(T)$, weighed by their $F'^*_k(T)$—i.e. $\ln \theta^*_{avg}(T) = \sum_{k \geq 1} F'^*_k(T) \ln \theta^*_k(T)$, as justified later in this chapter and in Chaps. 10 and 11—, T_{ref} is a reference temperature (for example, $T_{ref} = T_g \cong \frac{T_c}{1.2}$ [32, 33]) and T_0 another (iteratively chosen) reference temperature, the so-called *Vogel temperature*, at which $\theta^*_{avg}(T_0)$ is currently claimed to diverge.

Surprise!—Figs. 9.7, 9.8, 9.9, 9.10 and 9.11, for the same numerical example used in Chap. 8, show that volume does not need to change to obtain the known WLF and VTF type of plots. While Figs. 9.7, 9.8, 9.9 and 9.10 show that, at strictly constant volume, temperature changes of the CTMD-predicted dynamics follow a WLF relationship within the conventional limits of $T_{ref} \pm 50$ K—where T_{ref} was taken at what would be the approximate conventional glass transition temperature, T_g, of the assumed arbitrary material (given that its average response time turns of the order of 10^2 to 10^3 s, as shown in Fig. 9.2)—, Fig. 9.11 shows that CTMD also predicts two VTF relationships at different temperature ranges, as also experimentally found for a wide range of materials [32].

Figures 9.7, 9.8 and 9.9 also indicate how the two WLF parameters, C_1 and C_2, may easily be determined, whose values might arguably be similar above and below T_{ref} (cf. Fig. 9.9). While $C_1 \sim 20$ turns out reasonably close to its known average value when $T_{ref} \sim T_g$, C_2 does not, which may be the result of assuming strictly constant volume conditions. Finally, Fig. 9.10 plots the conventional shift factor, $a_{T/T_{ref}} = \theta^*_{avg}(T)/\theta^*_{avg}(T_{ref})$, much used in current reductions and extrapolations of linear viscoelastic behavior, as described in Chap. 6.

Similar plots have also been obtained and are shown in Chap. 11 for the same PMMA and PC materials studied in stress relaxation experiments. All of them invalidate classic claims that such type of behavior might be the direct and sole result

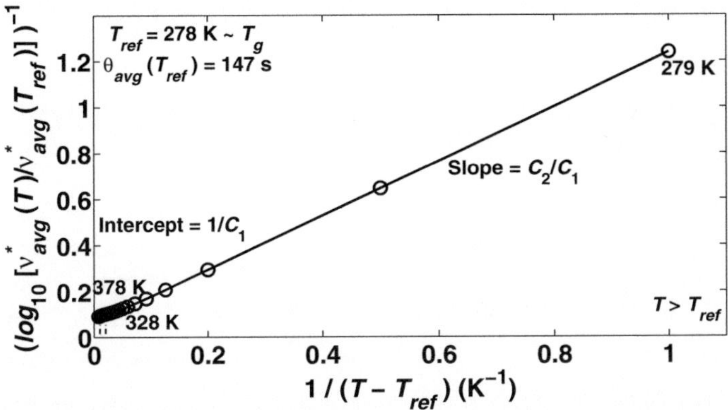

Fig. 9.7 WLF plot at temperatures higher than $T_{ref} \sim T_g$ for the Numerical Example specified in Chap. 8

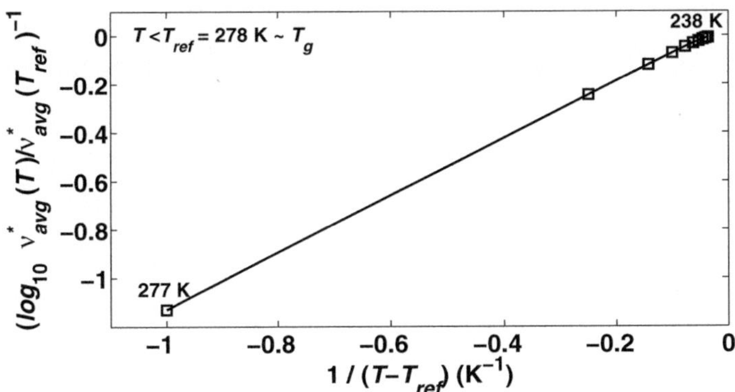

Fig. 9.8 WLF plot at temperatures lower than $T_{ref} \sim T_g$ for the Numerical Example specified in Chap. 8

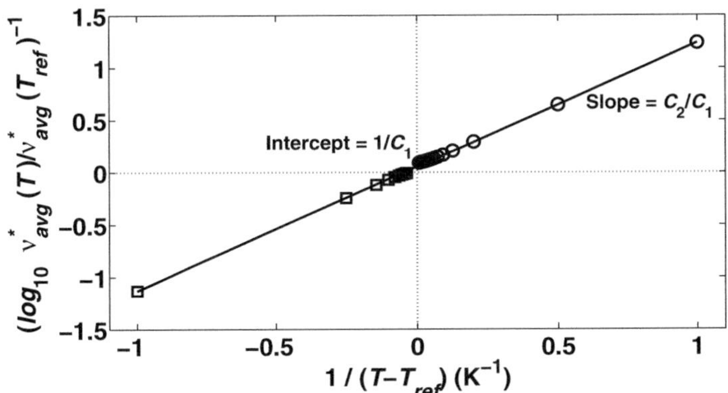

Fig. 9.9 Merged WLF plot from the previous two, for temperatures above and below T_{ref}

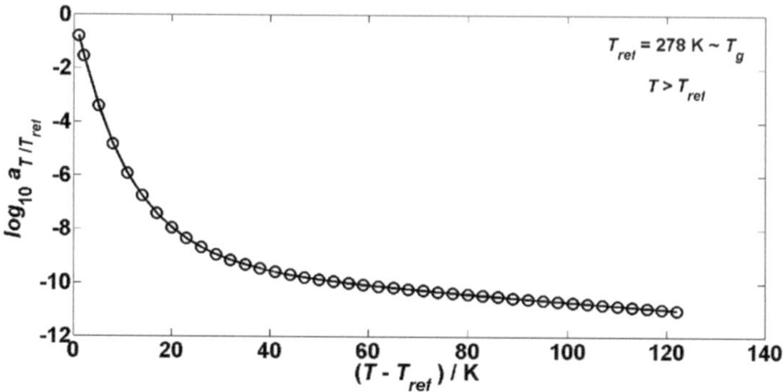

Fig. 9.10 Conventional shift factor at temperatures higher than $T_{ref} \sim T_g$ for the Numerical Example specified in Chap. 8

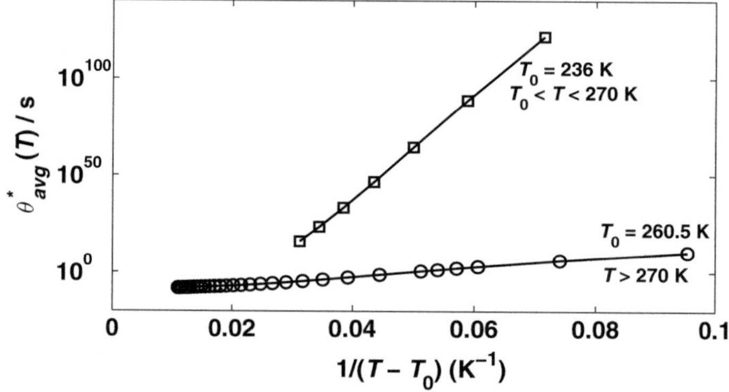

Fig. 9.11 VTF plots for the Numerical Example specified in Chap. 8, for the indicated temperature ranges and values of the Vogel temperature, T_0[67]

of volume changes, as abundantly described in textbooks and early literature and summarized in Sect. 6.10 of this book. Will this mean that CTMD reverses conventional wisdom by excluding any effect of molecular packing? Of course not, as will be seen in Sect. 9.11.

---

[67] Once more, note the extremely wide range of timescales of these calculations, much beyond any accessible (useful and even meaningful) range, given that the estimated age of the Universe stands at less than 5×10^{17} s! This just illustrates that *CTMD totally excludes any divergence whatsoever of response times except at 0 K*, which appears physically meaningful, not nonsense. The minimum temperature of the above calculations was 250 K, and lower temperatures and longer times could not be considered for this example system only because they could not be handled by the double precision numerical processor used (cf. Footnote 28 of Chap. 8). Cluster and average frequencies can be calculated down to $\sim 10^{-308}$ Hz, but obviously not average response times, $\theta^*_{avg}(T) = 1/\left[2\pi \nu^*_{avg}(T)\right]$, with 64-bit number storage.

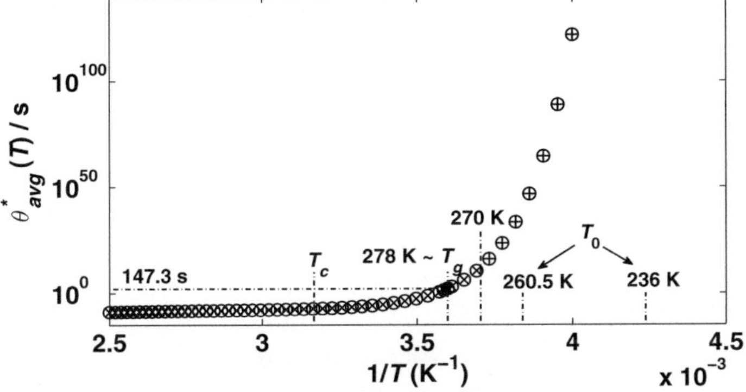

Fig. 9.12 Generalized geometric (or logarithmic) average equilibrium (or linear viscoelastic) response times for the Numerical Example specified in Chap. 8. The data within each of the VTF correlations of Fig. 9.11 are identified by $\otimes$ (< 270 K) and $\otimes$ (> 270 K)

The physical soundness of both WLF and VTF [34, 35][68] correlations may nevertheless be questioned. Do not they look like mere approximate fitting relationships, usable within rather limited temperature ranges? What may physically mean saying that WLF is only "valid" within $T_{ref} \pm 50$ K (or any other range) and that there are *two* VTF relationships—one above and another below some (or whatever) temperature, and *two* temperatures at which the average response time $\theta^*_{avg}(T)$ diverges, *i.e.* cannot be measured or calculated?! Also, the VTF line for $T > 270$ K is not exactly straight, but slightly sigmoid, meaning that the correlation fails both near its T_0, because response times do *not* diverge, and at high temperatures (above T_c), because the real behavior becomes Arrhenius-like.

Figure 9.12 should clarify this matter, where one confirms that average response times do not diverge at $T_0 = 260.5$ K, arguably not even at $T_0 = 236$ K, or at any other temperature (except 0 K).

The plot does not make clear that divergence may (or will) not occur at $T_0 = 236$ K, just because 250 K was the lowest temperature allowed by the numerical processor used in the calculations for this example system,[69] but CTMD predicts that $\theta^*_{avg}(T)$ is a continuous function and that all cluster frequencies become exactly zero *only at 0 K* (cf. (8.3) in Chap. 8 and Fig. 9.13), their approach to zero being faster for larger clusters. In other words, with VTF correlations, one cannot fit any experimental

---

[68] The authors of the first reference explicitly suggest that theories not predicting a dynamic divergence of the VTF form should be focused on.

[69] We again stress that CTMD, with most common numerical processors, allows to calculate and store frequencies down to $\nu \sim 10^{-308}$ Hz (in the absence of emerging intermediate overflows and similar numerical problems), but not the corresponding response times, $\theta = 1/(2\pi\nu)$. In addition to Footnote 67, the reader may here recall the discussion in Sec. 8.3.2.3 and its Footnote 28.

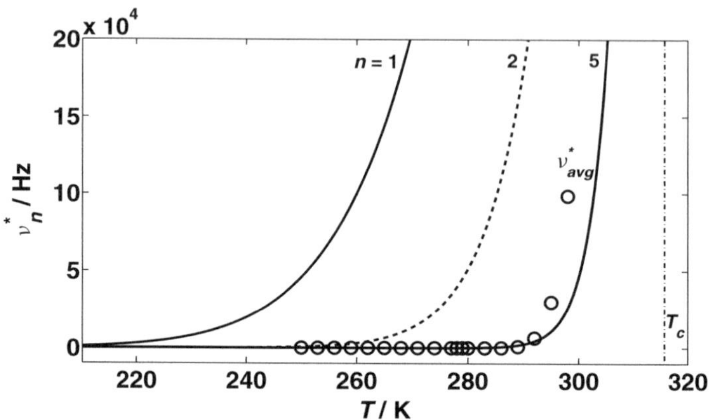

Fig. 9.13 Cluster and average calculated frequencies approaching zero at low temperature

or calculated[70] data following the above *smooth and non-divergent variation* with neither two nor a unique VTF equation (and even less with one with $T_0 = 0$ K, of the Arrhenius type, as Fig. 9.12 shows, except above T_c). This should speak for itself about the meaning and perhaps even the usefulness of VTF relationships.[71] The Figure locates all the above VTF-relevant temperatures, T_c and the approximate T_g at about $T_c/1.14$, at which the calculated average response time turns out 147.3 s. On a side note, in Chap. 11 the reader will find a detailed Matlab$^©$ code that will enable him/her (with the physical parameters of this Numerical Example, not PMMA's) to generate all the data plotted in Fig. 9.12, which he/she may then try and accurately fit to a single 5th order polynomial (with the same number of adjustable parameters as two VTF correlations), while CTMD needs only *three* specific physical property values to generate the same data: $E_{a,1}$, ν_c and T_c.[72] Chapter 11 shows how, from stress relaxation experimental data, one may calculate those properties in reasonable agreement with known values for PMMA, or estimate them for other materials, e.g. PC.

The $\theta^*_{avg}(T)$ calculated behavior of Fig. 9.12 is similar to what is experimentally found and published for a wide range of systems [21], and CTMD justifies it by significant increases in cooperativity upon cooling, as shown in Fig. 9.14, leading to the significantly growing activation energies responsible for the steepness of the $\theta^*_{avg}(T)$ curves. The definition of $F'^*_{n_{max}}$ used (and specified in the Figure) justifies the discrete and irregular values of n^*_{max}—the maximum cooperativities.

---

[70] The calculations behind Fig. 9.12 were intentionally exaggerated towards very long timescales to show that *no divergence* whatsoever of response times may be predicted by CTMD. What might be reasonable? That all motions completely freeze at some non-zero T, or only at $T = 0$ K?

[71] E.-J. Donth refers in Ref. [32] that the physical meaning of the VTF parameters is "*a long running issue and still a mystery*".

[72] VTF parameters do thus not add any extra (physically relevant) information to that of the polynomial's coefficients.

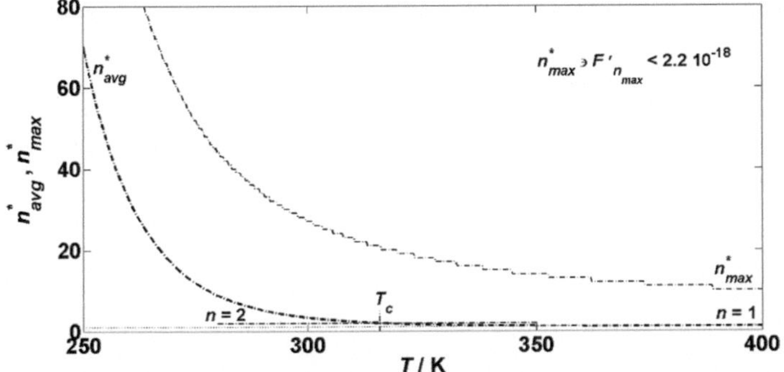

Fig. 9.14 Average and maximum cooperativities (number of primitive relaxors in each cluster) for the Numerical Example specified in Chap. 8

9.9 The Universality of the Behavior

The universality of the behavior of amorphous condensed matter is a often referred characteristic and a detail-independent theory has even explicitly called for [1]. The reader may recall and recognize that Figs. 8.1–8.4 and 9.15, 9.16 and 9.17, as many other diagrams of published data [32] (like Figs. 2.9, p. 39 and 6.1, p. 379 of Ref. [32]), suggest that materials in general show widely similar dynamic behavior, which ought to be characterized by a limited number of functional relationships among operating conditions (where temperature is of paramount relevance) and an also limited number of physically meaningful parameters. Figures 9.15, 9.16 and 9.17 and those of Ref. [32] just referred to do even suggest that relaxation maps for different materials might almost be reduced to a single one if one could identify adequate reducing parameters, one of which may be expected to involve the crossover temperature, T_c.

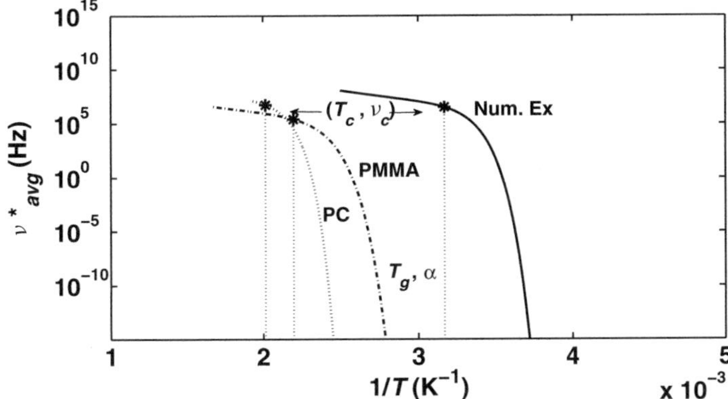

Fig. 9.15 Relaxation maps for the Numerical Example specified in Chap. 8 and the PMMA and PC studied in Chap. 11

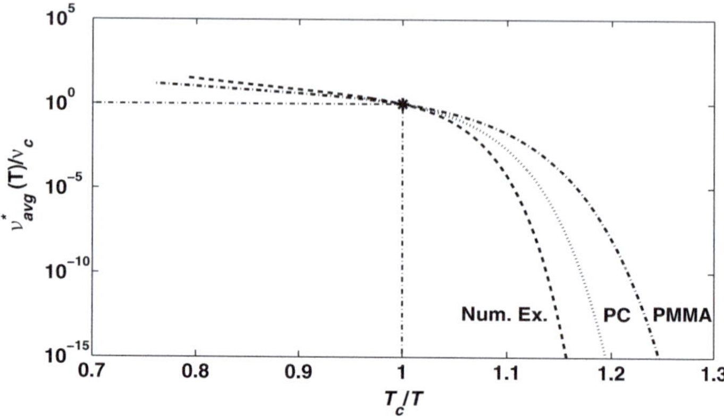

Fig. 9.16 The normalized relaxation maps of Fig. 9.15 relative to the crossover temperatures and frequencies

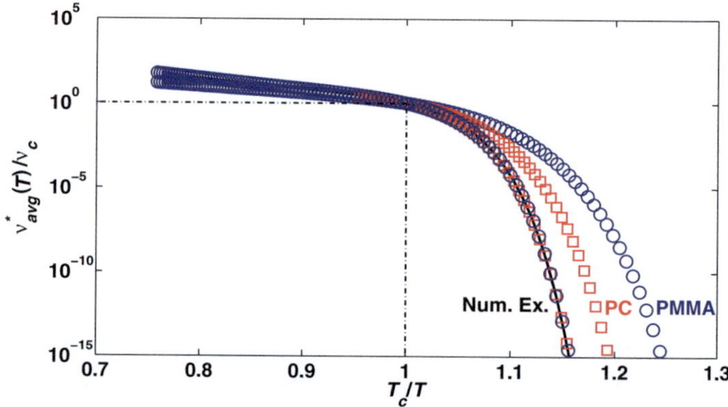

Fig. 9.17 The normalized relaxation maps of Figs. 9.15 and 9.16 relative to the crossover temperatures and frequencies (solid curve and symbols without curves) and corrected (by superposition of the three curves) for the same reduced minimum activation energy

From (9.16) (and (9.24a) for creep), using half of the values of ω_c in (9.18) as previously justified, one may see that, in the linear viscoelastic domain (i.e. for very low ε_0 and σ_0, where $E_{a,1}$ is not significantly altered by strain or stress), the response times of clusters of size k—ignoring for now any expected a priori differences between stress relaxation and creep—may be formulated as[73]

---

[73] In agreement with the *fluctuation–dissipation relationship* (cf. Refs. [24–26]), within the linear viscoelastic domain these response times turn out identical to those of the thermal fluctuations, $\theta_k^*(T)$.

$$\theta_k^*(T) = \frac{1}{\omega_c}\left\{\frac{\sinh\left(\frac{h\nu_c}{k_B T}\right)}{\sinh\left(\frac{h\nu_c}{k_B T_c}\right)}\exp\left[\frac{E_{a,1}-h\nu_c/2}{k_B T_c}\left(\frac{T_c}{T}-1\right)\right]\right\}^k \qquad (9.44)$$

and, as $\sinh\left(\frac{h\nu_c}{k_B T}\right)\approx\frac{h\nu_c}{k_B T}$ to within less than 0.5% for temperatures down to 10^{-4} K (for PC) or even 10^{-6} K (for PMMA),[74] the above equation may be replaced by

$$\theta_k^*(T) \approx \frac{1}{\omega_c}\left\{\frac{T_c}{T}\exp\left[\frac{E_{a,1}-h\nu_c/2}{k_B T_c}\left(\frac{T_c}{T}-1\right)\right]\right\}^k, \qquad (9.45)$$

confirming that normalization of the behavior of amorphous condensed matter appears possible if one uses the crossover temperature to define a *reduced temperature*, T/T_c, and the minimum activation energy, $E_{a,1}$ plus T_c with (or even without ν_c[75]) to define a *reduced minimum activation energy*, $\left(E_{a,1}-\frac{h\nu_c}{2}\right)/(k_B T_c)$, both pertaining to each primitive relaxor. A *reduced cluster response time* may likewise be defined by $\frac{\theta_k^*(T)}{\theta^*(T_c)}=\omega_c\theta_k^*(T)$, or a *reduced response frequency*, ν_k^*/ν_c, bearing in mind that all response times are in CTMD predicted to be identical and equal to ω_c^{-1} at T_c.

Figures 9.15, 9.16 and 9.17 compare and try to reduce the accurate relaxation maps (not the crude approximations of Figs. 8.3 and 8.4) calculated for the same numerical example specified in Chap. 8 and those of the PMMA and PC materials studied in Chap. 11 under stress relaxation, by $\nu_{avg}^*=1/(2\pi\theta_{avg}^*)$, with the θ_{avg}^* formulated in the preceding section. In addition to the obvious similarity of the three relaxation maps, the Figures also show how one may bring them to (1) almost coincidence by normalization by means of the reduced temperatures and frequencies defined just above—cf. Fig. 9.16, and (2) an exact single one *only* if one uses the same reduced minimum activation energy for the three curves—cf. Fig. 9.17, showing that such correction makes the three curves exactly coincident below and also above T_c. Without the latter artificial correction, there is of course no way of representing the behavior by a single two-dimensional curve.

---

[74] The respective ν_c values have been obtained from stress relaxation experiments and are given in Chap. 11.

[75] Cf. Footnote 17 of Chap. 8.

All amorphous materials at equilibrium (or within the linear viscoelastic domain, where the volume of the clusters does not explicitly appear in the expressions of $\theta_k^*(T)$—cf. (9.16) and (9.24a)), thus depend on three parameters—$E_{a,1}$, v_c and T_c—, given that $a_z = \left(z_{1,r}^{\#}/z_1\right)$ is a function of the other three (cf. (9.17)). However, the diagrams of Figs. 8.1 and 8.2 are oversimplified and incomplete, as amorphous materials may only be accurately represented within CTMD by points in tetra-dimensional space, where the fourth coordinate is whatever static or dynamic property one may choose as dependent variable, like θ_{avg}^*, for example. So, complete normalization aimed at illustrating universal behavior (if it were possible, or just approximate, as in Fig. 9.17) requires three reduced parameters, not just T/T_c and $\omega_c\theta_k^*(T) = v_c/v_k^*(T)$—or $\omega_c\theta_{avg}^*(T) = v_c/v_{avg}^*(T)$—, where the third parameter is the reduced activation energy considered in (9.44) and (9.45). The corresponding reducing parameter, $k_B T_c$, turns out to be proportional the basic thermal agitation energy at the crossover.

Using the three independent *reduced parameters* – temperature ($T_r = T/T_c$), frequency ($v_{k,r}^* = v_k^*(T)/v_c$, $v_{avg,r}^*(T) = v_{avg}^*(T)/v_c$) *or* response time ($\theta_{k,r}^*(T) = \omega_c\theta_k^*(T)$, $\theta_{avg,r}^*(T) = \omega_c\theta_{avg}^*(T)$), and activation energy ($E_{a,1,r} = \frac{E_{a,1}-hv_c/2}{k_B T_c}$), the formal relationships

$$\theta_{k,r}^*(T) \approx \left\{T_r \exp\left[E_{a,1,r}(T_r - 1)\right]\right\}^k \tag{9.46}$$

and

$$\theta_{avg,r}^*(T) = e^{\sum\limits_{k\geq 1} F_k'^* \ln\theta_{k,r}^*(T)} \approx \left[T_r e^{E_{a,1,r}(T_r-1)}\right]^{n_{avg}} = \left[T_r e^{E_{a,1,r}(T_r-1)}\right]^{1/\beta_\alpha}, \tag{9.47}$$

are indeed predicted by CTMD to be *universal* (for all amorphous condensed matter), at least below the crossover temperature, T_c, assuming that the nature and properties of a *primitive relaxor* will always be specifiable, where (as shown in Sect. 10.1) $\beta_\alpha = 1/n_{avg}$ stands for the most representative value of the quintessential KWW parameter—the reciprocal of the average cooperativity. And, as this average cooperativity, $n_{avg}(T) = \sum_{k\geq 1} k F_k'^*(T)$, is also a universal function $\mathcal{F}$ of T_r and $E_{a,1,r}$ [*Prove and obtain it from* (9.2)], the final universal relationship becomes

$$\theta_{avg,r}^*(T) = \left[T_r e^{E_{a,1,r}(T_r-1)}\right]^{\mathcal{F}(T_r, E_{a,1,r})}. \tag{9.48}$$

But is everything universal in the above relationships, excluding of course the actual values of the reduced variables? Of course not: the non-universal element must be the number of terms, k_{max}, necessary to obtain the average cooperativity by,

$$n_{avg} = \mathcal{F}\left(T_r, E_{a,1,r}\right) = \sum_{k\geq 1}^{k_{max}} k F_k'^*\left(k, T_r, E_{a,1,r}\right), \tag{9.48a}$$

with all $F'^*_{k>k_{max}} \sim 0$, where k_{max} is a measure of the *detectable maximum cooperativity* for the given material, at the precision level of the calculations,[76] which depends on the nature and properties of the material and its primitive relaxors. Apart from that, the above relationships predicted by CTMD do appear formally universal. Despite their apparent physical soundness, however, future more thorough evaluation will be necessary to establish or not their validity, depending on the physical reasonableness of the F'^*_k formulated at the beginning of this chapter or some future better alternatives.

9.10 CTMD as a Long-Sought Model of Amorphous Materials' Tunable Fragility

Taking the approximate general relationship $T_c/T_g \sim 1.2$ [32, 33], one may also see that CTMD is also consistent with the classical Laughlin-Uhlmann's [35] or Angell's [36–38] *fragility plots* traditionally used to characterize a wide range of liquids and glass formers, for each of which the characteristic viscosity, or an average response time, θ^*_{avg}, may be taken as a function of T_g/T.

The so-called *fragility index* is usually defined by $m = \left[d \log_{10}(\eta \text{ or } \theta^*_{avg})/d(T_g/T) \right]_{T=T_g}$, which increases significantly from *strong* to *fragile* glass formers [36–38], according to the magnitude of the activation energy barriers that control the response, and this is consistent with the role of $E_{a,1}$ in (9.45), coupled with the expected (and confirmed—cf. Fig. 9.2 and, even more clearly, Fig. 9.14 in this chapter) strong increase of the operating average relaxor size, n_{avg}, at decreasing temperatures, recalling that $E_{a,n} = nE_{a,1}$.

One may even anticipate that the above fragility index may turn out to be a weighed superposition of elementary contributions from each type and size of clusters, k, which may be obtained from (9.45) and the definition of θ^*_{avg} already used in Sect. 9.8 as

$$m = \sum_{k\geq 1} F'^*_k(T_g)m'_k, \text{ with } m'_k = m_k + \left[\frac{d \ln F'^*_k(T)}{d(T_g/T)} \log_{10} \theta^*_k(T) \right]_{T=T_g}$$

and

$$m_k = \left[d \log_{10}\theta^*_k(T)/d(T_g/T) \right]_{T=T_g} \approx 0.521k\left(0.833 + \frac{E_{a,1} - h\nu_c/2}{k_B T_c} \right)$$

$$= f\left(k, \nu_c, E_{a,1}, T_c\right) \tag{9.49}$$

---

[76] At non-equilibrium (dynamic) conditions, the maximum cooperativity may also depend on the timescale of the observation, experiment or calculation, as suggested in Chap. 14.

taking $T_c/T_g \sim 1.2$ and where $F_k^{'*}(T)$ and $\theta_k^*(T)$ are of course also functions of k, v_c, $E_{a,1}$ and T_c (cf. (9.2) and (9.45)), assuming the cluster weights formulated at the beginning of this chapter or some future better alternative. Any other definitions of θ_{avg}^* would of course also yield appropriately modified fragility indexes dependent on the same variables.

It may be seen that the contribution of each cluster size to the overall fragility index has two additive components—one due to the temperature dependence of its specific response time, yielding m_k, and another linked to the temperature dependence of the contribution to its response time spectrum, $F_k^{'*}$. This stems from CTMD's capability of predicting that, as temperature decreases towards and below T_g, greater contributions of larger clusters (with longer response times and higher activation energies) should result, as physically expected, leading to the high indexes known to characterize fragile glasses. The final closed form for the CTMD *fragility index, m,* becomes

$$ m \approx 0.521 \left(0.833 + \frac{E_{a,1} - hv_c/2}{k_B T_c} \right) \left(\sum_{k \geq 1} k F_k^{'*} \right)_{T=T_g} + \left[\sum_{k \geq 1} \frac{dF_k^{'*}}{d(T_g/T)} \log_{10}\theta_k^* \right]_{T=T_g}, \quad (9.50) $$

which looks very difficult to calculate, but not to CTMD (using a powerful processor like Matlab© to numerically differentiate the calculated $F_k^{'*}$ or, much better, directly the calculated θ_{avg}^*). In Chap. 11 it will be shown that from the T_c, $E_{a,1}$ and v_c values estimated for PMMA from stress relaxation experiments, we could obtain $m = 133.3$, in excellent agreement with the literature value of 136 [32].

Finally, an alternative but more fundamental and accurate, T_c-based, formulation of the fragility index of a glass, m^*, may and should now be suggested as $m^* = \left[d\ln \theta_{avg}^* / d(T_c/T) \right]_{T=T_c}$, to replace the traditional one (9.50) which is more difficult to calculate and inconveniently dependent on the measurement's timescale via T_g.[77] This new *fragility index* may then be expressed by

$$ m^* \approx \left(1 + \frac{E_{a,1} - hv_c/2}{k_B T_c} \right) \left(\sum_{k \geq 1} k F_k^{'*} \right)_{T=T_c} = 2 \left(1 + \frac{E_{a,1} - hv_c/2}{k_B T_c} \right), \quad (9.51) $$

a very simple result. This is because, at $T = T_c$, (1) all θ_k^* are identical, $\ln \theta_c$ may be factored out and (2) all $F_k^{'*}$ add up to 1, making null the second term in the T_c-analogue of (9.50), and (3) at $T = T_c$, $\left(\sum_{k \geq 1} k F_k^{'*} \right)_{T=T_c} = n_{avg}^*(T_c)$—the average number of primitive relaxors in each cluster, or average cooperativity, at the crossover temperature—is predicted by CTMD to be exactly 2 for *all materials*, as shown for the Numerical Example in Fig. 9.14[78] [*Prove it analytically as a simple exercise!*]. So, within CTMD, the crossover temperature may also be defined as the temperature

---

[77] S. A. Kivelson and G. Tarjus state in Ref. [1] that "*being time-scale dependent, T_g is clearly irrelevant from the standpoint of fundamental physics*" (in contrast with T_c, we would say).

[78] What is not universal is of course the nature and size of each primitive relaxor.

below which a material responds to thermal or other physical excitations by clusters of an average number of primitive relaxors higher than 2.

Equation (9.50) (or, preferably, (9.51)) should thus establish alternative definitions of a *universal fragility relationship* (for all materials), expressing it as a simple function of the previously defined primitive relaxor's reduced activation energy. What more appropriate definition! The much lower new fragility indexes, m^* (for example, 20.84 for PMMA as obtained in Chap. 11) should however stand as a much *better defined* and *fundamental, timescale-independent*, property of the system.

Therefore, CTMD also meets the specification explicitly set by G. Tarjus [39, 40] and other researchers, for example during the *5th International Workshop on Complex Systems* in Sendai, Japan, September 2007, for a promising theory of glass behavior—the ability to provide a *"model of tunable fragility"*. The relevant *tuning parameters* clearly emerge in CTMD as T_c, $E_{a,1}$ and v_c—the same that locate each specific material structure on the tridimensional crossover maps of Figs. 8.1 and 8.2.

The whole of CTMD, Figs. 9.15, 9.16 and 9.17, and the proposed new definition of fragility (normalized by the procedure of the preceding section), yielding $m^* = 2(1 + E_{a,1,r})$, where $E_{a,1,r}$ is the reduced activation energy defined after (9.45), all stand as excellent illustrations and proof of the until now puzzling similarity (or universality) of behavior of the whole range of glass formers.

9.11 Molecular Packing Effects Predictable by CTMD

The reader may have registered that, in stress relaxation, it was concluded in Sect. 9.4 that $\alpha_1 > 3|\alpha_2|$ and thus $(\alpha_1 - \alpha_2)$ and $(\alpha_1 + \alpha_2)$, both proportional to the total (occupied + *free*) volume of each cluster, are always positive, even when α_2 turns out negative as a possible result of direct transitions leading to states of negative (compressive) local stress. Therefore, from (9.16), one sees that all relaxation times, θ_k, are predicted to increase if free volume decreases by lowering the temperature. Likewise, as in creep (when related to stress relaxation of the same material) $\alpha \gg \alpha_1 > 0$, one concludes from (9.23) that all retardation times, τ_k, also increase when the material contracts as temperature is lowered, perhaps even more significantly than the corresponding relaxation times. To account for the effect of molecular packing within not too wide temperature ranges, one may thus (as a first approximation) take a simple linear variation of all cluster volumes with temperature, though at the expense of one more parameter—the average thermal expansion coefficient—whose value, however, will have to differ below and above some low temperature, T_g, depending on the timescale.

A more significant effect might however be that of the minimum activation energy, $E_{a,1}$, and therefore all clusters' activation energies, $E_{a,k} = kE_{a,1}$, which are expected to increase upon volume reductions by cooling.[79,80] How will that be? To our knowledge, a detailed theory of this effect on the potential energy landscape is still lacking, but one may semi-quantitatively guess that all $E_{a,k} \propto (free volume)^{-\delta}$ with $\delta < 2/3$, by analogy with, and contrast to, ideal gases, for which we know that their energy levels are exactly proportional to $(volume)^{-2/3}$. It thus appears reasonable to assume that temperature will determine the dynamic behavior through the *activation* of local structural transitions, whose activation energies are however expected to depend on molecular packing.[81]

CTMD is open to future improvements directed at accounting for variations in molecular packing, either (1) through the above semi-quantitative corrections and/or, in future more elaborate and accurate developments, (2) by trying to apply CTMD itself to the process of *volume relaxation* and *free volume redistribution*, because one expects that the same or similar types of single and clustered motions may be involved in molecular packing variations and free volume redistribution within any material. This might in the end lead, for example, to final small adjustments to the WLF parameters obtained by CTMD for each specific system at constant volume conditions, arguably without invalidating a predominant and direct effect of temperature on condensed matter dynamics, as quantified by this theory.

As, in stress relaxation and creep, we had to quantify the elastic stress and delayed elastic strain changes, respectively, imparted by each individual cluster transition, we will then need to find how those same transitions might yield local free and total volume changes and formulate them accordingly. Having in mind (and generalizing) what may be assumed to happen in the stress relaxation of polymers (where the uncoiling of polymer chains under constant total strain should be responsible for the relaxation of the stresses of the molecular skeleton), direct/reverse transitions in stress relaxation might be assumed to lead to negative/positive free volume changes proportional to the corresponding stress change. In the simplest possible formulation, one may assume the same absolute value per primitive relaxor for each of those elementary changes in free volume, possibly of the order of twice its activation volume, $v^{\#}$, as defined in Sect. 9.2.1. And, in line with the original stress relaxation formulation in that section, the mentioned proportionality constant of the free volume change to the stress change would turn out equal to v/E_0.

Likewise, having in mind the uncoiling of polymer chains under constant stress, direct/reverse transitions in creep might arguably also lead to negative/positive free volume changes and, in line with the original creep formulation of Sect. 9.3.1, the

---

[79] Cf. in Sect. 9.2.4 the text related to and including Footnote 44.

[80] The potential energy landscape can only be represented in multi-dimensional space. Nevertheless, imagine that the molecular energy landscape is "viewed" as "projected" onto a plane, thus "becoming" bi-dimensional, with energy represented along the orthogonal dimension, as might be "pictured" by a sheet of paper that you squeezed in your hands to make it full of wrinkles and folds. Place it on a table and compress it bi-dimensionally with both hands. Do the wrinkles and folds stay the same as before, attenuate, or increase in size?

[81] Why would chemical reactions be the only types of activated processes in the physical realm?

proportionality constant of the free volume change per primitive relaxor to the corresponding delayed elastic strain change might then turn out of the order of υ, its total volume.

Whether each direct/reverse transition in stress relaxation and creep of non-polymeric systems might also be assumed to lead to a local reduction/increase of free volume of the above magnitude, identically related to the stress and delayed elastic chain changes, is however less clear. How one could reach similar (or alternative) conclusions in terms of transitions from "solid-like" to "liquid-like" cells, following Cohen and Grest's nomenclature and formulation [41], or between "liquid-like" cells with different free volumes, seems to stand as a major outstanding problem in the expected and recommended future development of the free volume theories. One important thing that is clear, as recognized by Cohen and Grest [41] themselves, is that such "solid-like" to "liquid-like" cell transitions (or between different "liquid-like" cells, we would add), particularly at low temperatures, should be *activated*, and therefore directly and significantly dependent on temperature.

A more comprehensive future physical description and mathematical formulation of *dynamic heterogeneity* should thus ideally result, for both stress relaxation and creep situations, from the forementioned analysis, including the transition frequencies (as calculated in Chaps. 8 and 9), the effect of intermittency on clusters' weights, and the fluctuations of free volume.

The same strategy would be applicable to the formulation of simple thermal equilibrium (in the absence of any external physical excitation), by equating the total free volume changes per unit time resulting from the direct and the reverse transitions. As to the slow thermal expansion of the system (assumed at quasi-equilibrium) upon temperature increases at constant pressure, and adopting Cohen and Grest's concepts (if tried out), it could be accounted for by calculation of the renewed equilibrium of the "solid-like" to liquid-like" and reverse activated cell transitions, combined with the volume expansion of both types of cells. Within CTMD, the same strategy would be applicable with reference to the equilibrium between unrelaxed and relaxed primitive relaxors within clusters of the whole range of sizes. Under the simplifying assumption of equal direct and reverse activation energies, all equilibria, at any temperature, will be characterized by equal numbers of unrelaxed and relaxed structures (whose volumes will however change with temperature). In the absence of forced excitations, the mentioned equilibrium "*composition*" will only change with temperature when those activation energies will be different.

Chapter 15 in Part III sketches how CTMD might be extended to a first formulation of volume relaxation.

9.12 The Physical Parameters of CTMD

The text of Chaps. 8 and 9 already identified that the CTMD's specific parameters are $E_{a,1}$—the minimum activation energy, ν_c—the crossover frequency, T_c—the crossover temperature (where these three together determine the residual partition function ratio $z^{\#}_{1,r}/z_1$ by (9.17)), plus those of mechanical nature, E_0 – the instantaneous modulus (or D_0—the instantaneous creep compliance), E_∞—the relaxed modulus (or D_∞—the ultimate creep compliance), neglecting viscous flow, or the corresponding dielectric or other variables, and υ_1—the total (occupied and free) average volume of each primitive relaxor.

References

1. S.A. Kivelson, G. Tarjus, Nat. Mat. **7**, 831 (2008)
2. J.R.S. André, J.J.C. Cruz Pinto, Polym. Eng. Sci. **54**, 404 (2014)
3. J.R.S. André, Fluência de Polímeros—Fenomenologia e Modelação Dinâmica Molecular (Polymer Creep—Phenomenology and Modeling of Molecular Dynamics), Doctoral Thesis, Universidade de Aveiro, 2004
4. J.J.C. Cruz Pinto, J.R.S. André, Polym. Eng. Sci. **56**, 348 (2016)
5. J.L. Koenig, *Spectroscopy of Polymers*, in ACS Professional Reference Book (American Chemical Society, 1992)
6. Y. Lee, R.S. Bretzlaff, R.P. Wool, J. Polym. Sci. B: Polym. Phys. **22**, 681 (1984)
7. F. Guiu, P.L. Pratt, Phys. Stat. Sol. **6**, 111 (1964)
8. B. Escaig, in *Plastic Deformation of Amorphous and Semi-Crystalline Materials*, ed. by B. Escaig, C. G'Shell (Les Éditions de Physique Les Ulis, France, 1982), pp. 187–225
9. R.G. Larson, Rheol. Acta **24**, 327 (1985)
10. J. Sweeney, I.M. Ward, J. Mater. Sci. **25**, 697 (1990)
11. S. Houshyar, R.A. Shanks, A. Hodzic, Polym. Test. **24**, 257 (2005)
12. A.S. Krausz, H. Eyring, *Deformation Kinetics* (Wiley-Interscience, New York, 1975)
13. W.N. Findley, J.S. Lai, K. Onaran, *Creepand Relaxation of Non-Linear Viscoelastic Materials* (Dover Publications, New York, 1989)
14. J. Jäckle, R. Richert, Phys. Rev. E **77**, 031201 (2008)
15. G. Adam, J.H. Gibbs, J. Chem. Phys. **43**, 139 (1965)
16. K.L. Ngai, Comments Solid State Phys. **9**, 121 (1979)
17. K.L. Ngai, J. Chem. Phys. **109**, 6982 (1998)
18. K.L. Ngai, Phys. Cond. Matter **15**, S1107 (2003)
19. K.L. Ngai, M. Paluch, J. Chem. Phys. **120**, 857 (2004)
20. K.L. Ngai, J. Non-Cryst, Solids **353**, 709 (2007)
21. K.L. Ngai, *Relaxation and Diffusion in Complex Systems* (Springer, New York, 2011)
22. R. Kohlrausch, Ann. Phys. Chem. **91** (1), 56, 179 (1854)
23. G. Williams, D.C. Watts, Trans. Faraday Soc. **66**, 80 (1970)
24. H. Nyquist, Phys. Rev. **32**, 110 (1928)
25. E.-J. Donth, J. Phys. Condens. Matter: Condens. Matter **12**, 10371 (2000)
26. G.E. Crooks, Phys. Rev. E, **60**(3), 2721 (1999)
27. M.L. Ferrer, Ch. Lawrence, B.G. Demirjian, D. Kivelson, C. Alba-Simionesco, G. Tarjus, J. Chem. Phys., **109**, 8010 (1998)
28. G. Williams, in *Dielectric Spectroscopy of Polymeric Materials*, ed. by J.P. Runt, J.J. Fitzgerald (ACS, Washington DC, 1997), Chap. 1
29. M. Goldstein, J. Chem. Phys. **51**, 3728 (1969)
30. C.A. Angell, S. Borick, J. Non-Cryst, Solids **307–310**, 393 (2002)
31. D. Kivelson, G. Tarjus in *Physics of Glasses*, ed. by P. Jund, R. Jullien (AIP, Melville, NY, 1999)
32. E.-J. Donth, *The Glass Transition—Relaxation Dynamics in Liquids and Disordered Materials*, *Springer Series in Materials Science-48* (Springer, Berlin, 2001)
33. H.J. Sillescu, J. Non-Cryst, Solids **243**, 81 (1999)

34. T. Hecksher, A.I. Nielsen, N.B. Olsen, J.C. Dyre, Nat. Phys. (2008)
35. W.T. Laughlin, D.R. Uhlmann, J. Phys. Chem. **76**, 2317 (1972)
36. C.A. Angell, in *Relaxation in Complex Systems*, ed. by K.L. Ngai, G.B. Wright (U. S. Departmrnt of Commerce, Springfield, 1985), p. 1
37. C.A. Angell, J. Non-Cryst, Solids **131–133**, 13 (1991)
38. C.A. Angell, Science **267**, 1924 (1995)
39. G. Tarjus, *Invited Lecture*(not published), 5th International Worlshop on Complex Systems, Sendai, Japan (Sep 2007)
40. G. Tarjus, in *Dynamical Heterogeneities in Glasses, Colloids, and Granular Media*, ed. by L. Berthier et al. (OUP, Oxford, 2011)
41. M.H. Cohen, G.S. Grest, Phys. Rev. B **20**, 1077 (1979)

Chapter 10
Physical Comparison of CTMD with Other Theoretical Contributions and Dynamic Models

10.1 CTMD and Ngai's Coupling Model (CM)

As described in Sec. 7.1, *Ngai's coupling model* or *correlation* [1–6] started from the observation that most materials' responses are significantly extended in time—being well (even if only *approximately*) represented by an extended, KWW [7, 8], exponential—and the attempt to interpret them as the result of a transition, at some time, t_c, from the response of single, uncorrelated, *"primitive relaxors"* to the cooperative response of groups of such relaxors, in clusters whose sizes increase as the temperature is lowered, as also predicted by CTMD and shown in Sect. 9.1 of this book.

We recall that, taking the above assumed transition, at $t = t_c$, from a single-exponential (Debye) to an extended exponential (KWW) response, i.e. $e^{-t_c/\theta_{0\alpha}} = e^{-(t_c/\theta_\alpha)^{\beta_\alpha}}$, Ngai obtained the fundamental relationship of his *coupling model* (CM), $\theta_\alpha = \left[t_c^{-(1-\beta_\alpha)} \theta_{0\alpha} \right]^{1/\beta_\alpha}$, where $\theta_{0\alpha} = \theta_1$ is the characteristic response time of the primitive relaxors (expectedly the first that show their contribution to the measured response), θ_α is the average response time of the material during its cooperative, extended-exponential, overall time-dependent response, and $\beta_\alpha < 1$ is identical to Ngai's $(1 - n)$ parameter.

In the developed CTMD *analytical* theory of Chaps. 8 and 9, the details of the *random and intermittent* (and thus *chaotic*) *clustering* of the assumed primitive relaxors were considered by *statistical mechanical arguments*, with the main result that, for the clusters of k (≥ 1) primitive relaxors participating in any overall time-dependent response of the material (stress relaxation, creep, dynamic mechanical, etc.), one finds that their specific response times, $\theta_k(T)$, become, in the linear limit, or at thermal equilibrium,

$$\theta_k(T) = [\omega_c \theta_1(T)]^k / \omega_c, \tag{10.1}$$

as the reader may work out, where $\theta_1 = \theta_{0\alpha}$ is the *response time of the* assumed *primitive relaxors* and ω_c is the *crossover angular frequency*, in the exact sense used by e.g. Donth [9], also considered within the mode coupling, MCT, theory [10]. In terms of the retardation times, θ_k should be replaced by τ_k, and (9.16) by (9.24a).

As we have seen, the formulation turns out to potentially cover *any temperature*, T (with $0 \leq T < T_{degradation}$), *down to* 0 K, and *any timescales* (in principle, no matter how long), with predictions that appear here and in Chaps. 8, 9 and 11 to physically agree with most of the known experimental features of materials' responses (listed in Chap. 6). Though the behavior at such low temperatures is not experimentally accessible in a direct way, the theory's predictions make good physical sense, in addition to being at least approximate at experimentally accessible ones.

Within CTMD, the average response time, $\theta_{avg} = \theta_\alpha$, of the overall response may be best defined by[1]

$$\ln \theta_{avg}(T) = \sum_{k \geq 1} F_k'(T) \ln \theta_k(T), \quad \text{or}$$

$$\theta_{avg}(T) = \prod_{k \geq 1} [\theta_k(T)]^{F_k'(T)} \tag{10.2}$$

i.e. as a *logarithmic* or *generalized geometric mean* of the response times of the individual cluster sizes, where the $F_k'(T)$, with $\sum_{k \geq 1} F_k'(T) = 1$, are their relative weights, maybe as calculated in Sect. 9.1, such as to take into account the intermittent, and thus chaotic, nature of the primitive relaxors' participation in clusters of the various sizes (*dynamic heterogeneity*). This, we recall, was achieved by first (and tentatively) formulating them in Sect. 9.1, per primitive relaxor of the material, as

$$F_k'(T) \propto k\theta_k(T)/k!, \quad with \quad \sum_{k \geq 1} F_k'(T) = 1, \quad \text{to yield, as shown,}$$

$$F_k'(T) = \frac{1}{(k-1)!} \frac{\theta_k(T)}{\sum_{j \geq 1}\{\theta_j(T)/(j-1)!\}}. \tag{10.3}$$

Substituting the above $\theta_k(T)$ of (10.1) in CTMD's expression of $\theta_{avg} = \theta_\alpha$ (10.2), one obtains, with $\theta_1(T) = \theta_{0,\alpha}(T)$,

$$\ln \theta_\alpha(T) = \left[\sum_{k \geq 1}(k-1)F_k'(T)\right] \ln \omega_c + \left[\sum_{k \geq 1}k F_k'(T)\right] \ln \theta_{0\alpha}(T), \quad \text{or}$$

$$\theta_\alpha(T) = \theta_{avg}(T) = \frac{[\omega_c\theta_{0\alpha}(T)]^{n_{avg}(T)}}{\omega_c} = \frac{[\omega_c\theta_1(T)]^{n_{avg}(T)}}{\omega_c}, \tag{10.4}$$

---

[1] All CTMD-calculated $\ln \theta_k(T)$ turn out evenly spaced and, with reference to real experimental response (stress relaxation and creep or compliance) curves, their weighted average, $\ln\theta_{avg}(T)$, in addition to thus being a good average candidate, is generally coincident or close to the inflection point of the response curves.

where $n_{avg}(T) = \sum_{k \geq 1} k F_k'(T)$ stands as an *average cooperativity* of the material's response. We should note that 10.4 does not depend on the actual distribution of cluster sizes, $F_k'(T)$, and so it will remain valid in case the weights $F_k'(T)$ obtained in Chap. 9 might need modification.

Apart from the new identification and interpretation that may now be offered for the relevant parameters, the above result is identical to the one K. L. Ngai had originally obtained with his coupling model. This may be a logical consequence of both approaches (CM and CTMD)'s link to, and claimed attempt to account for, the *chaotic interaction of multiple relaxation centers* in varying numbers, now more explicitly dealt with by CTMD.

The main physical parameters of CTMD are thus the crossover frequency, ν_c or $\omega_c = 2\pi \nu_c$, the crossover temperature, T_c, and the average activation energy of the primitive relaxors's response, as already identified in Chap. 8, whose values are known for many materials (cf. Donth [9], and Beiner et al. [11]) or may be estimated. In Chap. 11, it will be shown that we are able to recover nearly those same known values (plus the correct estimate of the glass transition temperature, T_g) from 24-h stress relaxation experiments with PMMA and PC at 70 and 100 °C below their respective T_g—all this with relatively simple, analytical, mathematics and very fast computations.

So, the conclusions on the CTMD-CM relationships are:

(1)   The very well-named, but until now unidentified, *"quintessential parameter"* of Ngai's coupling model and of KWW extended exponential relationships may be taken as the *average cooperativity*, $n_{avg}(T)$, maybe better than $(1 - n)^{-1}$ or n_{avg}'s reciprocal, $\beta_\alpha(T)$;

(2)   Not just the average response time, $\theta_\alpha(T)$—(10.4)—, but also each and every one of the cluster size-specific response times, $\theta_k(T)$—(10.1)—, obey the basic relationship of Ngai's coupling model, their own specific "quintessential parameter" being the specific size of each type of clusters, k (what else?);

(3) t_c is just proportional to the reciprocal of the crossover angular frequency, ω_c, a *material-specific, time scale- and temperature-insensitive, property*;

(4)   The linkage of the above mentioned "quintessential parameter" with the *many-body relaxation* nature of the materials response now couldn't be clearer—according to CTMD, it is no less than its direct measure or, literally, *how many primitive relaxors take part in the various clusters*, both globally as an average, and in every one of the participating clusters as its specific size, k, at any given temperature, T.

As three closing notes on the KWW response function, one may now observe that CTMD also establishes the specific conditions that KWW's parameters must satisfy to adequately (even if only approximately) represent the experimental response behavior:

(a)   The function's average response time may well turn out to be CTMD's predicted *logarithmic* or *generalized geometric average*;

(b)   The "extending" parameter, $\beta_\alpha(T)$, turning out as the exact *reciprocal* of the predicted temperature-dependent *average cooperativity*, should therefore and does effectively decrease at lower temperatures, because of increased cooperativity;

(c)   CTMD's result that, not only the average response time, but also the response time spectrum width increase as the temperature is lowered (as illustrated in the examples of Figs. 9.2, 11.26 and 11.27 and experimentally found) explains the widespread recognition that the two KWW parameters are not independent, and thus not very adaptable to undisputable fits to experimental response curves.

In fact, as both parameters are directly and explicitly predicted by CTMD, and are necessarily related—because the width at the left of the average of the spectrum of $\ln(\omega_c\theta_k)$ may be calculated as $\ln\left[(\omega_c\theta_{avg})/(\omega_c\theta_1)\right] = \left(\frac{1}{\beta_\alpha} - 1\right)\ln(\omega_c\theta_1)$, which is equivalent to $\ln(\omega_c\theta_{avg}) = n_{avg}\ln(\omega_c\theta_1)$—, they should not be obtained by mere and independent numerical fitting to experimental data. Thus, CTMD's formulation is physically much better grounded than the empirical KWW one, in addition to (as it must) exposing itself to unequivocal falsification *or* validation of all its logical consequences by experimental measurements. The above relationships enable to quantify how the mentioned spectrum width, as the temperature is reduced, increases more than proportionately to the predicted increase of $n_{avg} \equiv 1/\beta_\alpha$, given that—as shown in Sect. 9.9—by (9.45) the factor $\ln(\omega_c\theta_1)$ above also increases more than proportionately to $(T_c - T)$. These facts also imply that *no exact*, only approximate, time–temperature equivalence or superposition should indeed be expected, except within a limited range of temperatures, as experimentally found and illustrated in Chaps. 9 and 11, which supports *rheological complexity*—not simplicity—for most materials.

10.2   CTMD and Götze's Mode Coupling Theory (MCT)

Section 7.2 of this book identified the essentials of the Mode Coupling Theory (MCT), and Chaps. 8 and 9 worked out the physical and mathematical foundations, details, and many of the predictions of this *cooperative theory of materials dynamics* (CTMD), while Chap. 11 and some appendixes expand on those and other theoretical predictions and compares them with what may be deduced from stress relaxation experiments on two amorphous polymers.

Based on the contents of the ·previous sections, chapters and appendixes, a preliminary qualitative comparison of both theories may be itemized as follows:

1.   MCT is a mean-field theory, which considers that the dynamic behavior of amorphous condensed matter is governed by density fluctuations. CTMD tries to delve more deeply into microscopic and mechanistic details of the thermal and other

physical responses of amorphous materials, considering specific local, thermally activated, motions of widely varying scales[2] at conditions of constant volume.

2. The MCT coupling parameters appropriate to any given system are variable with temperature, density, etc., and are not easily determined, their physical nature remaining quite opaque and unknown what universal principle(s) may determine their values, thus being so far difficult to distinguish them from "fitting constants". CTMD's parameters are all physically identifiable – specific properties, expressed in its own physical units—and at least approximately quantifiable, some of them being already known (and tabulated [9, 11]) for a range of systems.

3. MCT is mathematically very complex, even in the analysis of the simplest systems. CTMD uses simple and mostly analytical mathematics, only requiring the use of one or other conventional numerical algorithm (in very fast computations) to treat more complex cases.

4. MCT successfully predicts a transition at a temperature, T_c—initially wrongly identified as a glass transition, but then rightly as a crossover between different dynamics—, at which response times are predicted to diverge, unless some (phenomenological) ergodicity-restoring, hopping-like, activated mechanism is included in the analysis. CTMD predicts a crossover temperature, T_c, and frequency, ν_c, and quantitatively relates them to specific physical properties of the system, but predicts no divergence whatsoever of the response times, not even at very low and inaccessible temperatures (to both experiments, MCT calculations, and current molecular dynamics simulations)—strictly, and mathematically, only at 0 K. All CTMD response rates are *non-zero everywhere above* 0 K.

5. MCT appears physically and mathematically limited to (quasi-) equilibrium and linear behavior. CTMD starts with and formulates thermal equilibrium, but then extends to non-equilibrium, non-linear forced responses, no matter how slow they might be, within only numerical processor memory (not computing speed) limitations. In the linear, low excitation limit, CTMD's predictions comply with the fluctuation–dissipation relationship (FDR).

6. For the above reasons, MCT may only "visit" the vicinity of the crossover and higher temperatures. CTMD, although of doubtful or restricted applicability to temperatures higher than T_c, may at least approximately describe the behavior of amorphous (and arguably even crosslinked and/or semi-crystalline) condensed matter down to temperatures well below, not just T_c, but T_g.

7. With MCT it is difficult to account for and explain cooperativity, dynamic heterogeneity and the concept and values of fragility. CTMD incorporates and accounts for dynamic heterogeneity and ends up formulating, quantifying and generalizing the practical concept of the fragility of glasses, as a function of its main parameters, thus contributing to a model of tunable fragility.

---

[2] These motions surely lead to local density fluctuations, but they have not yet been directly formulated in the foregoing development of CTMD. However, it is pointed out in Chap. 11 and shown in Chap. 15, that changes in molecular packing might also be described in a future extension of the theory.

8. MCT succeeds in predicting the universality of the crossover temperature, the approximate two-step relaxation mechanism (one at high temperatures, followed by the extended, KWW, primary relaxation). CTMD quantitatively formulates and physically explains the universality of amorphous condensed matter behavior close to and well below T_c and T_g, and physically identifies both parameters of KWW-type of response functions. It also explains why the KWW correlations are only approximate and arguably not very useful fitting expressions, whose parameters are not independent.

10.3　Brief Reference to Specific Required Characteristics and Capabilities of Fruitful Theories

A significant variety of theoretical approaches have in the past been proposed to deal with the dynamics of condensed matter. In addition to CM and MCT, this book briefly describes free volume approaches in Sect. 6.10, in which thermal activation does not play the major role, but many other proposals would also deserve specific reference.

For brevity, though, we invite the reader to go through Chaps. 1 and 2 of the advanced work edited by Berthier et al. [12], where four prestigious researchers (J. Kurchan, J. S. Langer, Th. A. Witten and P. G. Wolynes) answer to a set series of questions formulated by the Editors (Chap. 1), and G. Tarjus gives an overview of the theories of the glass transition (Chap. 2). The following list of required (or sought) specific characteristics and capabilities of fruitful theories of glasses may be singled out from the mentioned two chapters (followed by a comment of ours within []), many of which were kept in mind and sought as references in the development of CTMD, as the reader is invited to recognize:

1. G. Tarjus, while giving the greatest emphasis to the growing collective and heterogeneous[3] dynamic behavior of amorphous materials, goes as far as recommending that "such collective features should even be exaggerated [12, 13], to explore" (and prove or disprove) "their asymptotic predictions", prior (one would say) to trying to tune them to the experimental behavior. [This specific suggestion was taken up in CTMD, as discussed in Chaps. 11 and 12.]
2. Some level of coarse graining (i.e. neglecting superfluous detail) is recommended to capture the universal features of the behavior. "Minimal models" are therefore what is needed. [Some or most of the atomistic and computer simulation models incorporate too many microscopic details, and so those models are not minimal. CTMD tries to strike a balance between microscopic insight and universal concepts and behavior.]
3. J. Kurchan stresses that "the internal logic and the tractability of a theory, *i.e.* the fact that the objects it invokes and relates have clear definitions and meanings,

---

[3] [Meaning the co-existence of fast and slow regions within the structure of the materials].

are what makes it satisfactory". [The physical concepts and objects as well as the physical properties used within CTMD are all well defined.]

4. J. S. Langer argues that there are very few theories that are comprehensive and sharp enough to make unambiguous predictions. They are mainly "models" that address limited specific features of the behavior of glasses. His "*excitation-chain mechanism*" [14–16] perhaps provides the closest picture and theoretical framework to CTMD, within which he was able to derive a Vogel-Fulcher formula for the transition rates. However, this excitation-chain formulation still predicts diverging relaxation times and length scales. [The reader is encouraged to identify the similitudes and the contrasting features and results of Langer's approach and CTMD.]

5. Numerical simulations by Glötzer [17] made chain-like motions to appear during thermally activated transitions between inherent structures. [In CTMD, the inherent structures involved in thermally activated transitions are the primitive relaxors and any of their clusters of varying sizes, which may be the result of chain-like synchronizations of primitive relaxors.]

6. Together with the excitation-chain mechanism, the Random First-Order Transition Theory (RFOT) [18–20] also share some general features with CTMD, as they also use activated dynamics and the transition state theory (TST). [By contrast with CTMD, J. S. Langer's excitation-chain formulation still predicts diverging relaxation times and length scales, and RFOT starts from a quasi-equilibrium assumption.]

7. P. G. Wolynes argues that a sound theory of glasses must explain (1) the non-Arrhenius temperature dependence, (2) the non-exponential time dependence, and (3) aging.

The above seven general characteristics and requirements were actively sought and at least partly met by CTMD. As to aging, specifically, the tentative strategy sketched in Chap. 14 might be a useful starting point for future work.

Closing Note

Be it with the CM, MCT, CTMD or any other existing or future theories, what matters in their relative evaluation is (1) their conceptual, physical and mathematical soundness, (2) their range of applicability, (3) the number, relevance and validity of their predictions (relative to real experimental behavior), (4) their computability, (5) the detailed physical insight provided, and, of course, (6) their potential for further improvement.

References

1. K.L. Ngai, Comments Solid State Phys. **9**, 121 (1979)
2. K.L. Ngai, J. Chem. Phys. **109**, 6982 (1998)
3. K.L. Ngai, Phys. Cond. Matter **15**, S1107 (2003)
4. K.L. Ngai, M. Paluch, J. Chem. Phys. **120**, 857 (2004)

5. K.L. Ngai, J. Non-Cryst, Solids **353**, 709 (2007)
6. K.L. Ngai, *Relaxation and Diffusion in Complex Systems* (Springer, New York, 2011)
7. R. Kohlrausch, Ann. Phys. Chem. **91**(1), 56, 179 (1854)
8. G. Williams, D.C. Watts, Trans. Faraday Soc. **66**, 80 (1970)
9. E.-J. Donth, *The Glass Transition—Relaxation Dynamics in Liquids and Disordered Materials, Springer Series in Materials Science - 48* (Springer, Berlin, Heidelberg, 2001)
10. W. Götze, *Complex Dynamics of Glass-Forming Liquids—A Mode-Coupling Model*, International Series of Monographs on Physics—143 (Oxford Science Publications, Oxford, 2009)
11. M. Beiner, H. Huth, K. Schröter, J. Non-Cryst, Solids **279**, 126 (2001)
12. L. Berthier, G. Biroli, J.-P. Bouchard, L. Cipelleti, W. Van Saarloos, *Dynamical Heterogeneities in Glasses, Colloids, and Granular Media* (OUP, Oxford, 2011)
13. S.A. Kivelson, G. Tarjus, Nat. Mat. **7**, 831 (2008)
14. J. S. Langer, Phys. Rev. E. **73**, 041504 (2006)
15. J. S. Langer, *Phys. Rev. Lett.* **97**. 115704 (2006)
16. J. S. Langer, *Phys. Rev. E*, **78**, 051115 (2006)
17. S.C. Glötzer, J. Non-Cryst, Solids **274**, 342 (2000)
18. V. Lubchenko, P.G. Wolynes, J. Chem. Phys. **121**, 2852 (2004)
19. V. Lubchenko, P.G. Wolynes, Adv. Chem. Phys. **136**, 95 (2007)
20. V. Lubchenko, P.G. Wolynes, Annu. Rev. Phys. Chem. **58**, 235 (2007)

Chapter 11
What Do CTMD and Experiments Say?

11.1 General Itemized Review of CTMD's Predictions

The reader is first invited to check that most of the eleven characteristics identified in Chap. 6 have been met as definite or at least expected future results of applying CTMD to the behavior of amorphous condensed matter. Where results are still very incomplete or uncertain is on the effects of molecular packing and on the hysteresis and aging effects, in addition to the persisting problem of the relationship between relaxation and retardation times and their spectra. The latter, as tentatively predicted by CTMD at the beginning of Chap. 9, will certainly require future modification, as discussed in Sect. 11.9.3 and Chap. 12.

It is however appropriate and useful to itemize in much greater detail, and locate in the text, all specific predictions of CTMD:

(1) Origin and nature of the cooperativity in materials' responses (Chap. 8).
(2) Super-Arrhenius behavior as an effect of the cooperativity in the responses (Chaps. 8 and 9, and this chapter).
(3) Formulation and calculation of the frequencies of motions of all types of clusters of primitive relaxors at constant volume, and the definition of the crossover frequency and crossover temperature (Chap. 8).
(4) Physical explanation and statistical mechanics formulation of the loss of independence of the primitive relaxors (Chaps. 8 and 14).
(5) Possible strategy for the future development of an equilibrium (ETG) and non-equilibrium (NETG) theory of glasses (Chaps. 8 and 14).
(6) Origin and nature of the dynamic heterogeneity, and its validity also under full thermodynamic equilibrium (Chap. 9).

J. J. Cruz Pinto and J. R. dos Santos André, *Analytical Molecular Dynamics of Amorphous Condensed Matter*, Springer Series in Materials Science 342, https://doi.org/10.1007/978-3-031-56517-5_11

(7) A first physically reasonable (though not yet accurate[1]) proposal for estimating the response time distributions (at equilibrium, or under forced excitations) and their variation with temperature, with clear evidence of rheological complexity. (Chap. 9, and this chapter).

(8) Improved dynamic modelling of linear and non-linear stress relaxation and creep, and of linear frequency response, while predicting non-affine stresses and strains relative to their overall average values. (Chap. 9 and this chapter).

(9) Formulation and calculation of the response times of each individual set of clusters of primitive relaxors and of their average relaxation and retardation times in linear and non-linear stress relaxation or creep, as functions of temperature and applied strain or stress, respectively (Chap. 9).

(10) Prediction of average response times of the order of 100 to 1000s at or near the conventional glass transition temperature (Chap. 9 and this chapter).

(11) Prediction of non-divergence of all individual and average response times at any temperature, except at 0K (Chaps. 8 and 9 and this chapter).

(12) Average response times approximately following WLF and VTF correlations even at constant volume (Chap. 9 and this chapter).

(13) Approximate time–temperature superposition or equivalence and additional support to rheological complexity (Chap. 9 and this chapter).

(14) Universal behavior of amorphous condensed matter (Chap. 9).

(15) Calculation of fragilities and formulation of a model of tunable fragility (Chap. 9 and this chapter).

(16) Possibility of future inclusion of molecular packing effects (Chaps. 9 and 15).

(17) Reduced number of parameters, all of them specific physical properties of each system (Chaps. 8, 9 and 10 and this chapter).

(18) Possibility of estimating the above parameters from 24-h stress relaxation and other experiments, so far from a simplified version of the theory, awaiting the future application of its full version (Chaps. 4, 5 and this chapter)

(19) Agreement of the individual and average cluster response times with Ngai's CM relationship and physical identification of its parameters, as well as those of the approximate extended exponential KWW experimental relationships (Chap. 10).

(20) A contribution to a possible clarification of the relationship between relaxation and retardation times and their spectra (Chap. 9).

(21) Very wide range of accessible timescales at any temperature (within mere numerical precision, or processor word-length, limits), near and well below the crossover temperature. (Chaps. 8, 9 and this chapter).

(22) Relatively simple analytical formulation and equally fast computations irrespective of timescale (Chaps. 8, 9, 10 and this chapter).

(23) Possible generalization to very simple molecular systems, as well as crosslinked amorphous and semi-crystalline materials. (Chaps. 9 and 13).

---

[1] [As discussed in Sect. 11.9.3 and Chap. 12].

As the above listing specifies, this chapter will further discuss and illustrate some of the above results in greater detail.

11.2 More Accurate Relaxation Maps

As mentioned in Chap. 8, the relaxation maps of Figs. 8.3 and 8.4 were very crude first estimates, arbitrarily assuming symmetrical absorption peaks. They played the role of convincing one of the authors that the initial idea of accounting for cooperativity by means of clusters' formation, activation and transition could be pursued with good prospects of obtaining the super-Arrhenius behavior found in experiments. Once convinced, the development of CTMD was continued until its formulation yielded the generalized geometric or logarithmic average as the likely most representative value of average response times, as $\ln\theta^*_{avg}(T) = \sum_{k\geq 1} F_k'^*(T)\ln\theta_k^*(T)$, where the $*$ superscript applies to equilibrium conditions, with the $F_k'^*(T)$ as tentatively formulated in Chap. 9.

Figures 11.1 and 11.3 show the corresponding more accurate relaxation maps for the three systems studied (Numerical Example, PMMA and PC), calculated for the latter two cases with the CTMD parameters estimated from 24-h stress relaxation experiments following the procedure described in Chap. 4, where the average frequencies plotted were obtained from $v^*_{avg}(T) = 1/[2\pi\theta^*_{avg}(T)]$. While for the Numerical Example the physical properties v_c, T_c and $E_{a,1}$ have the values arbitrarily assumed in Chap. 8, for the PMMA and the PC their values, obtained by treating the mentioned stress relaxation experimental data, were ($v_c = 2.4888 \times 10^5$ Hz, $T_c = 455.24$ K, $E_{a,1} = 35.655$ kJ^2) and ($v_c = 5.3449 \times 10^6$ Hz, $T_c = 496.24$ K, $E_{a,1} = 49.957$ kJ), respectively.

It becomes apparent that all relaxation maps are significantly displaced to higher temperatures or lower frequencies relative to the initial crude ones and, as previously illustrated in Fig. 9.2 and now also in Fig. 11.2, the range of active cluster sizes and response times increases as temperature is lowered. As mentioned in Chap. 8, these relaxation maps may represent how the glass transition temperature will vary in measurements at the timescales corresponding to horizontal lines at any chosen frequency, v, and Fig. 11.2 also suggests that absorption peaks at fixed timescales will be skewed towards low temperatures, while at fixed temperature they may be skewed either way, towards low or high frequencies (cf. loss modulus and tanδ peaks in Sect. 11.4).

---

[2] These values agree with those tabulated by Donth, and Beiner et al., in [1, 2], respectively, for PMMA: 2.52×10^5 Hz and 455K, though not with $E_{a,\beta} = 79$ kJ/mol, possibly implying clusters with twice the size predicted in our calculations (or a different nature of the β relaxation). The values for PC are first ever estimates, as far as we know.

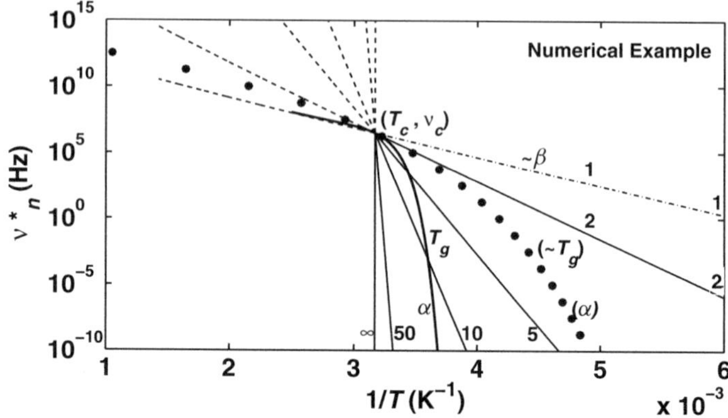

Fig. 11.1 Approximate (•) and accurate (–) CTMD-predicted relaxation map for the numerical example specified in Chap. 8

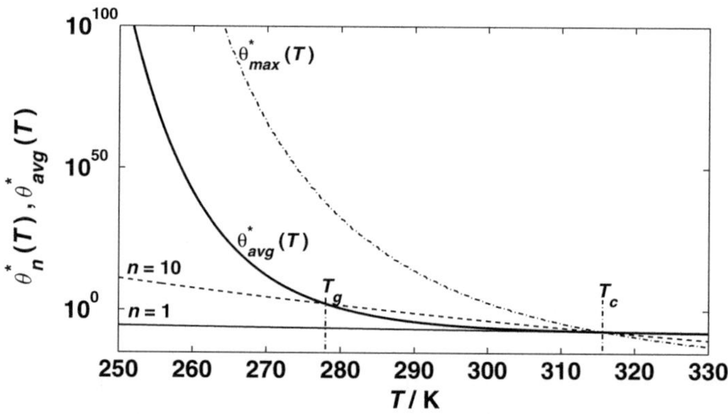

Fig. 11.2 Cluster size-dependent [$n = 1, 10$], maximum and logarithmic average CTMD-predicted relaxation times as functions of temperature in linear scale, for the numerical example specified in Chap. 8

It is again noted that, pending deeper evaluation, the data for $T > T_c$ may have to be discarded, at least in the absence of multiple transition and activation paths for each primitive relaxor. As previously mentioned, the Figures also illustrate that very wide ranges of timescales are indeed accessible to CTMD.

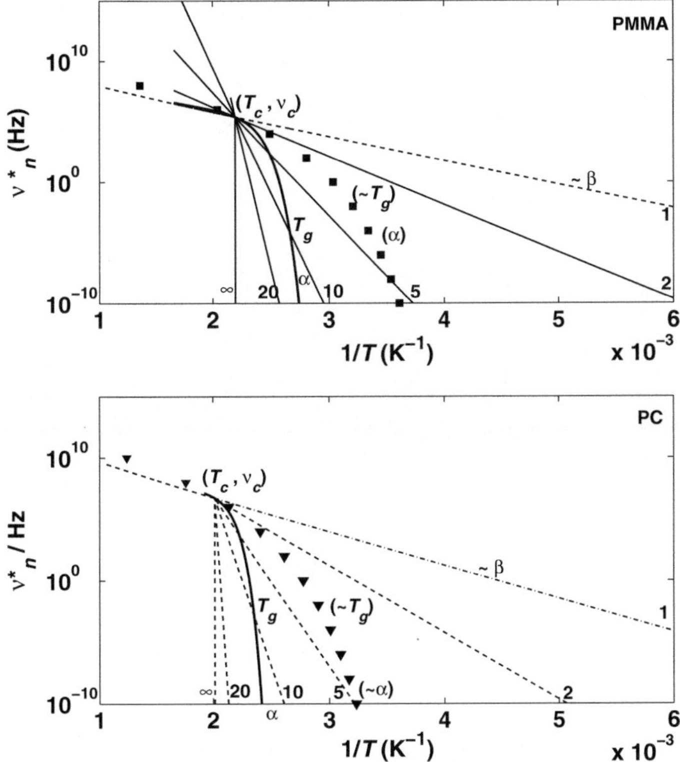

Fig. 11.3 Approximate (■, ▼) and accurate (–) CTMD-predicted relaxation maps for PMMA (top) and PC (bottom) calculated from stress relaxation experimental data

Figures 11.4 and 11.5 plot the cluster size-dependent and average relaxation times at equilibrium or linear viscoelastic conditions for the same three systems, with indication of the conventional glass transition region (at $T \sim T_c/1.2$ [1, 3]), where relaxation times are generally within 100–1000 s or so. A relevant prediction is that, within that range, the glass transition temperature varies by just a few degrees per power of ten of timescale, as experimentally found. It may also be noted that the tangent to the average relaxation time curve at T_c is the line representing the behavior or clusters of just two primitive relaxors, which is what was in Chap. 9 shown to be the exact predicted average size of the clusters at that temperature, according to CTMD.

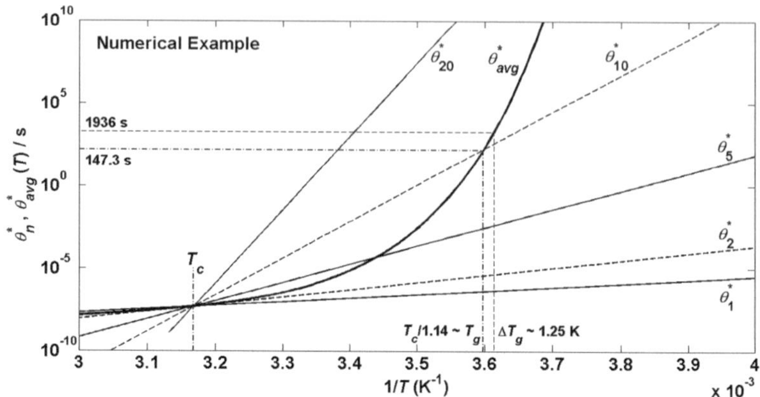

Fig. 11.4 Cluster size-dependent and logarithmic average CTMD-predicted relaxation times at equilibrium as functions of reciprocal temperature for the numerical example specified in Chap. 8

Other interesting and encouraging observations are that the estimated T_g of PMMA and PC turned out 106°C and 151°C, or 140°C, if in the latter case one divides T_c by 1.2 instead of 1.17,[3] respectively, which are close to known experimental values.

11.3   CTMD-Predicted Linear Uniaxial Tensile Stress Relaxation Moduli and Creep Compliances

Figures 11.6 and 11.7 plot the relative linear uniaxial tensile stress relaxation moduli at varying temperatures near T_g, as well as near T_c just for reference, assuming no viscous flow. They were calculated by Eqs. 9.2, 9.16 (with $\epsilon_0 \to 0$) and 9.19, and were complemented by the corresponding relative linear creep compliance at $\sim T_g$ (from Eq. 9.24 in the same linear limit), taking all corresponding retardation times as $\tau_n = (E_0/E_\infty)\theta_n$, with $E_0 = 1$ GPa and $E_\infty = 1$ MPa) for the Numerical Example, and the corresponding optimized values for PMMA and PC as determined from the stress relaxation experimental data presented and plotted in Sect. 11.9 ($E_0 = 1.2641$ GPa, $E_\infty = 1.0000$ MPa) and ($E_0 = 1.0014$ GPa, 0.55626 MPa), respectively.

The lower part of Fig. 11.6 shows a zoomed-in plot of the same curves for the Numerical Example, suggesting that whole curve time–temperature superposition or equivalence may turn out only a crude approximation.

---

[3] The calculations were obviously performed for a discrete set of temperatures, and $T_c/1.17$ turned out to be yielding an average relaxation time closest to 100s (which is a rather strict, perhaps unnecessary condition).

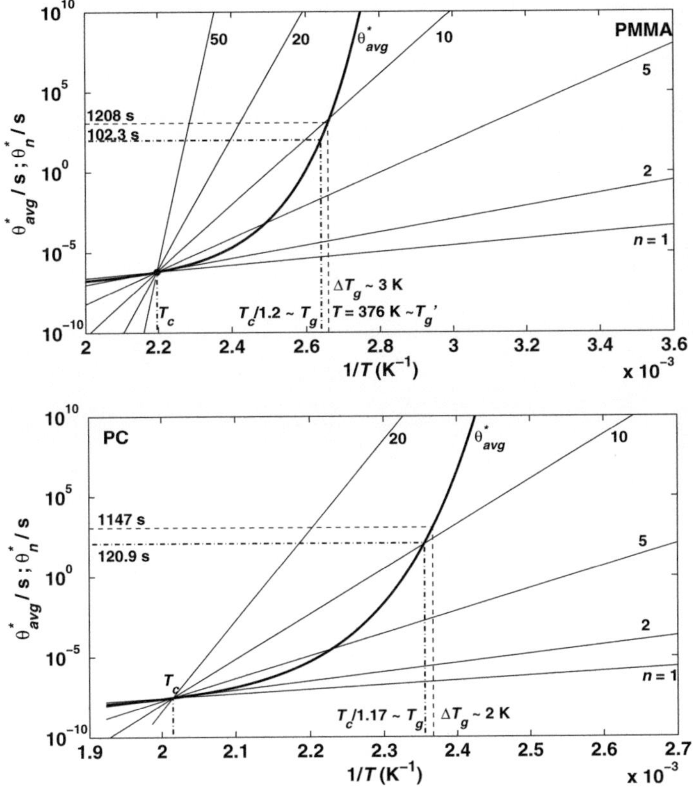

Fig. 11.5 Cluster size-dependent and logarithmic average CTMD-predicted relaxation times at equilibrium as functions of reciprocal temperature for PMMA (top[4] [4]) and PC (bottom)

The wide range of timescales necessary to register the whole stress relaxation or creep processes at or below T_g is clearly documented. Other very important qualitative and quantitative features of the behavior will be considered further below, but one pertaining to the universality issue is documented in Fig.11.8: the behavior of different amorphous materials appears not only similar (largely universal) at or near T_c, but also at or near T_g, which is not surprising in view of the approximate relationship between T_g and T_c. This means that microscopic phenomena at the vicinity of the glass transition are, in themselves and in their macroscopic consequences, also relatively universal.

Figures 11.9 and 11.10 follow the pattern of Figs. 11.6 and 11.7 but now for the relative linear uniaxial tensile creep compliance of the same three systems, keeping the above same relationships between the retardation and relaxation times, and they

---

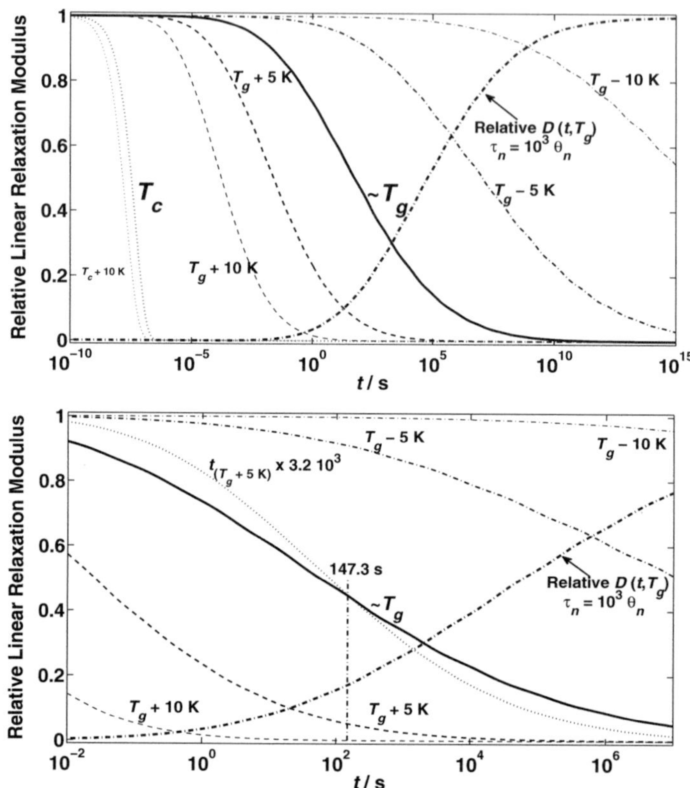

Fig. 11.6 Temperature- and time-dependent relative linear non-viscous uniaxial tensile stress relaxation modulus $[E_r(t, T) - E_\infty]/(E_0 - E_\infty)$ for the Numerical Example specified in Chap. 8 with the superposition of the relative linear non-viscous uniaxial tensile creep compliance at $\sim T_g$ with $\tau_n = 10^3\theta_n$ (top) and their zoomed-in plots for $10^{-2} \le t/s \le 10^7$ (bottom)

make apparent that before creep is half-way through, stress relaxation is already approaching its final stages. Again, time–temperature superposition or equivalence might only be an approximation, at least considering whole response curves, from short to long times. The gradual transition from super-Arrhenius to Arrhenius, low activation, regime as temperature increases is also apparent, as experimentally found.

11.4   Dynamic Mechanical Responses

This section presents and discusses the results obtained by applying the formulations of Sect. 9.7, with reference to the more extended set of Figs. 11.11, 11.12, 11.13, 11.14, 11.15, 11.16, 11.17, 11.18, 11.19 and 11.20, to consider all the relevant dynamic properties.

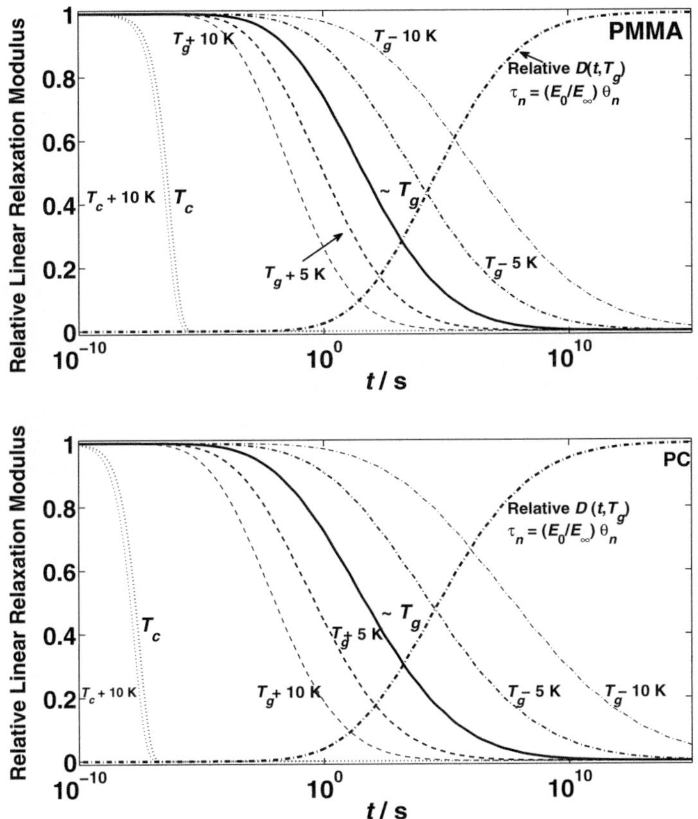

Fig. 11.7 Temperature- and time-dependent relative linear non-viscous uniaxial tensile stress relaxation modulus $[E_r(t, T) - E_\infty]/(E_0 - E_\infty)$ for PMMA (top) and PC (bottom) with the superposition of the relative linear non-viscous tensile creep compliance at uniaxial $\sim T_g$ with $\tau_n = (E_0/E_\infty)\theta_n$

Figures 11.11, 11.12 and 11.13 plot the storage properties (E' and some D' data), Figs. 11.14, 11.15 and 11.16 the loss properties (E'' and some D'' data), and Figs. 11.17, 11.18, 11.19 and 11.20 the tan δ values (calculated mainly by E''/E' and some by D''/D').

It may be seen that the E' and D' curves, being a superposition of the responses of independent (though intermittent) sets of clusters, are slightly wavy at each of their characteristic frequencies, but that effect does not show up with temperature as the main independent variable because a range of cluster frequencies will be active to variable extents at each temperature. As physically expected, the storage modulus increases with frequency at constant temperature and decreases with temperature at constant frequency, because the clusters will have less time to respond at higher frequencies of testing and respond faster at higher temperatures, respectively.

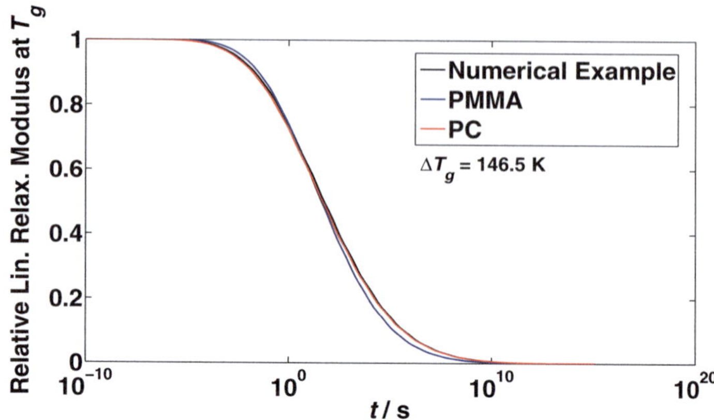

Fig. 11.8 Near superposition of relative linear non-viscous uniaxial tensile stress relaxation modulus curves for three different amorphous systems at their specific, widely different, T_g (where $\theta^*_{avg} \sim 10^2 s$)

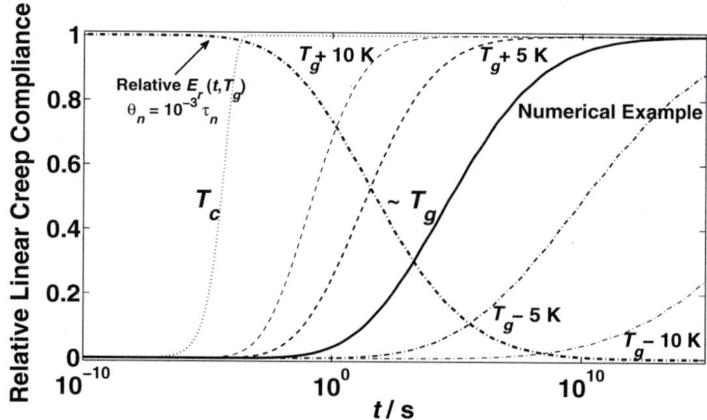

Fig. 11.9 Temperature- and time-dependent relative linear non-viscous uniaxial tensile creep compliance $[D(t, T) - D_0]/(D_\infty - D_0)$ for the Numerical Example specified in Chap. 8 with the superposition of the relative linear non-viscous uniaxial tensile stress relaxation modulus at $\sim T_g$ with $\theta_n = 10^{-3} \tau_n$

It may also be recognized that $\nu = 0.01$ Hz (corresponding to θ of the order of 100 s) will not be very different from the average frequency at which the primitive relaxors and clusters will be moving near T_g, if one notes that the maximum negative slopes of the $E'(\nu = 0.01\,\text{Hz}, T)$ curves occur around each respective estimated T_g—278 K (Num. Ex.), 379 K (PMMA) and 424 K (PC), for which the D' curves are also shown. The next three Figs.11.14, 11.15 and 11.16 show the loss modulus, E'', and compliance, D'', curves.

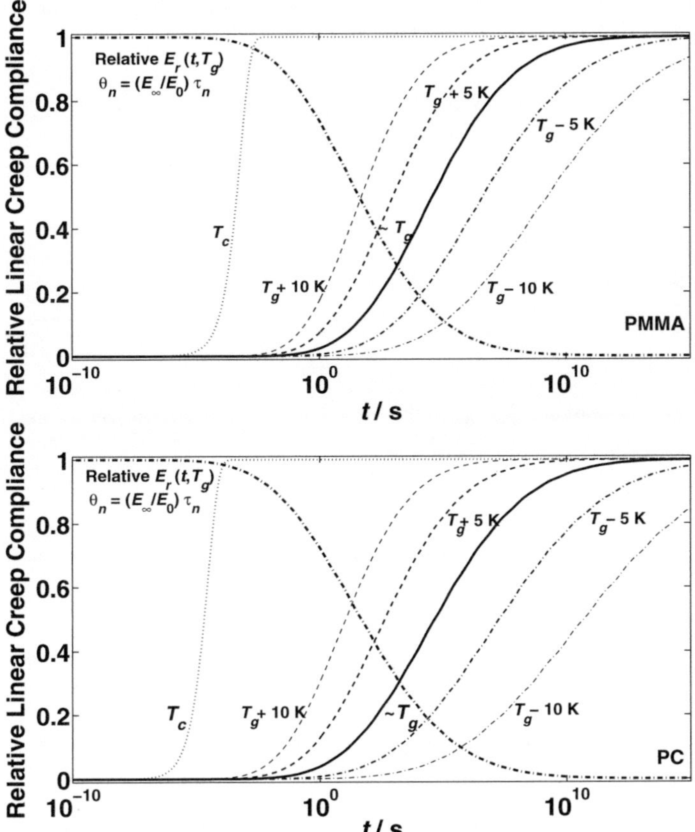

Fig. 11.10 Temperature- and time-dependent relative linear non-viscous uniaxial tensile creep compliance with the superposition of the relative linear non-viscous uniaxial tensile stress relaxation modulus at $\sim T_g$ with $\theta_n = (E_\infty/E_0)\tau_n$ for PMMA (top) and PC (bottom)

The loss curves as functions of frequency now expectedly show a much more pronounced wavy nature, with each of the "micro-peaks" exactly located at specific cluster frequencies, as explicitly indicated in Fig.11.14-top with the corresponding cluster sizes specified, and each curve mirrors somewhat the corresponding cluster frequency distributions after moderate spreading and smoothing. Their skewness is towards low frequencies or long response times, in agreement with approximately truncated (near ν_1^* at equilibrium) log-normal relaxation spectra and, less clearly, towards low temperatures. Some "shoulders" in temperature-dependent curves are here found, as in many experimental data.

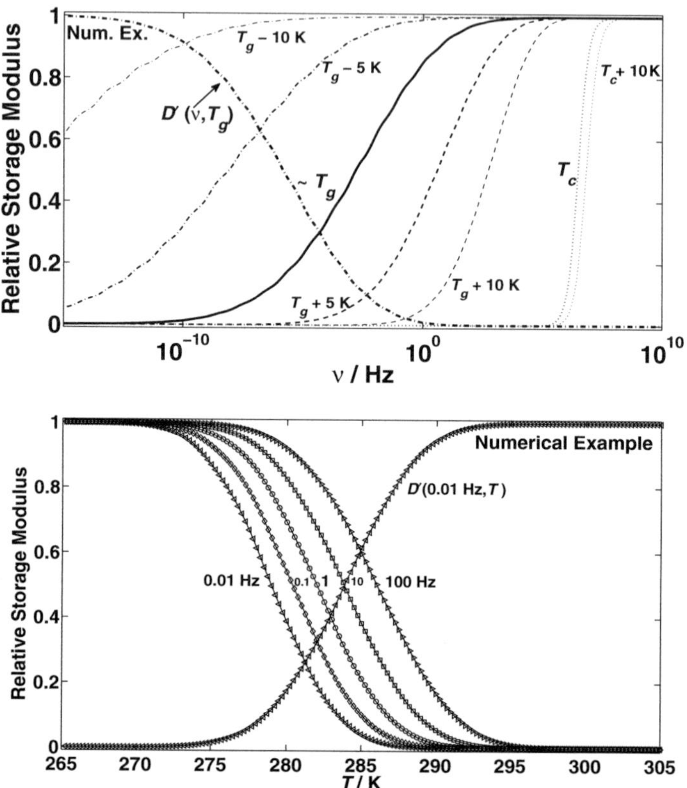

Fig. 11.11 Temperature- and frequency-dependent relative linear non-viscous uniaxial tensile storage modulus $\left[E'(\nu, T) - E_\infty\right]/(E_0 - E_\infty)$ for the Numerical Example specified in Chap. 8 with the superposition of the relative linear uniaxial tensile storage creep compliance at $\sim T_g$ (top) and at 0.01 Hz (bottom), assuming $\tau_n = 10^3 \theta_n$

Though often experimentally inaccessible (in case of excessive softening, melting or eventual thermal or mechanical degradation), the predicted peak at T_c is large, irrespective of the occurrence or not of relaxations by large clusters, and narrow, as all primitive relaxors in the system will be vibrating at the crossover frequency, ν_c.

The frequencies selected for the lower parts of the Figures (here and in the tanδ ones) cover the range accessible to the most performing dynamic analyzers, and the plots rightly show that the absorption peaks shift to higher temperatures and increase in size as the frequency is increased, in line with the increased mobility of all relaxors and clusters at higher temperatures.

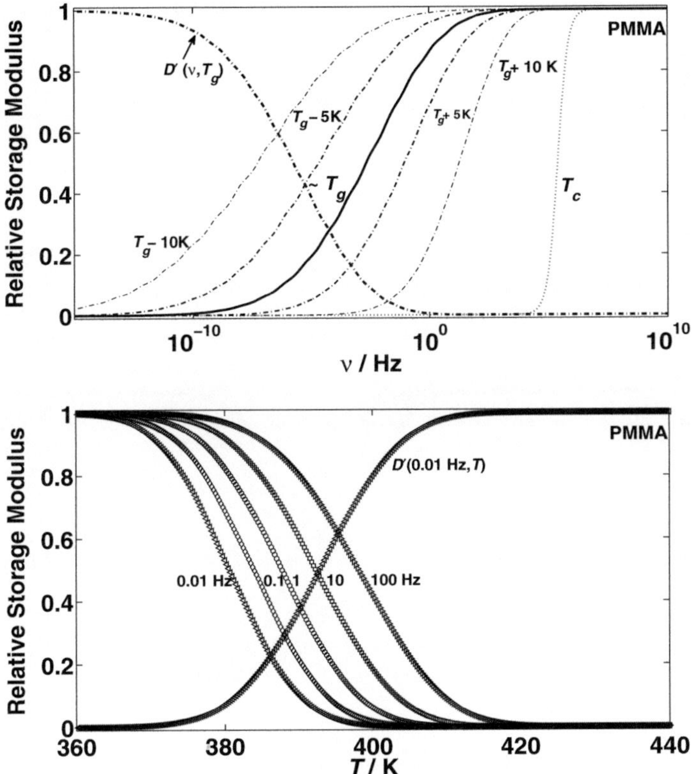

Fig. 11.12 Temperature- and frequency-dependent relative linear non-viscous uniaxial tensile storage modulus $[E'(\nu, T) - E_\infty]/(E_0 - E_\infty)$ for PMMA with the superposition of the relative linear uniaxial tensile storage creep compliance at $\sim T_g$ (top) and at 0.01 Hz (bottom), assuming $\tau_n = (E_0/E_\infty)\theta_n$

An important observation is the coincidence of the relative loss creep compliances at 0.01 Hz with the loss moduli at 10 Hz, and it must be related to the fact that the mentioned ratio of frequencies almost coincides with the reciprocal of that of the (real or assumed) individual retardation and relaxation times.

As to Figs. 11.17, 11.18 and 11.19, they first show that, despite the known equality of E''/E' and D''/D' for each individual relaxor or cluster (behaving as a standard linear solid), as shown in Chap. 3 and in Fig. 11.20, they are widely different when we have a wide distribution of response times.

Study Question 11.1 Interpret Fig. 11.20 and physically justify the fact that, for a real material, one generally has $\tan\delta_E \neq \tan\delta_D$ from their definition and with reference to the curves of top Figs. 11.11 , 11.12 , 11.13, 11.14, 11.15 and 11.16.

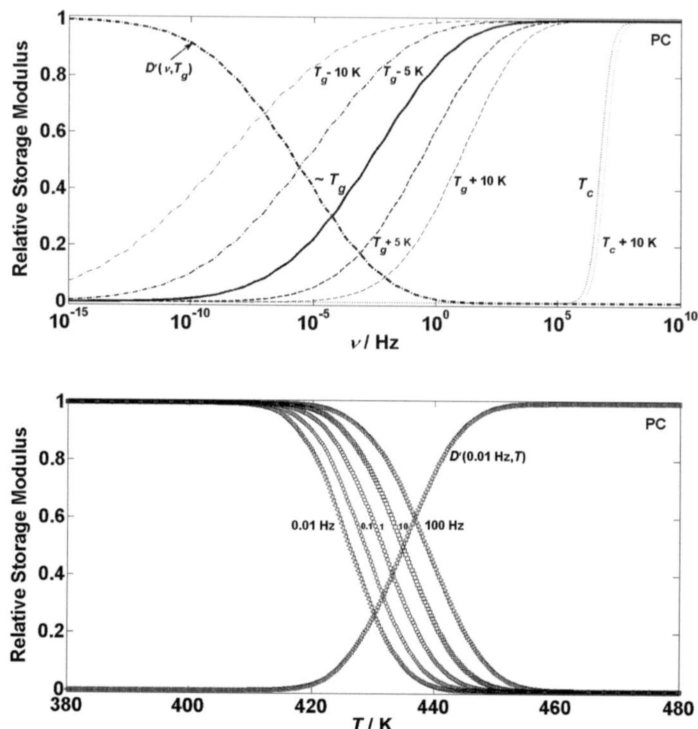

Fig. 11.13 Temperature- and frequency-dependent relative linear non-viscous uniaxial tensile storage modulus $\left[E'(v,T) - E_\infty\right]/(E_0 - E_\infty)$ for PC with the superposition of the relative linear uniaxial tensile storage creep compliance at $\sim T_g$ (top) and at 0.01 Hz (bottom), assuming $\tau_n = (E_0/E_\infty)\theta_n$

Study Question 11.2 Recall Sect. 2.3.6 and its Footnote 23, where it is noted that there is no good justification for the widespread use of $\tan\delta$ as a dynamic characterizing parameter of materials.

(1)   Observe the locations of the crossover frequencies, v_c, at the top Figs. 11.17, 11.18 and 11.19 and the top Figs. 11.14, 11.15 and 11.16 relative to the corresponding theoretical absorptions at T_c (assuming it can, or could, be detected).

(2)   Discuss their relation to, and justification for, the abovementioned footnote.

(3)   Extend your discussion to the locations of the E'' and D'' micro-peaks at the top Figs. 11.14, 11.15 and 11.16 (explicitly identified for the Numerical Example) and those of $\tan\delta$ at the top Figs. 11.17, 11.18 and 11.19.

(4)   What relationship would you guess as existing between the locations of the E'' and D'' micro-peaks (the latter shown only at $\sim T_g$).

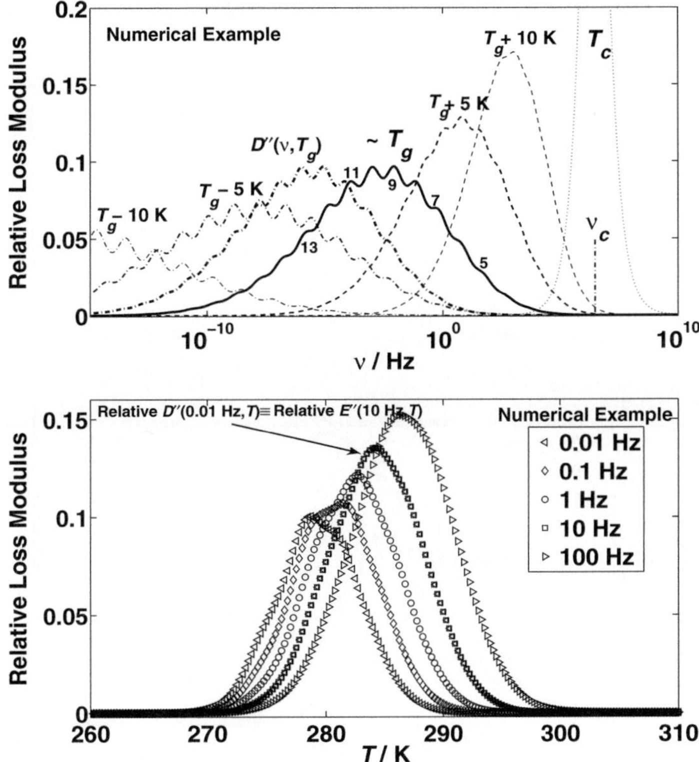

Fig. 11.14 Temperature- and frequency-dependent relative linear non-viscous uniaxial tensile loss modulus $[E''(\nu, T) - E_\infty]/(E_0 - E_\infty)$ with the superposition of the relative linear non-viscous uniaxial tensile loss compliance at $\sim T_g$ (top) and at 0.01 Hz (bottom) for the Numerical Example specified in Chap. 8, assuming $\tau_n = 10^3 \theta_n$

All the above Figures plotting the dynamic properties show that very similar curves are obtained at "physically analogous" temperatures for each system, e.g. at their respective T_g. This is apparent throughout, for example, between the $\tan\delta_D$ curves at 0.01 Hz, or between the $E''(\nu, T_g)$ and $D''(\nu, T_g)$, where the latter are only shifted in ν in inverse proportionality to τ_n and θ_n, even suggesting possible reducibility to identical curves by appropriate normalization. This may also be related to what is implied by Fig. 11.8. The shifts of the curves with temperature or frequency are as for all other dynamic properties, but the skewness of the $\tan\delta$ traces are reversed relative to those of E'' and D'', because of the different shapes of the storage curves.

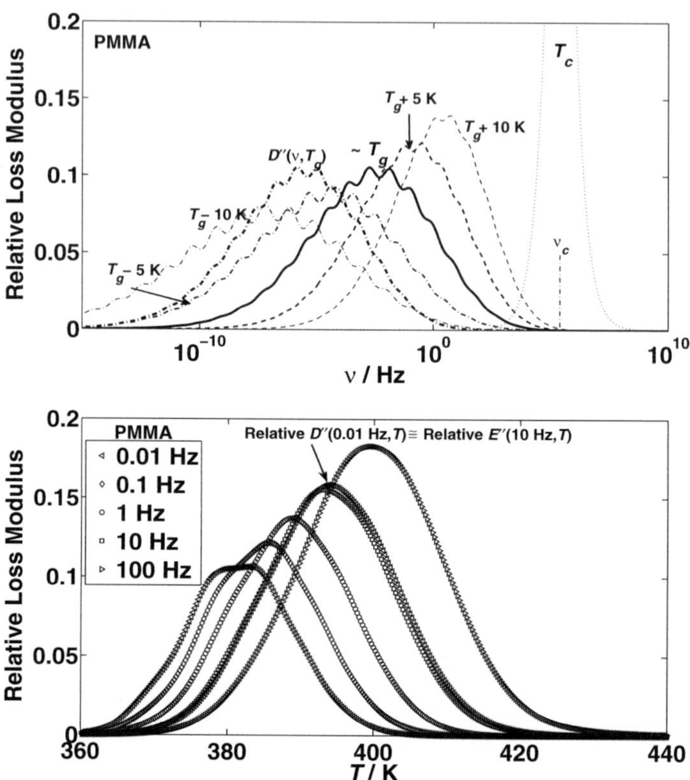

Fig. 11.15 Temperature- and frequency-dependent relative linear non-viscous uniaxial tensile loss modulus $[E''(\nu, T) - E_\infty]/(E_0 - E_\infty)$ with the superposition of the relative linear non-viscous uniaxial tensile loss compliance at $\sim T_g$ (top) and at 0.01 Hz (bottom) for PMMA, assuming $\tau_n = (E_0/E_\infty)\theta_n$

A final note should be made to stress that all the above calculations and plots assume one single type or nature of primitive relaxors in the systems, such as to yield monomodal spectra and loss peaks (after smoothing out the micro-peaks, of course).

11.5   Time–Temperature Equivalence or Superposition

In Chap. 9 it was already shown for the Numerical Example specified in Chap. 8 that WLF and VTF relationships are satisfied by the CTMD-calculated average relaxation times at equilibrium or within the linear viscoelastic domain, $\theta^*_{avg}(T)$, even at constant volume. In this section we further analyze the subject of time–temperature

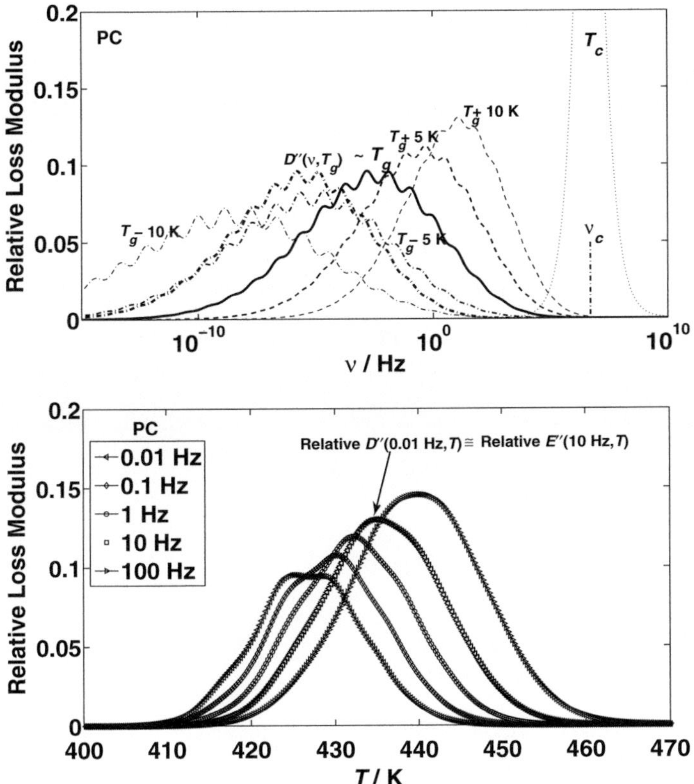

Fig. 11.16 Temperature- and frequency-dependent relative linear non-viscous uniaxial tensile loss modulus $[E''(\nu, T) - E_\infty]/(E_0 - E_\infty)$ with the superposition of the relative linear non-viscous uniaxial tensile loss compliance at $\sim T_g$ (top) and at 0.01 Hz (bottom) for PC, assuming $\tau_n = (E_0/E_\infty)\theta_n$

equivalence, not just for the mentioned example but for the other two systems-PMMA and PC-with the help of Figs. 11.21, 11.22, 11.23, 11.24 and 11.25.

The procedure adopted to prepare this discussion involved calculating by CTMD the stress relaxation moduli for the three systems at ± 5 K increments within a range of times not greatly exceeding those easily accessible to measurements, namely 10^4 seconds,[5] and then extrapolating to lower and higher temperatures to estimate the complete stress relaxation modulus curve at each respective approximate T_g. Additionally, we also assess the validity of WLF and VTF relationships for PMMA and PC.

---

[5] In fact, in various such experimental studies, the range of times did not exceed three powers of ten.

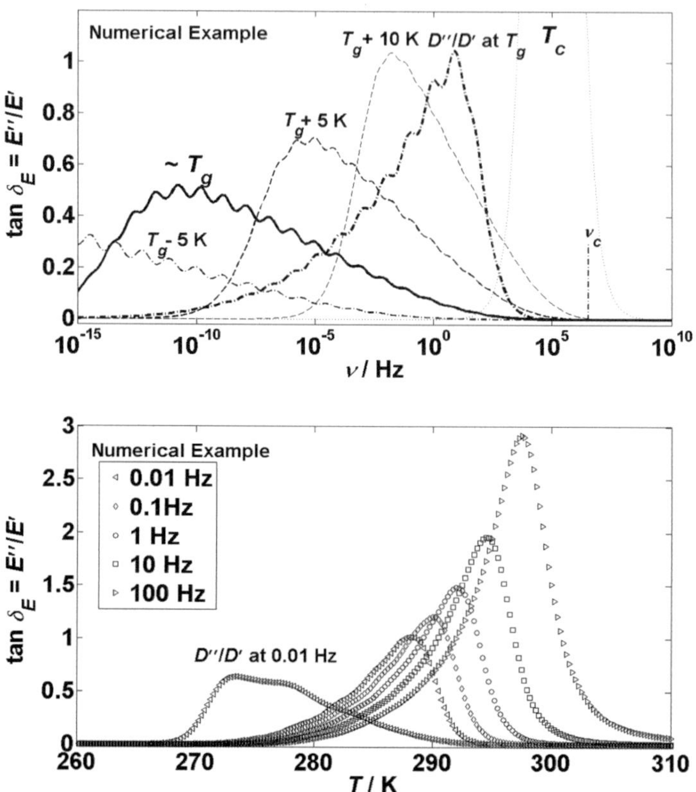

Fig. 11.17 Temperature- and frequency-dependent non-viscous uniaxial tensile $\tan\delta_E = E''/E'$ with the superposition of $\tan\delta_D = D''/D'$ at $\sim T_g$ (top) and at 0.01 Hz (bottom) for the Numerical Example specified in Chap. 8, assuming $\tau_n = (E_0/E_\infty)\theta_n$

Shifting the curves at low temperatures to the left (shorter times) and the ones at higher temperatures to the right (longer times), we obtain almost perfect superposition to a single curve at T_g, like in many other past similar works using actual experimental data [5–8], but the final reduced curve does not exactly fit the entire predicted trace (solid curve) at T_g spanning a much wider range of timescales.

It may be suspected, and guessed, that the same would have happened in the above classical works if experiments could have been conducted over 10 or more decades of timescale at the final extrapolated temperature. From the present calculations, it is suggested that possible "inaccuracies" of the reduced curve might be concentrated at very short and very long timescales, and not significantly in the main transition region.

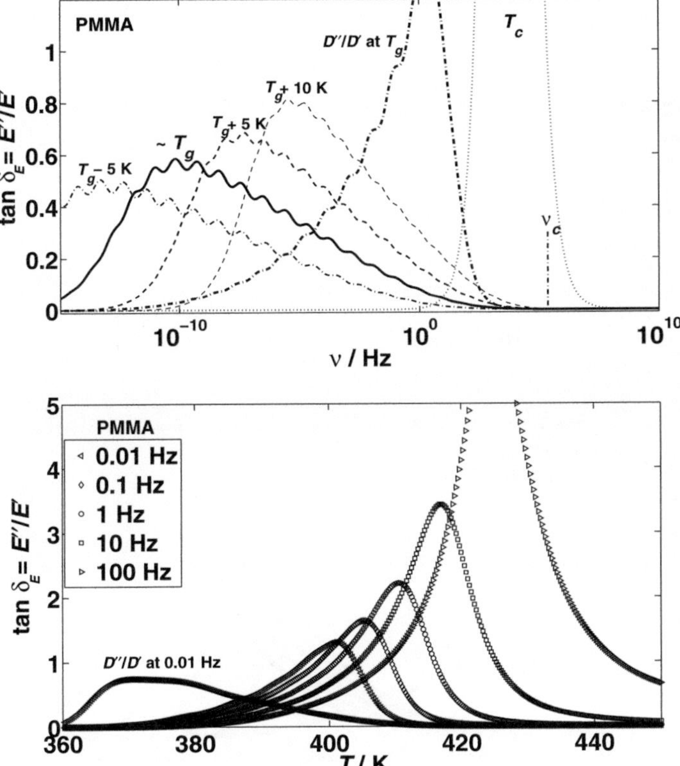

Fig. 11.18 Temperature- and frequency-dependent non-viscous uniaxial tensile $\tan\delta_E = E''/E'$ with the superposition of $\tan\delta_D = D''/D'$ at $\sim T_g$ (top) and at 0.01 Hz (bottom) for PMMA, assuming $\tau_n = 10^3\theta_n$

At the present constant volume conditions, what seems to happen in the calculations of the full response traces at T_g (relative to a hypothetical more perfect time–temperature superposition, if it were physically valid) is some excessive, non-uniform, narrowing of the response spectra for very short response times (i.e. slower response), and some also non-uniform widening at long relaxation times (i.e. also slower response), *or* that perfect superposition should not be expected. In the first case, it could mean that the spectra formulated in Chap. 9 would need modification, either in the weighing of the various clusters according to their intermittency, their a priori relative abundance, or both.[6]

---

[6] This could result in the small *non-uniform apparent vertical displacements* shown in the plots at very short and very long times.

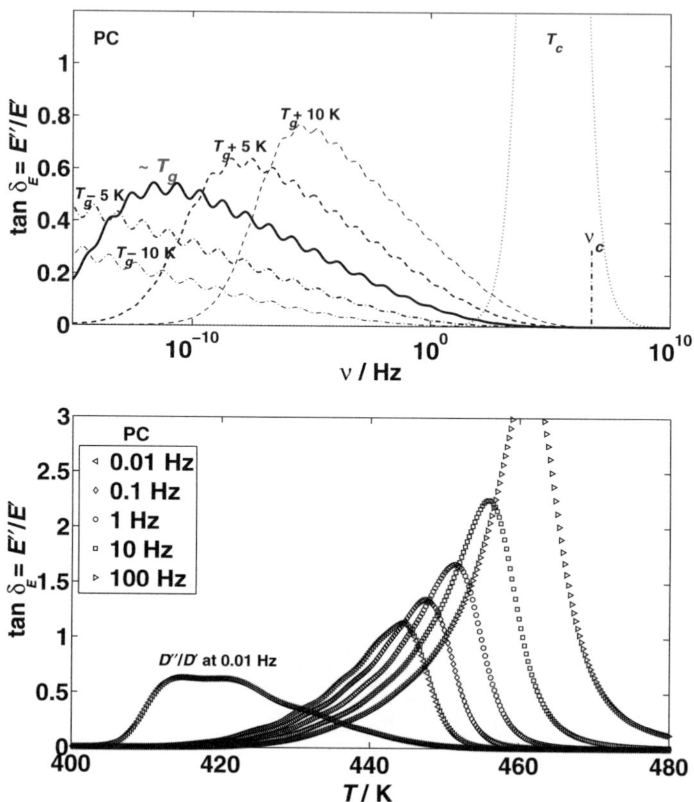

Fig. 11.19 Temperature- and frequency-dependent non-viscous uniaxial tensile $\tan\delta_E = E''/E'$ with the superposition of $\tan\delta_D = D''/D'$ at ~ T_g (top) and at 0.01 Hz (bottom) for PC, assuming $\tau_n = 10^3 \theta_n$

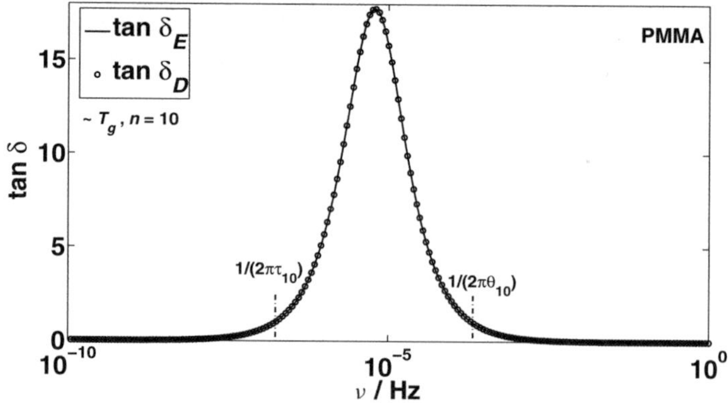

Fig. 11.20 Example ($n = 10$ at $T \sim T_g$) of the coincidence of the non-viscous uniaxial tensile $\tan\delta_E$ and $\tan\delta_D$ for any individual set of clusters of primitive relaxors

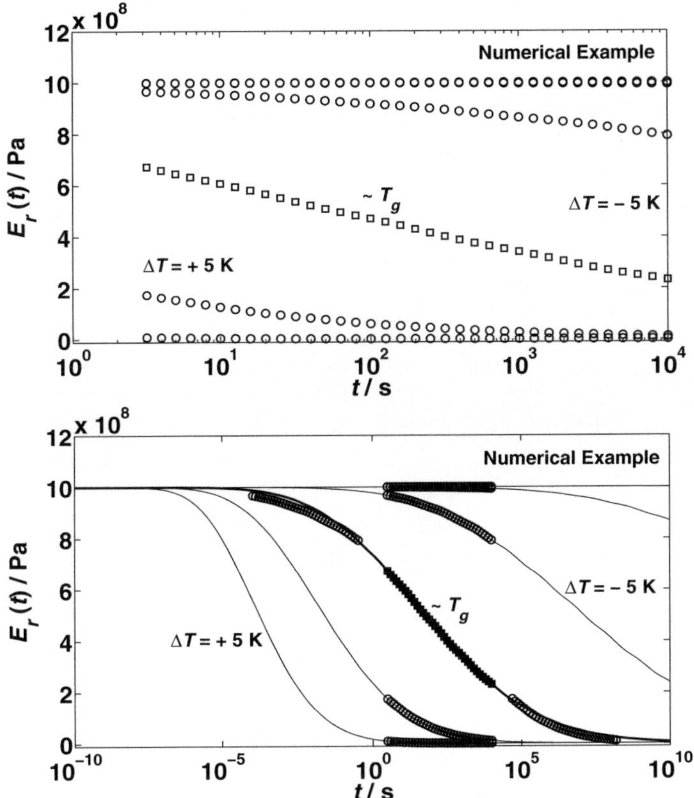

Fig. 11.21 Limited time–temperature superposition (or equivalence) of CTMD-predicted uniaxial tensile stress relaxation moduli for the Numerical Example specified in Chap. 8

Nevertheless, whatever the reason for the mentioned small deviations from perfect superposition, what matters is that such approximate, or an eventual more accurate, superposition would not be achieved by temperature-dependent displacements of complete curves, but only of small (ideally, differential) sections of them, according to significantly varying local operating activation energies among different sections of the curves, corresponding to contributions by different clusters of primitive relaxors (recall Chap. 8 and the wavy nature of the relaxation, compliance, storage and loss curves in this chapter). This is true of the present CTMD calculations, but we again guess that the same might have happened in the extrapolations of the above classical experimental data, often spanning a maximum of three decades of timescale. Therefore, the evidence more strongly appears to be in favor of *rheological complexity* (changing active cluster sizes, response times and activation energies), rather than simplicity. The way of disproving this and falsifying CTMD altogether would be by impossible (temperature-controlled) measurement of full response curves over 10 or more decades of timescale.

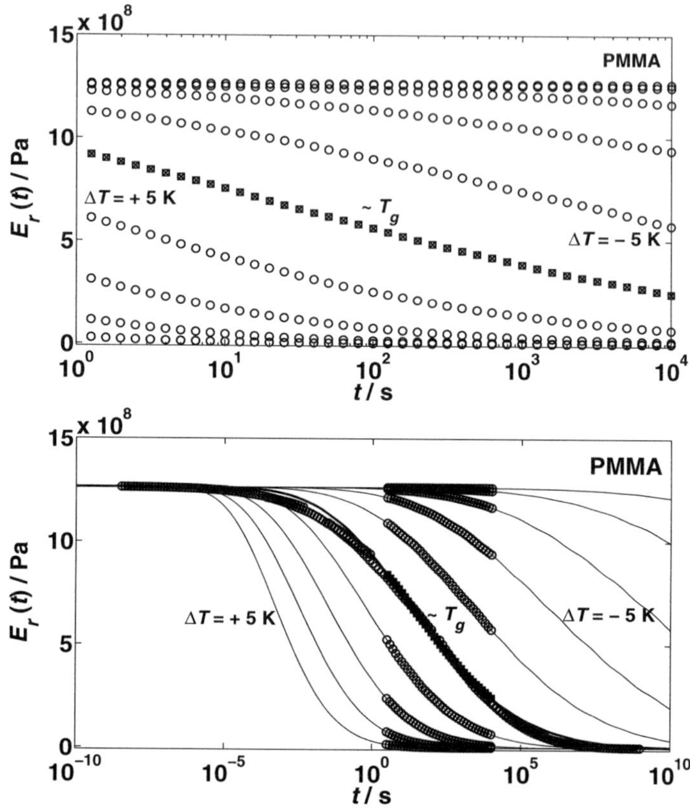

Fig. 11.22 Limited time–temperature superposition (or equivalence) of CTMD-predicted uniaxial tensile stress relaxation moduli for PMMA, with physical parameters derived from ~ 24-h stress relaxation measurements at ~ 60 K below T_g in the linear viscoelastic domain[7] [4]

Finally, Figs. 11.23 and 11.25 show that CTMD-calculated average relaxation time data for PMMA and PC at constant volume also almost accurately obey WLF and VTF correlations, thus strengthening the rejection of classical claims that molecular packing changes might alone explain that type of behavior. But what matters most is that CTMD does not assign a clear physical meaning (only a mild interpolation and extrapolation utility) to such correlations.

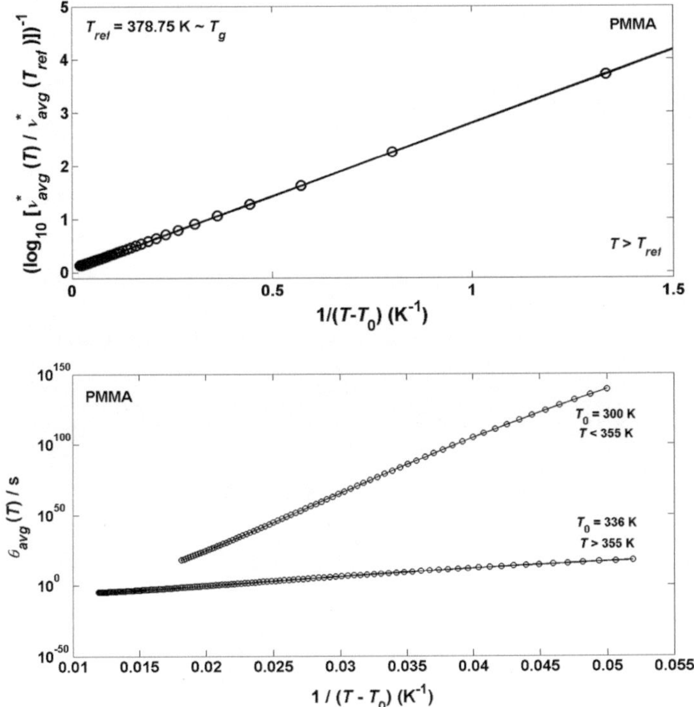

Fig. 11.23 CTMD-predicted WLF and VTF relationships for PMMA, with physical parameters derived from ~ 24-h uniaxial tensile stress relaxation measurements at ~ 60 K below T_g in the linear viscoelastic domain

11.6 Additional Highlight of the Ability of CTMD to Predict Dynamic Heterogeneity, Widening Response Spectra at Decreasing Temperatures, and Rheological Complexity

All three features in the title have been treated in depth in the book, the first of which determines the remaining two, but samples of the CTMD-predicted cluster response time distributions have only previously been shown for the Numerical Example of Chap. 8. Given their relevance, Figs. 11.26 and 11.27 show for completeness several calculated spectra at a range of temperatures for PMMA and PC.

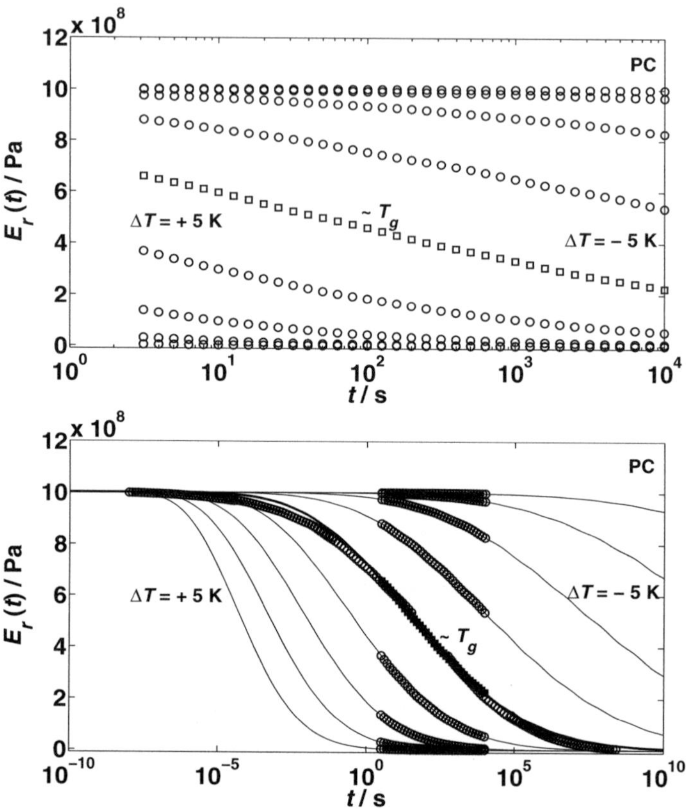

Fig. 11.24 Limited time–temperature superposition (or equivalence) of CTMD-predicted uniaxial tensile stress relaxation moduli for PC, with physical parameters derived from ~ 24-h uniaxial tensile stress relaxation measurements at ~ 100 K below T_g in the linear viscoelastic domain

11.7   CTMD-Predicted "Fragilities"

One of the stringiest general tests of CTMD must involve evaluating what it says about the so-called and classical concept or property of amorphous materials. We have seen in Chap. 9 that a model of tunable fragility appears to naturally emerge from the theory. How do such *fragility plots* [9–12] look like within CTMD, and what are its quantitative predictions?

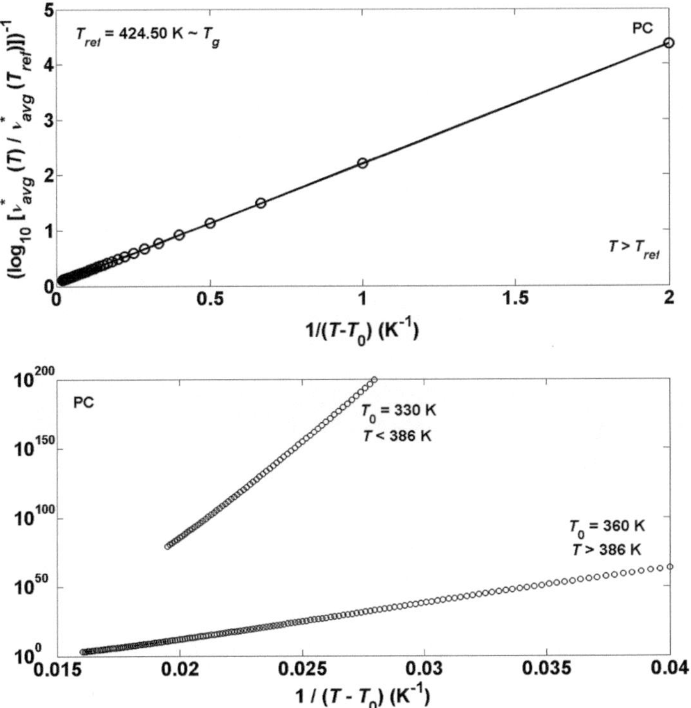

Fig. 11.25 CTMD-predicted WLF and VTF relationships for PC with physical parameters derived from ~ 24-h uniaxial tensile stress relaxation measurements at ~ 100 K below T_g in the linear viscoelastic domain

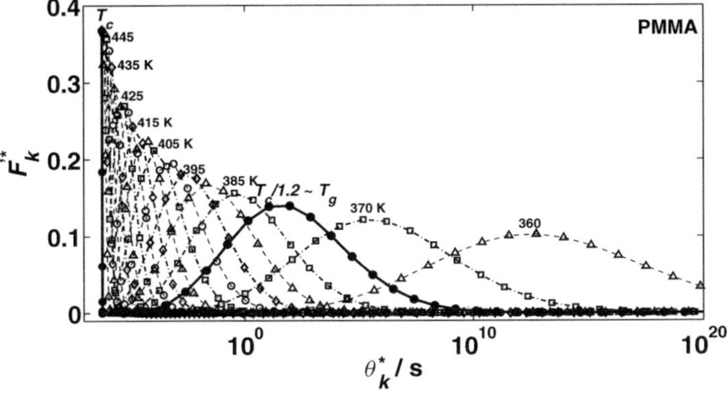

Fig. 11.26 PMMA's CTMD-predicted equilibrium response time distributions

Figure 11.28 shows the zoomed-in fragility plots in the vicinity of their respective T_g for the three systems studied. As those T_g values and the corresponding average

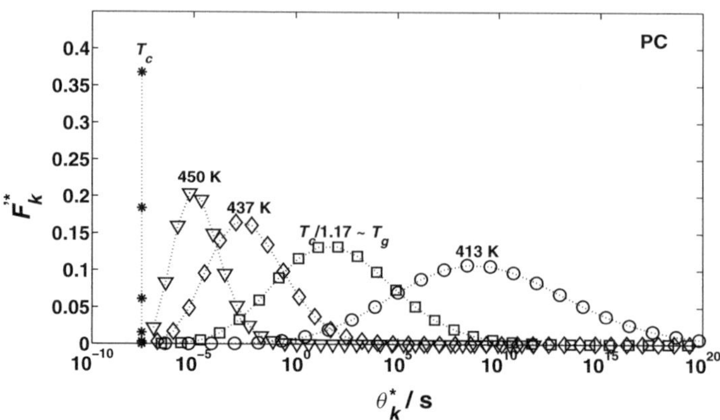

Fig. 11.27 PC's CTMD-predicted equilibrium response time distributions[8] [4]

relaxation times differ, it makes perfect sense to normalize the plots to each of those relaxation times, by plotting $\theta_{avg}^*(T)/\theta_{avg}^*(T_g)$, as well as extending the plots below T_g and up towards T_c,[9] as shown in the lower part of the Figure and in Fig. 11.32.

The different slopes of the curves at T_g are easily determined, not just graphically, but by accurate calculation within CTMD itself, either by Eq. 9.50 or, in a very straightforward way, by direct numerical differentiation of the high-resolution $\ln\theta_{avg}^*(T)$ data, as plotted in Fig. 11.29, for the Numerical Example. Figures 11.30 and 11.31 shows the results for the other two systems, where the lower part of Fig. 11.30 documents (for the PMMA example) the inconvenience of the classical definition of the fragility index, as it depends on the timescale of the measurements through the value of T_g obtained or chosen. A mere 3K decrease in T_g causes a 15% increase in the fragility index, which is a poor physical characterization of a material.

---

[9] Where available theories mostly fail is at any temperature below T_c (not just T_g, the lowest the worse, of course). The markers shown represent actual high resolution ($\Delta T = 0.25$K) CTMD-calculated data in a very narrow temperature range, but the available data cover a much wider range than any published fragility plots, as the reader may recognize by inspection of all previous relaxation times and temperatures plotted in the many Figures of this book.

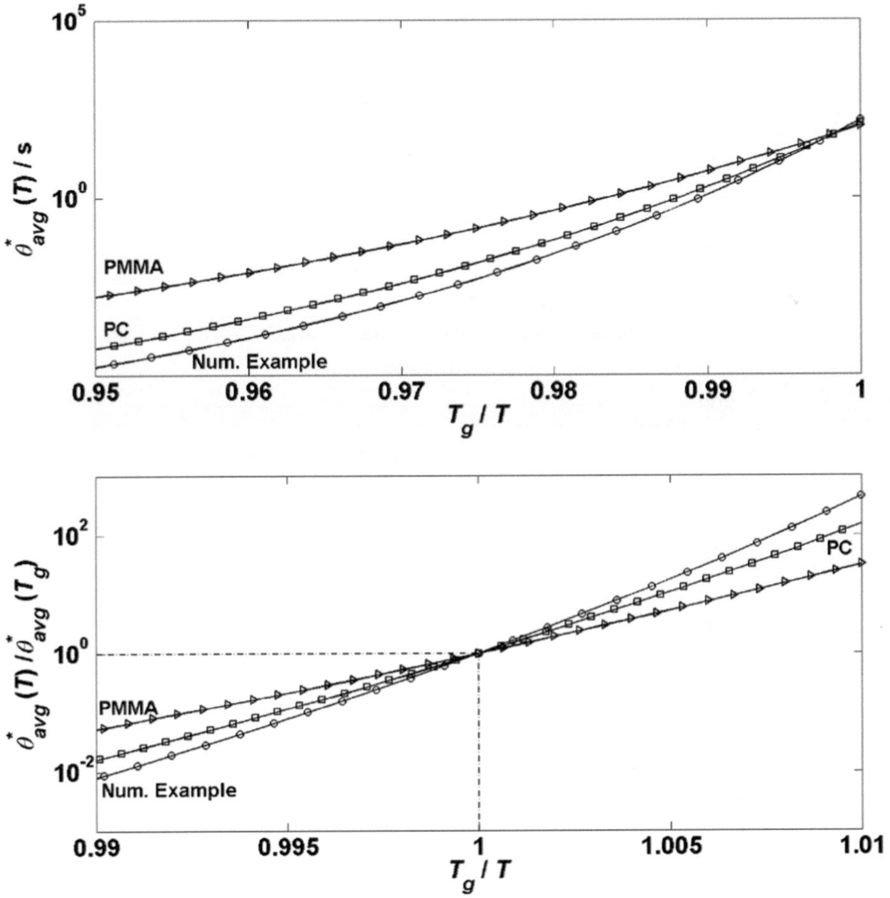

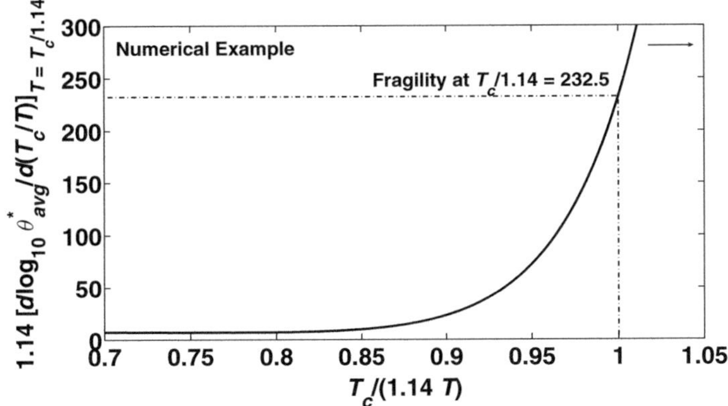

Fig. 11.28 Fragility plots for the three systems studied

Fig. 11.29 Calculation of the fragility index of the Numerical Example specified in Chap. 8

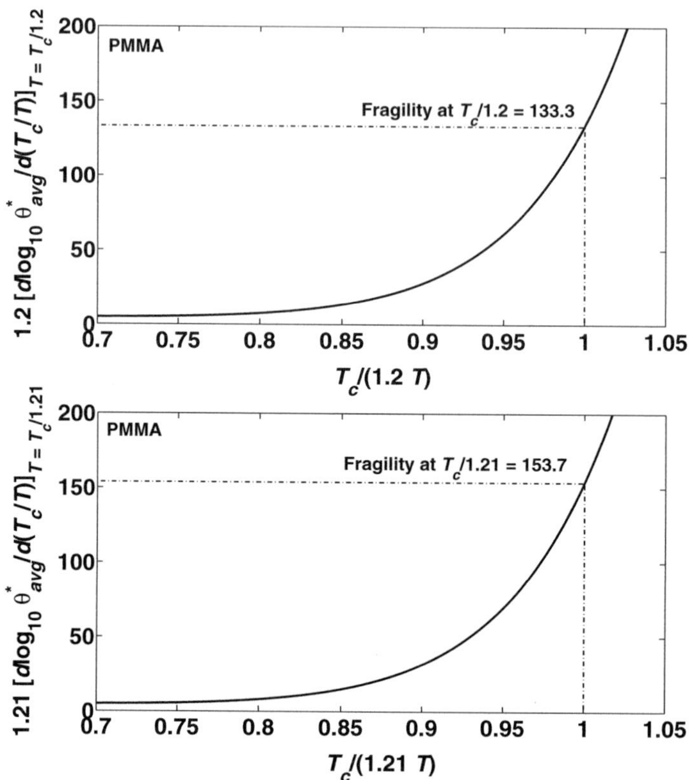

Fig. 11.30 Calculation of the fragility index of PMMA (top[10] [4]) and its dependence on the glass transition temperature measured or selected (bottom)

For PMMA, the value calculated at $T_g = T_c/1.2$, $m = 133.3$, agrees with the tabulated value of 136 [1, 2], at which the average relaxation time is of the order of 100s.[11] As for PC, the CTMD-calculated value at $T_g = T_c/1.17$, where its average relaxation time is closest to 100s, is $m = 198.5$, a first ever estimate (cf. Figure 11.31) as far as we know, like its crossover frequency and crossover temperature given in Sect. 11.2.

The higher fragility index of PC relative to PMMA is the most eloquent demonstration that "fragility" is a totally inadequate name for the property in question, as implicitly suggested by Donth [1]. Worse than the latter two cases is perhaps the one taken as our Numerical Example of a very flexible structure, for which the glass

---

¹⁰ Copyright © 2019 From *Processing and Characterization of Multicomponent Polymer Systems: New Insights* by Jose James et al. (eds.)/J. J. C. Cruz Pinto and J. R. S. André (auths.). Adapted with expanded data by permission of Taylor and Francis Group, LLC, a division of Informa plc.

¹¹ Recall that the CTMD-calculated values of PMMA's crossover frequency, ν_c, and temperature, T_c, also very closely agree with values tabulated in [1, 2] (cf. Footnote 2).

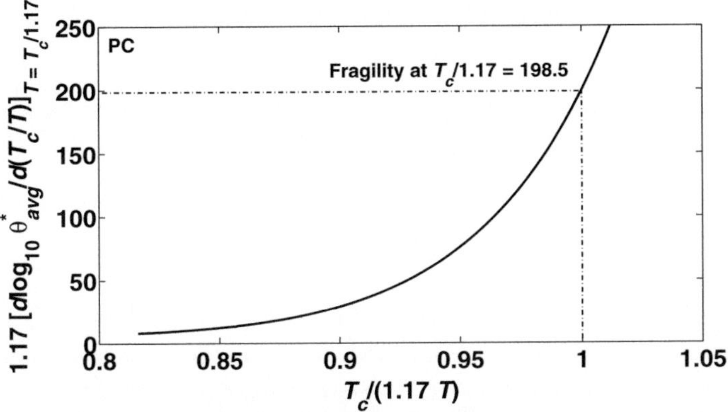

Fig. 11.31 Calculation of the fragility index of PC

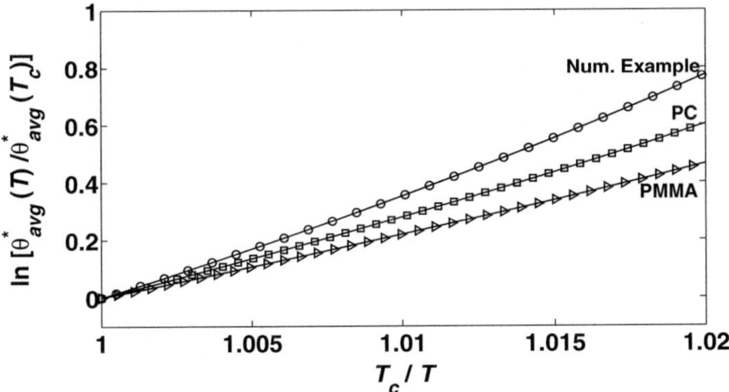

Fig. 11.32 More fundamental normalized "fragility" plots

transition temperature is only $\sim 5\,°C$ with no steric hindrances $\left(z^{\#}_{1,r}/z_1\right) = 1$. Such material might even not exist but would not be "fragile" at all.

However, as proposed in Chap. 9, it would make better sense to calculate fragility (or whatever) indexes at some truly *fundamental* temperature, like T_c, to physically characterize any given structure without depending on the timescale of the measurement. In addition, as shown, the calculation turns out straightforward from basic dynamic properties of the system. Doing that, more fundamental, normalized, "fragility" plots would be obtained, as in Fig. 11.32, whose accurate, more physically significant slopes at $T = T_c$ could be calculated by $m^* = 2\left(1 + E_{a,1,r}\right)$, to yield 32.48 (Numerical Example), 10.84 (PMMA) and 26.22 (PC).

A relevant final note in this section is that, while most predictions by CTMD imparting on features or properties dependent on the average relaxation times turn out physically very reasonable,[12] the same will not yet apply to the actual relaxation spectra, except near T_g.[13] This matter is given full recognition and deeper analysis in Sect. 11.9 and Chap. 12, as one of the persisting loose ends and future update requirements of CTMD.

11.8   Computability of CTMD

What if, for each set of physical properties needed and used within a theory, we could compute all the specific results shown in this book for any given material in a matter of *seconds* on an ordinary personal computer, laptop or tablet, or even a simple smartphone with access to an off-site powerful server? Pending more complete testing, significant future update and substantial validation,[14] that is exactly what we can do with CTMD with just 1 2/3 A4-pages of simple computer code,[15] as the one below for the PMMA data in this book.[16]

```
format long e;
Ea1_PMMA=35655.48316306835;
nuc_PMMA=2.488750545812028e+05; h=6.6260755e-34;
kB=1.3806568e-23;
NA=6.0221367e23; R=8.31451;
Tc_PMMA=4.552364787204446e2;          T=[[320:0.25:455]          Tc_PMMA
[456:0.25:550]];
sum_E=0.0*T;sum_D=0.0*T;sum_s_E=0.0*T;sum_s_D=0.0*T;sum_n_
E=0.0*T;sum_n_D=9.0*T;
thet_avg_T_E=0.0*T;nM=0.0*T;lthet_avg_s_T_E=0.0*T;
thet_avg_s_T_E=0.0*T;thet_avg_s_T_D=0.0*T;thet_avg_T_D=0.0*T;
del_thet_E=0.0*T;del_lthet_s_T_E=0.0*T;del_n_E=0.0*T;r_lthet_
E=0.0*T;r_n_E=0.0*T;
r_lthet_D=0.0*T;r_n_D=0.0*T;del_n_D=0.0*T;
del_thet_D=0.0*T;del_lthet_s_T_D=0.0*T;lthet_avg_s_T_D=0.0*T;
lt=[-10:0.1:15];t=10.0.^lt;nt=size(t,2);nT=size(T,2);
lnu=[-15:0.05:10];nu=10.0.^lnu;nnu=size(nu,2);
E_r_sls_rel=zeros(nt,nT); D_sls_rel=E_r_sls_rel;
az_PMMA=sinh(h*nuc_PMMA/2/kB/Tc_PMMA)*exp((Ea1_PMMA/R-h*nuc_
PMMA/2/kB)/Tc_PMMA);
E0_PMMA=1.264144717979428e+09;Einf_PMMA=1.000000000000941e+06;
a=1.0/exp(1);

for j=1:nT
```

---

[12] [Like the compliance to Ngai's coupling relationship(s), the T_g estimates, or the fragility relationships and its actual values].

[13] Nevertheless, the increase of the spectral width at lower temperatures is qualitatively well described.

[14] Falsification of CTMD is of course also, not just possible, but wide open.

[15] [Matlab® entirely open-source code, which interested readers are welcome to explore.] Full meanings of the variables may be requested from one of the authors (JJCP).

[16] [Only excluding the separate but straightforward calculation of fragility indexes].

```
sum(j)=1.0;
del_thet_E(j)=1.0;del_thet_D(j)=1.0;
nu_f=az_PMMA*exp(-(Ea1_PMMA/R-h*nuc_PMMA/2/kB)/T(j))/
sinh(h*nuc_PMMA/2/kB/T(j));
i=0;

while del_thet_E(j)/sum_E(j) >= 1.0e-2*eps & del_thet_D(j)/sum_
D(j) >= 1.0e-2*eps;
i=i+1;
F_i(i,j)=a/factorial(i-1);
nu_i_s_T(i,j)=nuc_PMMA*(nu_f)^i;
thet_i_s_T_E(i,j)=1.0/(2*pi*nu_i_s_T(i,j));
thet_i_s_T_D(i,j)=(E0_PMMA/Einf_PMMA)*thet_i_s_T_E(i,j);
del_thet_E(j)=F_i(i,j)*thet_i_s_T_E(i,j);
del_thet_D(j)=F_i(i,j)*thet_i_s_T_D(i,j);
thet_avg_T_E(j)=thet_avg_T_E(j)+del_thet_E(j);
thet_avg_T_D(j)=thet_avg_T_D(j)+del_thet_D(j);
sum_E(j)=thet_avg_T_E(j);
sum_D(j)=thet_avg_T_D(j);
nM(j)=i;
end

for i=1:nM(j)
F_i_p_E(i,j)=F_i(i,j)*thet_i_s_T_E(i,j)/sum_E(j);
F_i_p_D(i,j)=F_i(i,j)*thet_i_s_T_D(i,j)/sum_D(j);
del_lthet_s_T_E(j)=F_i_p_E(i,j)*log(thet_i_s_T_E(i,j));
del_lthet_s_T_D(j)=F_i_p_D(i,j)*log(thet_i_s_T_D(i,j));
del_n_E(j)=i*F_i_p_E(i,j);
del_n_D(j)=i*F_i_p_D(i,j);
sum_s_E(j)=sum_s_E(j)+del_lthet_s_T_E(j);
sum_s_D(j)=sum_s_D(j)+del_lthet_s_T_D(j);
sum_n_E(j)=sum_n_E(j)+del_n_E(j);
sum_n_D(j)=sum_n_D(j)+del_n_D(j);
lthet_avg_s_T_E(j)=sum_s_E(j);
lthet_avg_s_T_D(j)=sum_s_D(j);
n_avg_T_E(j)=sum_n_E(j);
n_avg_T_D(j)=sum_n_D(j);
z_s_coop_i(i,j)=(nu_i_s_T(i,j)/nuc_PMMA)^(nu_    i_s_T(i,j)/nuc_
PMMA);
end

thet_avg_s_T_E(j)=exp(lthet_avg_s_T_E(j));
thet_avg_s_T_D(j)=exp(lthet_avg_s_T_D(j));
nu_avg_s_T_E(j)=1.0/(2*pi*thet_avg_s_T_E(j));
nu_avg_s_T_D(j)=1.0/(2*pi*thet_avg_s_T_D(j));
r_lthet_E(j)=del_lthet_s_T_E(j)/sum_s_E(j);
r_lthet_D(j)=del_lthet_s_T_D(j)/sum_s_D(j);
r_n_E(j)=del_n_E(j)/sum_n_E(j);
r_n_D(j)=del_n_D(j)/sum_n_D(j);

for it=1:nt
sum_E_r_sls_rel=0.0;sum_D_sls_rel=0.0;
```

```
for i=1:nM(j)
sum_E_r_sls_rel=sum_E_r_sls_rel+F_i_p_E(i,j)*exp(-t(it)/thet_
i_s_T_E(i,j));
sum_D_sls_rel=sum_D_sls_rel+F_i_p_D(i,j)*(1.0-exp(-t(it)/thet_
i_s_T_D(i,j)));
end

E_r_sls_rel(it,j)=sum_E_r_sls_rel;
E_r_sls(it,j)=Einf_PMMA+(E0_PMMA-Einf_PMMA)*E_r_sls_rel(it,j);
D_sls_rel(it,j)=sum_D_sls_rel;
D_sls(it,j)=1/E0_PMMA+(1/Einf_PMMA-1/E0_PMMA)*D_sls_rel(it,j);
end

for inu=1:nnu
sum_Ep_rel=0.0;sum_E2p_rel=0.0;
sum_Dp_rel=0.0;sum_D2p_rel=0.0;

for i=1:nM(j)
XE=2*pi*nu(inu)*thet_i_s_T_E(i,j);
Y1E=F_i_p_E(i,j)*XE/(1+XE^2);
Y2E=Y1E*XE;
XD=2*pi*nu(inu)*thet_i_s_T_D(i,j);
Y1D=F_i_p_D(i,j)*1/(1+XD^2);
Y2D=Y1D*XD;
sum_Ep_rel=sum_Ep_rel+Y2E;
sum_E2p_rel=sum_E2p_rel+Y1E;
sum_Dp_rel=sum_Dp_rel+Y1D;
sum_D2p_rel=sum_D2p_rel+Y2D;
end

Ep_rel(inu,j)=sum_Ep_rel;
E2p_rel(inu,j)=sum_E2p_rel;
Dp_rel(inu,j)=sum_Dp_rel;
D2p_rel(inu,j)=sum_D2p_rel;
tandelta_E(inu,j)=(E0_PMMA-Einf_PMMA)*E2p_rel(inu,j)/(Einf_
PMMA+(E0_PMMA-Einf_PMMA)*Ep_rel(inu,j));
tandelta_D(inu,j)=((1/Einf_PMMA)-(1/E0_PMMA))*D2p_rel(inu,j)/
((1/E0_PMMA)+((1/Einf_PMMA)-(1/E0_PMMA))*Dp_rel(inu,j));
end

end
```

The first third of the above listing is made up of mere variables assignment[17] and array dimensions statements. The execution time of this entire procedure for wide temperature ranges and time and frequency ranges of 10^{-10} to 10^{15} s and 10^{-15} to 10^{10} Hz, respectively, along grids of 0.25 K temperature, 0.1 $\log_{10}$(time/s) and 0.05 $\log_{10}$(frequency/Hz) steps on a HP Spectre x360 14 [®], is about 50 *s*, and is

---

[17] [To the values separately calculated from experimental stress relaxation data, using an approximate version of CTMD and a separate Fortran direct global search optimization procedure].

almost "instantaneous" using Matlab® online (MathWorks Cloud) from an Android smartphone or tablet via fiber optics communications. This may prove relevant when pursuing the aim of CTMD's direct future accurate use[18] within an appropriate optimization algorithm to more accurately calculate all physical properties, including v_c, T_c and $E_{a,1}$, from stress relaxation or other mechanical response data, without approximations like Alfrey's integration one,[19] other than those explicit and implicit in CTMD itself.

11.9 Estimate of Physical Dynamic Data (v_c, T_c and $E_{a,1}$) from Uniaxial Tensile Stress Relaxation Measurements

As explained in Chap. 4, the option taken in our previous creep and stress relaxation work [13–15] was that of assuming a simple truncated log-normal distribution of response times, bearing in mind their expected final continuous character, such as to yield for the final relaxation modulus, $E_r(t)$,

$$E_r(t) = E_\infty + (E_0 - E_\infty)\frac{\int_{\ln\theta_1}^{+\infty} e^{-t/\theta} e^{-[b\ln(\theta/\theta^*)]^2} d\ln\theta}{\int_{\ln\theta_1}^{+\infty} e^{-[b\ln(\theta/\theta^*)]^2} d\ln\theta}, \tag{11.1}$$

or

$$E_r(t) = E_\infty + (E_0 - E_\infty)\frac{\text{erf}(b_0) + \text{erf}[b\ln(\theta^*/t)]}{1 + \text{erf}(b_0)}, \tag{11.2}$$

after making the classical Alfrey's approximation [13–15] on the numerator's integrand of Eq.11.1, where $b = b_0/\ln(\theta^*/\theta_1)$, θ_1 is a minimum relaxation time (of the order of the one characterizing uncorrelated single primitive relaxors), $\theta^* = \theta_{avg}$ an average relaxation time, and imposing $E_r(t \leq \theta_1) = E_0$, to avoid inconsistent values in excess of E_0 at very short times, resulting from the assumed truncated nature of the relaxation spectra.

The above approximations and procedure were also adopted here with the practical purpose of estimating physically reasonable values of the basic dynamic properties v_c, T_c and $E_{a,1}$, in addition to E_0 and E_∞, to be then used in the subsequent analysis of CTMD. For that, both θ_1 and θ^* were formulated according to Chap. 9 as functions of the above properties, the applied strain, and an average cluster size,

---

[18] Nevertheless, it was reasonable to first test CTMD from an approximate estimate of its main physical parameters obtained by a simplified version of the theory, where the still conjectural distributions of response times of Chap. 9 were replaced by the continuous truncated log-normal distributions specified in Chap. 4 (fitted to the actual experimental data), using Alfrey's integration approximation.[19]

[19] Cf. Chap. 4.

n_{avg}, and their values numerically optimized[20] to obtain stress relaxation moduli in the closest possible agreement with the experimentally measured values for PMMA and PC [14, 15].

11.9.1   Experimental Results

Totals of 114 (PMMA) and 248 (PC) time-dependent stress relaxation modulus values were measured [15] in uniaxial tension for the specified amorphous polymer materials, for times up to 9×10^4 s at 40 and 50°C (to within $\pm$ 1°C), and strains of 3 (only in some PC experiments), 4 and 5%, using a Zwick Z100 universal testing machine fitted with a 2.5 kN load cell. Crazing has been visually checked and avoided on all 11 different test specimens (including during some additional experiments lasting a full week, on separate test specimens), as both the experiments and the models were intended to cover conditions far below crazing, yield and fracture. The procedure physically justified in Chap. 5 was followed to correct the results for the initial strain ramp, to determine the *ramp time*, as the time at which the strain reaches its exact intended value. Only stress and modulus values for times above that value were actually used in the calculations, and the time values collected by the testing instrument were subtracted by that ramp time.[21]

11.9.2   Approximate Theoretical Results and Their Physical Discussion

The results of the measurements and Alfrey's-approximate model calculations agreed to within relative errors of only 0.9% for PMMA (Fig.11.33) and 0.3% for PC (Fig.11.34). These results do not include calculations for PMMA at 3% strain, because at this strain stress relaxation for this material was limited and experimentally uncertain (likely due to lower reproducibility among different test specimens[22]), and thus difficult to measure and model with comparable precision.[23] In the case of PC, two of

---

[20] A separate global optimization, direct search, routine in Fortran was written and used for that purpose. Execution times were of less than 1s per 10^4 function calls in the same HP Spectre x360 14[®] to compute the sum of the squares of the deviations between the entire sets (114 and 248) of calculated and experimental values of the stress relaxation modulus for PMMA and PC, respectively.

[21] Assuming that, during the initial strain ramp, the behavior may be very close to linear viscoelastic, Chap. 5 explains that, in future calculations, the measured time values should be best subtracted by 2/3 of the ramp time.

[22] One test specimen was of course spent per single test at each temperature and applied strain.

[23] This difficulty was not met in the PC tests, for which better reproducibility and lower relative errors were obtained between experimental and calculated values, and so some limited reasonable quality data at 3% strain could be obtained and treated for this material, despite its slower stress relaxation.

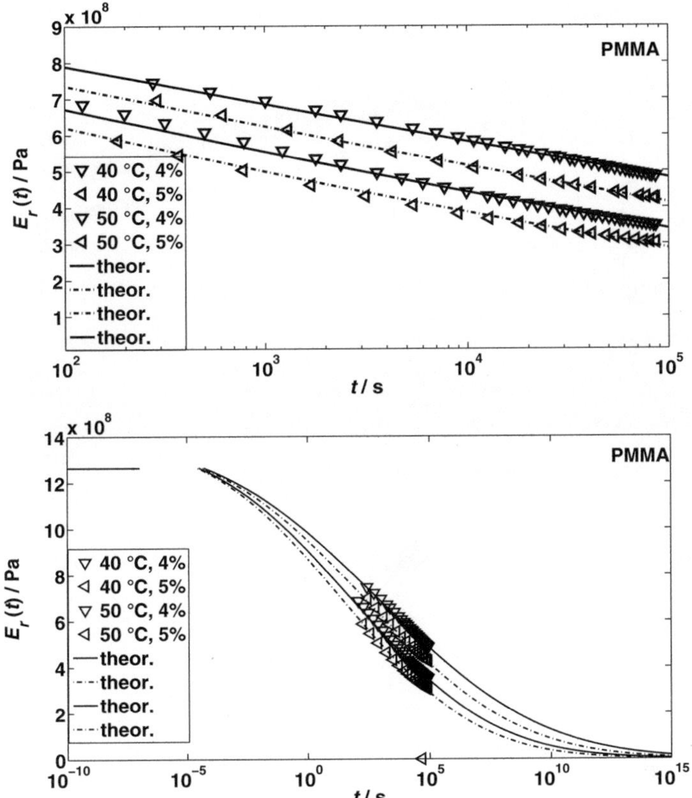

Fig. 11.33 Zoomed-in (top[24] [4]) and zoomed-out (bottom) experimental uniaxial tensile stress relaxation data for PMMA (markers) and their theoretical predictions (lines) by the simplified version of CTMD of Chap. 4

the curves in the top plot (for 4% strain at 40 and 50 °C) illustrate the data reproducibility that could be achieved among different test specimens (light and dark markers in the curves), which the approximate theoretical model and numerical optimization algorithm could interpret as corresponding to different material specimens with just very small differences in the average cluster sizes and average relaxation times.

The ability of the model to predict the about 10^5 s long experiments and extrapolate to what may reasonably be expected to be the very short and very long-time behavior (up to and including the seemingly correct instantaneous and infinite time modulus plateaus, E_0 and E_∞) should be highlighted. For PMMA, *ca.* 25 decades become necessary at 40–50 °C (i.e. 60 or more degrees below its T_g) for the full stress relaxation process to develop (cf. Figure 11.33), in line with available, textbook-classical,

---

[24] Copyright © 2019 From *Processing and Characterization of Multicomponent Polymer Systems: New Insights* by Jose James et al. (eds.)/J. J. C. Cruz Pinto and J. R. S. André (auths.). Adapted with re-scaled data by permission of Taylor and Francis Group, LLC, a division of Informa plc.

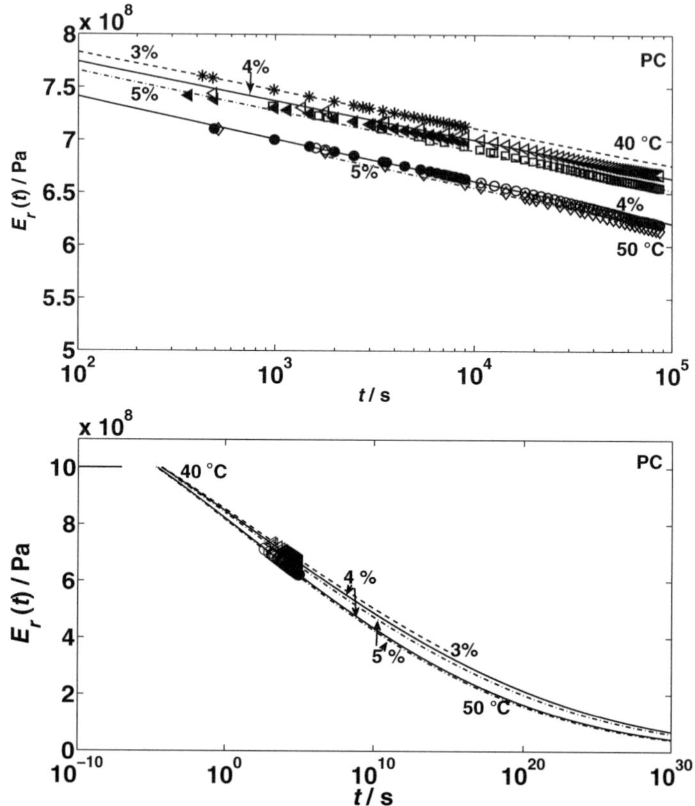

Fig. 11.34 Zoomed-in (top[25] [4]) and zoomed-out (bottom) experimental uniaxial tensile stress relaxation data for PC (markers) and their theoretical predictions (lines) by the simplified version of CTMD of Chap. 4 (open and solid markers are for repeat experiments on different test specimens)

data [16] at varying temperatures near and above T_g for this material. We remind that the present calculations do not consider the effect of simultaneous (though extremely slow) viscous flow, which would be responsible for somewhat shortened response times and a drift of $E_r(t)$ below E_∞, experimentally noticeable at extremely long times, as also theoretically justified and illustrated in Chap. 3.

In the case of PC (Fig. 11.34), stress relaxation turns out much slower, resulting now in the estimation of over 55 decades in the time scale (completely inaccessible to any conceivable experiment!), also neglecting viscous flow, as being necessary to approach E_∞ about 100 degrees below the material's T_g. Of course, similarly to other extremely long time (or low frequency) and low temperature predictions of CTMD in this book, these also cannot possibly ever be directly checked, but they should not be dismissed as completely unreasonable, assuming of course that such test, or the material itself, would not fail for many other possible reasons. We cannot be sure at this stage, but it remains open the possibility of the approximate and full future CTMD application to precisely controlled experiments with PC, at higher temperatures during significantly shorter times, indirectly validating the present predictions, with allowance for only crude (not strict) time–temperature superposition below T_g, as experimentally known and supported by the calculations illustrated previously in this chapter.

Also apparent in Figs. 11.33 and 11.34 is the fact that an increase in temperature, as predicted by Eq.9.8 (as by Eq.9.14 of the improved model), attenuates the non-linearity of the materials, *i.e.* the effect of increasing the strain is reduced.

The optimized values obtained for the various physical parameters of both materials, which yielded the continuous curves of Figs. 11.33 and 11.34 (partly listed in Sects. 11.2 and 11.3) were ($v_c = 2.4888 \times 10^5$ Hz, $T_c = 455.24$ K, $E_{a,1} = 35.655$ kJ/mol, $v_1(40°C) = 1.0460 \times 10^{-3}$ m^3/mol, $E_0 = 1.2641$ GPa, $E_\infty = 1.0000$ MPa) and ($v_c = 5.3449 \times 10^6$ Hz, $T_c = 496.24$ K, $E_{a,1} = 49.957$ kJ/mol, $v_1(40°C) = 1.1659 \times 10^{-3}$ m^3/mol, $E_0 = 1.0014$ GPa, $E_\infty = 0.55626$ MPa) for PMMA and PC, respectively, v_1 being their respective estimated volumes of 1 mol of primitive relaxors at 40°C.[26] From the estimated optimum values of v_c, T_c and $E_{a,1}$, the corresponding values of the structural parameter $a_z = \left(z_{1,r}^{\#}/z_1\right)$ were calculated by Eq.9.17 as 1.6179×10^{-4} and 4.6856×10^{-2} for PMMA and PC, respectively. It should however be noted that these results would have been affected by any significant level of viscous flow that might have been present.[27]

Published data for PC [17] is limited and not so well studied, covering temperatures only near and above T_g but, even at these conditions, the whole process requires a high number of powers of ten of timescale to complete or reveal any viscous flow. A detailed discussion was already provided [14] of the experimental and theoretically expected behavior of the maximum absolute rate of change of the stress relaxation modulus, $-\frac{dE_r(t)}{d\ln t}$, and suggests that the much quoted, apparently universal, value of $(0.10 \pm 0.01)(E_0 - E_\infty)$ may be valid only near and above the material's T_g, gradually but significantly decreasing as the temperature is lowered, and this contributes to the extremely long timescales predicted for the process at temperatures below T_g.

The present theoretical results and corresponding experimental data cannot be directly compared with data and predictions that span a maximum of only six decades of timescale [18–20], in the case of fast relaxing materials (like lithium fluoride, cadmium, polybutylene, low-density polyethylene, etc., close to room temperature), nor with the seemingly unrepresentative results with a maximum of only *two* decades (even less than for an *ideal standard linear solid*!) of some earlier molecular dynamics simulations [21, 22]. Other molecular dynamics simulations [23] appear much more realistic but, like most if not all simulations, have not been directly compared with experimental data for real materials and do not provide specific

---

[26] The calculations also allowed for, and estimated, a small linear volume expansion between 40 and 50°C.

[27] The reader should relate this to the remark made in the last paragraph of Sect. 3.2.2.

(quantified) timescales and numerical values of the relevant physical parameters (with specific units, not dimensionless).

The calculations using the analogue of Eq. 4.12 for a truncated spectrum at θ_1 also yield the approximate smoothed or continuous relaxation spectra of the materials at the various temperatures and strains

$$\frac{H(\theta)}{E_0 - E_\infty} = \frac{b_0/\sqrt{\pi}}{[1 + \mathrm{erf}(b_0)]\ln(\theta*/\theta_1)} e^{-[b_0 \ln(\theta/\theta*)/\ln(\theta*/\theta_1)]^2}, \qquad (11.3)$$

as illustrated by Fig. 11.35 (top). As expected, increasing temperatures and strains gradually narrow the spectra, and shift them to shorter times and very slightly lower the average cooperativity levels, as measured by the calculated values of the average cluster sizes, $n*$ (4.85 to 4.71 for PMMA and 3.25 to 3.21 for PC from 40 to 50°C and 4 to 5% strains). The predicted spectrum of PC at 50 °C and 5% strain is also shown for comparison, turning out very much wider.

In addition to the reasonable values of both plateau moduli (neglecting the effect of viscous flow), the other values are physically very significant. As previously pointed out, calculated crossover frequencies and temperatures, v_c and T_c, very closely agree with the published experimental data for PMMA [1, 2], the best characterized of the materials considered—0.25 MHz $\pm$ 1 decade and 455 $\pm$ 10 K, based on the extrapolated crossings of α and β traces in an Arrhenius plot from far below. This should lend additional support to the present formulations, combined with the estimate of PMMA's $T_g \sim T_c/1.2$ at around 106°C. The same happens with the T_c value calculated for PC, from which one obtains the estimate of just over 140°C for its T_g. As for PC's crossover frequency, to our knowledge, 5.3449 MHz is its first ever estimate, as no direct or indirect measurement has apparently been reported yet. This result, which obviously calls for future validation, should however not lead to confusion—it points to a much *faster jump from* the activated state in PC than in PMMA despite its *slower activation*, as clearly results from the minimum activation energy values, $E_{a,1}$, about 50% higher in PC than in PMMA.

The reader is reminded that the above crossover data have already been marked in Fig. 8.1, showing that they are close to, but not exactly onto, the crossover surface calculated for the $\frac{z_{1,r}^{\#}}{z_1} = 1$ example, in line with their respective optimum calculated values for $z_{1,r}^{\#}/z_1$. Bearing in mind the physical meaning of $z_{1,r}^{\#}/z_1$ (explained in Chap. 8), one might reasonably associate the above values to greater steric hindrances opposing PMMA's transitions *at* the activated state, due to its long and bulky methyl ester groups, despite the lower $E_{a,1}$ value—35.655 kJ/mol versus 49.957 kJ/mol in PC. These activation energies compare reasonably well with the published experimental data for the minimum (PMMA) or typical (PC) activation barriers of β-processes [24, 25], which should be like those expected to control the dynamic behavior of the primitive relaxors.

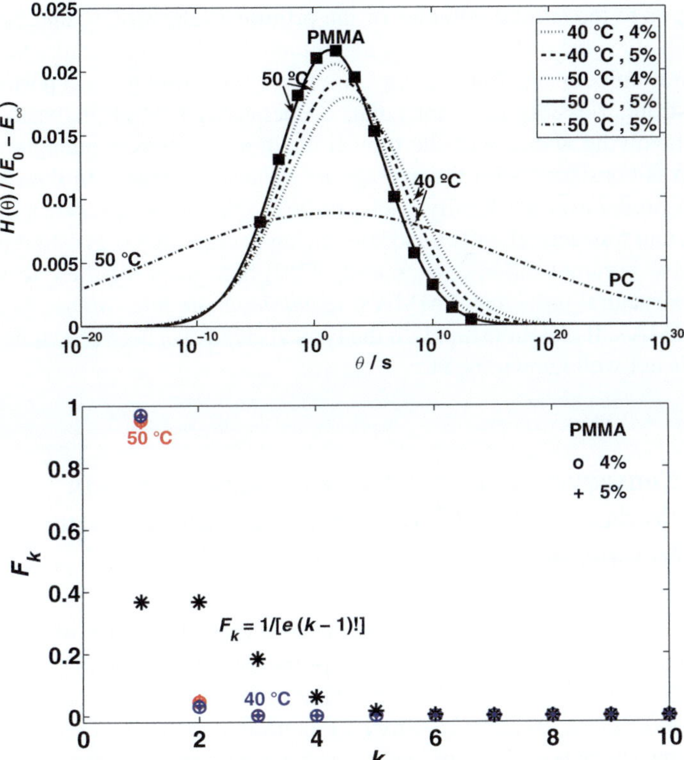

Fig. 11.35 Calculated truncated log-normal relaxation spectra of PMMA at the conditions indicated, plus one example for PC (top), and the comparison between the CTMD's initial F_k(that overestimate cooperativity-∗ markers) with those directly estimated from the continuous spectra obtained from the experiments, coincident with the highly conjectural values formulated in Chap. 12 for lower clustering probabilities (bottom)

The above minimum activation energy values might predominantly be the result of greater friction and dynamic arrest, or "slowness", *all along* each individual primitive relaxor transition in PC (in line with its greater toughness), and its higher T_c must also necessarily mean and entail a higher effective supercooling at any given temperature below the crossover relative to PMMA, while a higher localized skeleton stress in PMMA, exactly *at* each activated state, might be the critical factor determining the $z_{1,r}^{\#}/z_1$ ratio. However, all this should, at this stage, be deemed still conjectural, therefore requiring more detailed and comprehensive future analysis before a definite conclusion, given that the available T_c and ν_c experimental data taken as reference for PMMA may themselves have uncertainties of $\pm$ 10 K and $\pm$ 1 decade, respectively [1, 2], and these (combined with the uncertainty in $E_{a,1}$) may affect the corresponding $z_{1,r}^{\#}/z_1$, though perhaps not to the extent of bringing the values for PMMA and PC significantly closer.

Finally, as to the similar volumes of the primitive relaxors, υ_1, and based on an average density value of 1.2 g/cm^3, its magnitude for PMMA is similar to just under 14 structural units of this polymer (quantitatively supporting the reported suggestion of a strong, possibly intramolecular, cooperativity [25] of the basic motions), therefore involving about ¼ of the typical chain size between entanglements [14, 26], which is consistent with the average n^* values calculated. In the case of PC, based on a similar average density, each primitive relaxor should correspond to only between 5 and 6 structural units (also supporting the reported suggestion of a lower, possibly also intramolecular, cooperativity [25]) but, given the three to four times larger PC structural units than PMMA's, its *minimum mobile volume*, υ_1, turns out near to PMMA's. Its relationship with the typical chain size between entanglements, however, is not well known for PC.

11.9.3   Consequences of the Above Dynamic Property Predictions to CTMD's Evaluation and Ensuing Development

Noting the two simplifying assumptions made in the calculations of the preceding section to estimate the main three dynamic properties of the system (v_c, T_c and $E_{a,1}$) [14] and [15] (1) truncated log-normal distributions of relaxation times (replacing the CTMD ones of Chap. 9) and (2) Alfrey's integration approximation (Cf. Chap. 4), one must now check how they affected the predictions relative to the full application of CTMD, by relaxing just one of the above assumptions (Alfrey's), but keeping and using discretized substitutes-replacing the $H(\theta)/(E_0 - E_\infty)$ by the corresponding (non-CTMD) F_k'-of the same truncated log-normal distributions that were adjusted to the experimental stress relaxation data.

Why would we do that? For two reasons: (1) because Alfrey's approximation, as illustrated in Chap. 4 and again further below, often turns out very crude, and (2) the greatest doubt that clouds CTMD at its present stage is in the F_k' formulated in Chap. 9, which may in fact exaggerate the collective or cooperative nature of the behavior[28] by assuming that adjacency and large numbers ensure easy or equally probable clustering. The latter likelihood, more than mere possibility, was suspected from the large deviations obtained toward much longer times of the non-linear, strain-dependent (full original CTMD), relaxation modulus predictions for PMMA and PC at the same temperature and strain conditions of the experiments.

---

[28] [Following a corresponding and explicit suggestion by G. Tarjus in [27, 28]].

The conversion of the $H'(\theta) = H(\theta)/(E_0 - E_\infty)$ to an F'_k histogram is straightforward, given that the $\ln\theta_k$ are equally spaced (exactly in the case of Eq.9.8), such that the H'_k are as those shown by the ■ marks in Fig.11.35 (top) for the particular case of PMMA at 50°C and 5% strain. Assuming that dynamic heterogeneity might still be reasonably portrayed by the proportionally of the F'_k to the θ_k as proposed in Chap. 9, the greatest doubts may then lay in the combinatorial formulation developed for the F_k that are likely to overestimate the collective behavior. So, it is important to apply CTMD to the same PMMA and PC experimental conditions using as substitutes of the F_k those given by $F_k = H'_k(\Delta\ln\theta_k)/\theta_k\left[\sum_{j\geq 1}H'_j(\Delta\ln\theta_j)/\theta_j\right]$, as may be concluded from Eqs. 9.2.

As the lower part of Fig.11.35 makes clear, CTMD's original, purely combinatorial, F_k values overestimate the weights of clusters of two and more primitive relaxors and underestimates that of single, uncorrelated, primitive relaxors. The result of recalculating the stress relaxation moduli for PMMA with CTMD (which, though avoiding Alfrey's approximation, uses parameters separately optimized under that approximation), with the above modified F_k[29] and the new corresponding F'_k, are shown in Fig.11.36.

It may be recognized that the CTMD "exact" but unsmoothed wavy curves at the bottom of the Fig.11.36 illustrate the kind and magnitude of the deviations relative to the corresponding Alfrey-approximate smooth curves that result from the calculations of Sects. 11.9.1 and 11.9.2. The same kind of deviations are also illustrated at the top of the figure for a hypothetical normal distribution of relaxors around an average relaxation time of 1s, with the exact curve calculated in Matlab® by a high precision numerical integration function. One thus expects that subsequent accurate calculations assuming truncated log-normal spectra *without* Alfrey's approximation might lead to adjustments to the $E_{a,1}$ values and the resulting spectra, for nearly the same v_c and T_c, because the activation energies are expected to have greater influence on the position of the response curves along the time axis.

---

[29] [Seemingly more consistent with the experimental data].

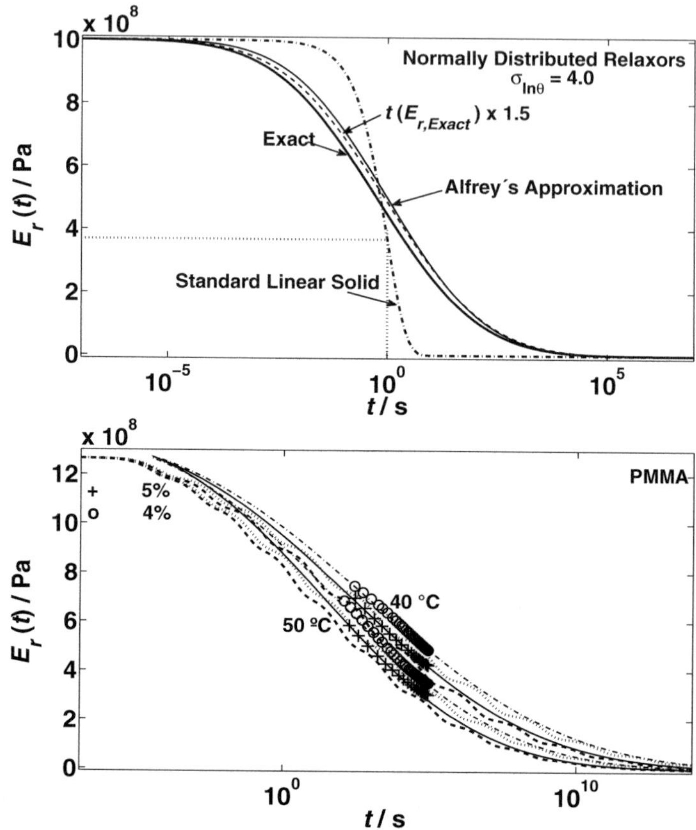

Fig. 11.36 Effect of Alfrey's approximation on predicted normally distributed uniaxial tensile stress relaxation modulus curves (top) and its influence on the CTMD-predicted uniaxial tensile relaxation moduli for PMMA at the two temperatures and strains of the experiments (wavy curves, bottom)[30] [4]

References

1. E.-J. Donth, *The Glass Transition—Relaxation Dynamics in Liquids and Disordered Materials*, Springer Series in Materials Science, vol 48 (Springer, Berlin, 2001)
2. M. Beiner, H. Huth, K. Schröter, J. Non-Cryst, Solids **279**, 126 (2001)
3. H.J. Sillescu, J. Non-Cryst, Solids **243**, 81 (1999)
4. J.J.C. Cruz Pinto, J.R.S. André, Cooperative molecular and macroscopic dynamics of amorphous polymers—a new analytical view and its potential, in *Processing and Characterization of Multicomponent Polymer Systems—New Insights*, eds. by J. James, S. Thomas, N. Kalarikkal, Y. Weimin, K. Pal, Chap. 13 (Apple Academic Press, Taylor & Francis, 2019), pp. 237–260
5. E. Catsiff, A.V. Tobolsky, J. Colloid Sci. **10**, 375 (1955), J. Polym. Sci. **19**, 111 (1956)

---

6. A.V. Tobolsky, J. Polym. Sci. Lett. **2**, 103 (1964)
7. J.R. McLoughlin, A.V. Tobolsky, J. Colloid Sci. **7**, 555 (1952)
8. J.D. Ferry, *Viscoelastic Properties of Polymers*, 3rd edn. (Wiley, New York, 1980)
9. W.T. Laughlin, D.R. Uhlmann, J. Phys. Chem. **76**, 2317 (1972)
10. C. A. Angell, in *Relaxation in Complex Systems*, eds. by K.L. Ngai, G.B. Wright (U. S. Dept. Commerce, Springfield, 1985), p. 1
11. C.A. Angell, J. Non-Cryst, Solids **131–133**, 13 (1991)
12. C.A. Angell, Science **267**, 1924 (1995)
13. J.R.S. André, *Fluência de Polímeros-Fenomenologia e Modelação Dinâmica Molecular* (Polymer Creep-Phenomenology and Modeling of Molecular Dynamics), Doctoral Thesis (Universidade de Aveiro, 2004)
14. J.R.S. André, J.J.C. Cruz Pinto, Polym. Eng. Sci. **54**, 404 (2014)
15. J.J.C. Cruz Pinto, J. R. S. André, Polym. Eng. Sci. **56**, 348 (2016)
16. A.V. Tobolsky, *Properties and Structure of Polymers* (Wiley, New Tork, 1960)
17. J.P. Mercier, J.J. Aklonis, M. Litt, A.V. Tobolsky, J. Appl. Polym. Sci. **9**, 447 (1965)
18. J. Kubát, Makromol. Chem. Suppl. **3**, 233 (1979)
19. Ch. Högfors, J. Kubát, M. Rigdahl, Phys. Stat. Sol. **107**, 147 (1981)
20. J. Kubát, Phys. Stat. Sol. **111**, 599 (1982)
21. W. Brostow, J. Kubát, Phys. Rev. B **47**, 7659 (1993)
22. S. Blonski, W. Brostow, J. Kubát, Phys. Rev. B **49**, 6494 (1994)
23. M. Warren, J. Rottler, Phys. Rev. Lett. **104**, 205501 (2010)
24. B. Brûlé, Relations entre Caractéristiques Moléculaires et Propriétés Mécaniques de Polyamides Amorphes et de Polycarbonates, Doctoral Thesis (University of Pierre et Marie Curie, France 1999)
25. J.L. Halary, F. Lauprêtre, L. Monnerie, *Mécanique des Matériaux Polymères* (Belin, Paris, 2008)
26. W.W. Graessly, Adv. Polym. Sci. **16**, 1 (1974)
27. S.A. Kivelson, G. Tarjus, Nat. Mat **7**, 831 (2008)
28. G. Tarjus, in *Dynamical Heterogeneities in Glasses, Colloids and Granular Media*, eds. by L. Berthier et al (OUP, Oxford, 2011).

Chapter 12
Open Questions and the Way Forward

12.1 A Conjecture on More Accurate Formulations of Relaxation Spectra

As anticipated, the discussion in Sect. 11.9.3 confirms that the weakest point of CTMD at its present stage is that collective behavior is being exaggerated, and so needs to be downscaled (or the cooperativity reduced) by accepting that adjacency of primitive relaxors does not guarantee clustering, not even a constant overall clustering probability irrespective of cluster size. A first, but highly conjectural, tentative proposal on how to tackle this outstanding weakness of CTMD is offered below.

We may conjecture that one way of accounting for a low average local *clustering probability*, p_c, of any two adjacent primitive relaxors, depending on local structural entanglements of variable nature, might be by considering a *chain-like process*[1] from any given reference or initiating primitive relaxor, with due consideration of an average number, n_c, of first neighbors. This will in all cases be a *tridimensional geometrical and topological problem* of the greatest complexity,[2] but let us simplify it as much as possible for stress relaxation, considering how clustering might be roughly described across de resisting area, *i.e.* in just two dimensions, as done in the simplest formulations of stress relaxation of Chap. 9. For each primitive relaxor of the system within or crossing the resisting area, taken as the reference in our calculation, the objective must then be to count all possible distinguishable sequences of clustering events of further $k-1$ primitive relaxors, for a given assumed number of first neighbors, for example $n_c = 6$, for a hypothetical hexagonal compact packing.

Such formulation should yield an expression for $f_k(k, n_c, p_c)$ proportional to the total number of clusters of k strictly adjacent primitive relaxors that include

---

[1] [Somewhat analogous to locally competing chain polymerizations, of propagation probability, p_c, without active termination].

[2] The reader may here recall the discussion starting at the 9th paragraph of Sect. 9.4, where it is made clear the tridimensional nature of clustering.

© The Author(s), under exclusive license to Springer Nature Switzerland AG 2024

J. J. Cruz Pinto and J. R. dos Santos André, *Analytical Molecular Dynamics of Amorphous Condensed Matter*, Springer Series in Materials Science 342, https://doi.org/10.1007/978-3-031-56517-5_12

the reference one. However, the clusters thus counted would include many others that, through the same procedure, would be generated from any of the other $k - 1$ primitive relaxors, and so one must divide the result by the factor responsible for that overestimate. As the overestimate would correspond to the inclusion of the same total k primitive relaxors in different clusters by any possible order, the factor sought should be $k!$, and so the F_k, before considering the clusters' intermittency that leads to the dynamic heterogeneity considered in Chap. 9, but including the proportionality of their responses to their sizes, would become $F_k = f_k(k, n_c, p_c)/(k - 1)!$.

To obtain $f_k(k, n_c, p_c)$, with the meaning given, conscious that the problem's complexity calls for much simplification, try and (1) take an ordinary checkered sheet of paper (replacing a resisting area in stress relaxation), where one can easily mark points defining an almost perfect hexagonal array ($n_c = 6$) and, starting at any given point, (2) initiate and continue growing an hexagonal cluster of points, by selecting at random one point after the other from any of the first neighbor positions of any of the preceding points, (3) counting the total number of different growing possibilities at each stage and (4) multiplying them by p_c also at each stage, while (5) registering the results.[3] Repeat a few times, building different hexagonal arrays of points. You might (or not[4]) reach an at least similar result to the one that follows for $f_k(k, n_c, p_c)$: $f_1 = 1$; $f_2 = n_c p_c$; $f_{k\geq3} = f_{k-1}(n_c + k - 1)p_c$. From these f_k (assuming they would all be correct), dividing them by their corresponding $(k - 1)!$ as justified in the preceding paragraph, the resulting weights F_k, before normalization by division by $S = \sum_{k=1}^{k_{max}} F_k$ (calculated within full machine precision up to an adequate k_{max}) will be

$$F_1 \propto 1; \quad F_2 = F_1 n_c p_c; \quad F_{k\geq3} = F_{k-1}(n_c + k - 1)p_c/(k - 1). \qquad (12.1)$$

It is recognized that, when, $p_c = 0$ (no clustering), $F_1 = S = 1$, $F_{k>1} = 0$ and all primitive relaxors behave independently and, when $p_c > 0$, $F_1 < 1$, with decreasing values of $F_{k>1} \geq 0$, corresponding to greater cooperativities.

Figure 12.1 shows the best adjustments obtained to the experimental stress relaxation data for PMMA by applying the full CTMD formulation with the above conjectured (bidimensional) F_k, used to calculate the final cluster weights, F'_k, that include the effect of dynamic heterogeneity according to Eq. 9.2. It may be seen that, with $n_c = 6$ and the probability values specified in the Figure for each of the curves, the agreement of the wavy, unsmoothed, CTMD-calculated curves with the experimental data would turn out "acceptable", as far as the conjectural formulation of the new cluster weights could be proven correct.

---

[3] This puzzling exercise may make one feel dizzy (and possibly give up), but keep trying, [… or ask some patient and curious junior to help].

[4] [As most of the authors' junior family members live across the Atlantic Ocean, and probably wouldn't be bothered with such "crazy game", we did not benefit from much help].

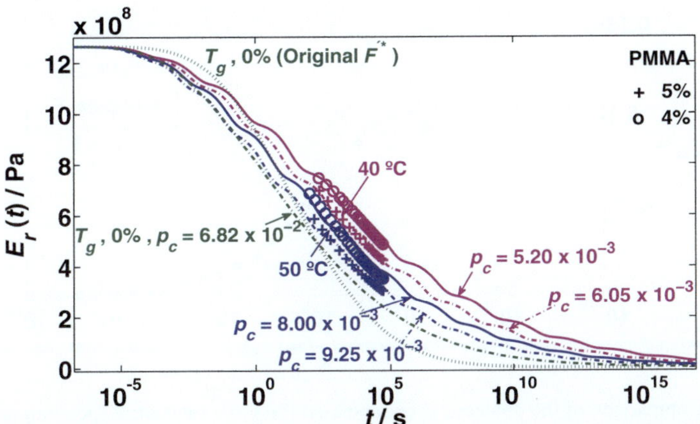

Fig. 12.1 Conjectural adjustment of the CTMD-predicted uniaxial tensile stress relaxation moduli for PMMA to the experimental data at 4% and 5% strains, with scaled-down clustering to the levels indicated by the p_c values, including a comparison of the original (pointed green) and new (green - ·) predictions at T_g and 0% strain[5] [1]

To add to the complexity of the problem, the specified conjectural probabilities, p_c, appear to show temperature and strain dependencies of the type $p_c \sim \exp\left(\frac{E - \alpha \varepsilon_0}{k_B T}\right)$,[6] where E turns out of the order of the minimum activation energy, $E_{a,1}$. It appears easy to understand why strain might help clustering, but the increase with temperature is very doubtful and would warrant deeper analysis. It may also be noted that, as shown in Fig. 12.2, assuming the above $p_c(T, \varepsilon_0)$ and extrapolating them to $T = T_g = T_c/1.2$ and $\varepsilon_0 = 0$ (thus yielding an approximate $p_c = 6.82 \times 10^{-2}$ at these conditions), leads to a relaxation spectrum ($F_k^{'*}$) not widely different from the original CTMD one as estimated in Chap. 9. However, that extrapolation is a big "jump" to the unknown, and the significance of this observation (or mere coincidence) is therefore very far from clear.

Another related observation, now back from Fig. 12.1, is that the new predicted relaxation moduli at T_g and 0% strain are not widely different from the original results at the same conditions; in particular, the predicted average relaxation times are very similar, as also implied in Fig. 12.2. This could be related to the comments made in the closing paragraphs of Sects. 11.3 and 11.4, concerning the physical reasonableness of most of the original CTMD predictions dependent on the average relaxation times near and above the "conventional" T_g, including the main dynamic physical parameters (ν_c, T_c, $E_{a,1}$) and properties such as fragilities. At much lower temperatures, we nevertheless expect that the "would be more accurate" relaxation

---

[6] It must however be recognized that two temperatures × two strains provide too limited data to draw any safe conclusions.

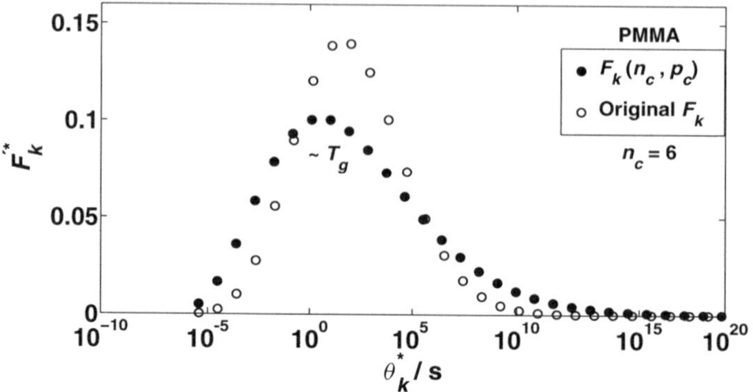

Fig. 12.2 Comparison of the conjectural extrapolated PMMA's relaxation spectrum (at the linear viscoelastic limit at T_g and with scaled-down clustering) with the original CTMD spectrum at the same conditions

maps of Figs. 11.1 and 11.3 should after all fall closer to the original ones,[7] such as to yield shorter average response times at those low temperatures. This of course reduces even further any likelihood of response time divergence.

A clustering probability, p_c, increasing with temperature as referred above, in addition to being difficult to explain, would attenuate or even erase the effect of dynamic heterogeneity on the F_k'[8] as formulated in Sect. 9.1.2 and the present one. Will that point to the possibility of significantly higher minimum activation energies affecting θ_k, as at least suggested by published data for PMMA-$E_{a,1} \sim 79$ kJ/mol [2, 3]? This should warrant closer analysis.

As made clear at the beginning of Chap. 8, a complete solution of the many problems and "puzzles" (even mysteries) of amorphous condensed matter are still very far from approximate success, and the present developments and proposals may add little to our understanding. However, most of the predictions of CTMD appear at least qualitatively reasonable, and even quantitatively approximate in a range of conditions, namely from T_c down to "conventional" T_g. Its main weaknesses have been clearly marked-the identification and quantitative description of structural and topological limitations to clustering, its tridimensional character, and the outstanding problem of the relationships between relaxation and retardation times, to which the crude and highly conjectural proposals of the foregoing chapters might contribute still too little.

Of course, any advances on the above issues will bear on future better predictions by CTMD of the details of the equilibrium and non-equilibrium thermal behavior (hysteresis) and of the various time- and frequency-dependent response curves presented in this book. Nevertheless, we think that the essentials of the nature

---

[7] Progress in this field is not really expected to follow a smooth and straight path.

[8] With $p_c \propto e^{-E_{a,1}/(k_B T)}$ and $\theta_k \propto e^{kE_{a,1}/(k_B T)}$, the $F_k' \propto \theta_k p_c^k$ would turn out independent of temperature!

and effects (even if exaggerated) of the collective or clustering behavior in amorphous materials, with emphasis on the identification and evaluation of dynamic (ν_c, T_c and $E_{a,1}$) and other physical properties (like the conventional values of T_g and fragility) may have been captured.

12.2 CTMD Updating Priorities

The recognized outstanding weaknesses of CTMD call for the following list of needed further studies:

1. New stress relaxation and other calculations assuming a truncated log-normal relaxation spectra, but relaxing Alfrey's approximation by accurate time integration within the same original optimization procedure used to estimate CTMD's physical parameters from the experimental data for PMMA and PC. [Small adjustments will be expected to the parameters, most probably to the minimum activation energy, $E_{a,1}$, as well as to the resulting smooth spectra.]
2. Formulation of more accurate CTMD discrete cluster relaxation time distributions (to improve on those of Sect. 9.1 and the conjectured ones of Sect. 12.1), with guidance from the adjusted continuous truncated log-normal spectra obtained in 1 and possible account for limited probabilities of clustering of adjacent primitive relaxors.
3. Attempt at the inclusion in the various CTMD formulations (equilibrium and non-equilibrium thermodynamics, stress relaxation, creep, and frequency response) of the true tridimensional character of clustering.
4. Further analyses of the relationships between relaxation and retardation times.
5. Approximate allowance for the effects of molecular packing, followed by direct and more accurate dynamic formulation of the actual volume changes, including free volume redistribution,[9] by CTMD itself, according to the proposals in Sect. 9.11 and Chap. 15 in Part III.
6. Inclusion of the effects of viscous flow in the CTMD formulations of stress relaxation, creep and frequency response, following a strategy analogous to the one outlined for linear viscoelasticity in Chap. 3, based on the definition of a viscous flow time.

If CTMD exaggerates collective behavior and the effect of temperature by assuming that all motions and relaxations are activated (even though molecular packing is also given a role), the solution might be to scale down (not eliminate) clustering and formulate free volume redistribution by the same, or adequately modified, transition frequencies. On the other hand, if free volume theories do not explicitly

---

[9] As explained in Sect. 9.11, free volume redistribution will most likely require primitive relaxor and cluster motions of identical nature to those formulated in Chap. 8.

handle clustering, and recognizably neglect single as well as clustered activated transitions [4],[10] the solution might be to allow for single and clustered activated changes among "solid-like" and "liquid-like" cells of varying free volumes, and only then perhaps relate them to changes in communal entropy and percolation concepts.

With the objectives and criteria listed in the Closing Note of Chap. 10, be it with CTMD or any other theories (just from first principles, or guided by at least some sound phenomenology), there are thus many open challenges to keep researchers motivated and busy for the foreseeable future, in evaluating and improving each of the theoretical options and proposals in this field of the equilibrium and non-equilibrium dynamic behavior of amorphous condensed matter. So far, we have only scratched the surface of the problem.

References

1. J.J.C. Cruz Pinto, J.R.S. André, Cooperative molecular and macroscopic dynamics of amorphous polymers—a new analytical view and its potential, in *Processing and Characterization of Multicomponent Polymer Systems—New Insights*, eds. by J. James, S. Thomas, N. Kalarikkal, Y. Weimin, K. Pal, Chap. 13 (Apple Academic Press, Taylor & Francis Group, 2019), pp. 237–260
2. E.-J. Donth, *The Glass Transition—Relaxation Dynamics in Liquids and Disordered Materials*, Springer Series in Materials Science, vol 48 (Springer, Heidelberg, 2001)
3. M. Beiner, H. Huth, K. Schröter, J. Non-Cryst, Solids **279**, 126 (2001)
4. M.H. Cohen, G.S. Grest, Phys. Rev. B **20**, 1077 (1979)
5. M. Paluch, R. Casalini, C.M. Roland, Phys. Rev. E **67**, 021508 (2003)

---

[10] [By viewing and formulating the effect of temperature only in an indirect way, through its effect on volume, and then finding that three of the theory's four parameters (A, B, and C in Eq. 6.7) may themselves have to depend on temperature [5]].

Chapter 13
The Dynamics of a Highly Flexible Chemical Structure—The Example of Cyclohexane

13.1 Introduction

Cyclohexane may be taken as an example of a simple substance with considerable molecular flexibility within a wide range of temperatures, due to its characteristic boat $\rightleftharpoons$ chair *conformational transitions*. On the other hand, its not too low melting point (280 K) provides significant but not complete motion freezing as the temperature is lowered down to solid state conditions. It is thus interesting and relevant to test CTMD's predictions with such simple system.

As the above transitions involve synchronized associations of simple rotations about C–C bonds, and the corresponding activated state's excess energy results from the electrostatic repulsion between the electronic clouds of the bonds to other C and H atoms (which stretch the bond angles in a way similar to the scissor-like motion responsible for the *ca.* 1240 cm^{-1} infrared absorption in methane and cyclohexane itself), 1240 cm^{-1} may perhaps be taken as representative of the characteristic decomposition frequency of the activated (half-chair) state, $\nu_1^{\#} \sim 3.72 \times 10^{13}$ Hz, as defined in Chap. 8. Also, all the above bonds that must rotate in coordination during each transition (i.e. the whole molecule) may together be taken as what is being called a *primitive relaxor* within the CTMD theory, so that each of the transitions involving single molecules could be characterized by a value $n = 1$ in calculating their corresponding frequencies, $\nu_1^{*}(T)$, as functions of temperature, by Eq. 8.3 of that same chapter.

13.2 Frequency Calculations by CTMD

Knowing the accepted values for the relevant activation energies—ca. 22 (boat $\rightarrow$ chair) and 43 [1] (chair $\rightarrow$ boat) kJ/mol—and assuming the partition functions ratio parameter ($z_{1,r}^{\#}/z_1$, defined in Chap. 7) equal to 1 (possibly not exact, lower values

J. J. Cruz Pinto and J. R. dos Santos André, *Analytical Molecular Dynamics of Amorphous Condensed Matter*, Springer Series in Materials Science 342, https://doi.org/10.1007/978-3-031-56517-5_13

allowing for lower molecular flexibilities), the following approximate estimates for $v_1^*(T)$ may be obtained, in Hz: (1) boat $\rightarrow$ chair—1.33×10^8 at 200 K (with cyclohexane very much solid), 4.94×10^7 Hz at 186 K, 2.29×10^5 at 135 K, and 2.85×10^{-3} at 70 K (i.e. 1 transition in about 5.8 min) and (2) just as contrasting data, chair $\rightarrow$ boat—438 Hz at 200 K and 62.5 at 186 K. If only one possibility or path is considered for each of the above transitions, the corresponding frequencies are halved. We could also conjecture on the possibility of higher order, cooperative, motions involving molecular clusters of n (> 1) cyclohexane molecules, all of which will have lower frequencies.

13.3   Discussion and Further Future Applications

Could chemists and chemical-physicists comment, validate, or invalidate these predictions,[1] with possible reference and comparison with previous or new estimates, maybe by using RRKM (Rice-Ramsperger-Kassel-Markus theory [2]) or NMR measurements? Also, what modern (and so popular) molecular dynamics simulations have to say as to both (1) actual results (at the above or any other temperatures) and (2) computation times?

With the numerical physical data specified (or any proven better alternative), and Planck's and Boltzmann's constants, the above calculations, at *any high or low temperature* we may choose, is (in human terms) almost instantaneous,[2] involving the execution of one single line command within a numerical processor such as Matlab® in whatever computing platform, or even in a simple scientific calculator. Within Matlab®, being a matrix processor, the same single line command may simultaneously process and yield data for any chosen array of different temperature values in similar short times.

Despite the possible uncertainties in the above IR-estimate of $v_1^{\#}$ and the assumed $(z_{1,r}^{\#}/z_1)$ value, those calculated orders of magnitude of $v_1^*(T)$ for cyclohexane may not prove unreasonable, from comparison with measured NMR reorientation frequencies, which currently refer to unspecified mean values [3]. For example, in [3], it is reported that "just above 186 K, the *mean reorientation frequency* exceeds 3×10^7 Hz". As yet, we do not know of any actual data on specific direct and reverse (boat $\rightleftharpoons$ chair) experimental transition frequencies but, for the boat $\rightarrow$ chair transition, CTMD gives (as quoted above) 1.33×10^8 Hz at 200 K and 4.94×10^7 Hz at 186 K (but 438 Hz at 200 K and 62.5 Hz at 186 K for the chair $\rightarrow$ boat one) which, combined with the above calculated results at 135 and 70 K, points to a sharp dynamic

---

[1] [Bearing in mind that $z_{1,r}^{\#}/z_1$, having the precise physical meaning specified in Sect. 8.2.1, is not known and has here just been assumed close to 1 (meaning not too strong steric straining against, nor facilitation of, the relevant transitions)].

[2] Further, calculation time does not depend on the chosen temperature. Very low temperatures (even approaching 0 K) may be freely chosen, subject only to the same number representation limitations implied in Footnote 28 of Chap. 8. It is here recalled that all CTMD-predicted frequencies only become 0 at 0 K!

transition somewhere between 200 and 100 K. Also, given the quite different activation energies and frequencies involved between the two types of transitions, most detected transitions in reported experimental work should have been of the boat $\rightarrow$ chair type.

If CTMD theory and procedures will resist scrutiny, (1) evaluating the equilibrium conformational composition of a sample as a function of temperature, or (2) following the dynamics of the conformational composition changes as a function of time and temperature, in controlled isothermal or slow scanning experiments, may become manageable tasks,[3] involving either (13.1) the straightforward analytical solution of the equation

$$N_c v^*_{1,cb}(T) = N_b v^*_{1,bc}(T) \tag{13.1}$$

for N_c/N_b, , or (13.2) the numerical solution of

$$dN_c/dt = N_b(T,t)v^*_{1,bc}(T,t) - N_c(T,t)v^*_{1,cb}(T,t) \text{ and } dN_b/dt = -dN_c/dt, \tag{13.2}$$

for $N_c(T,t)$ and $N_b(T,t)$, with the appropriate initial conditions, where the subscripts c, b, bc and cb mean "chair", "boat", "boat-to-chair" and "chair-to-boat", and t is real physical time, no matter how long.

It might (or not) be reasonable to expect that the first and simplest of the above tasks might be within the capabilities of molecular dynamics simulations in acceptably "short" computation times, but the second task may turn out beyond present capabilities of such type of simulations. Where molecular dynamics simulations may indeed be expected to shed valuable light into these matters may well be in attempting to determine (by whatever long computations may prove necessary – computation time would then not count) more trustworthy values for $v^\#_1$ and $(z^\#_{1,r}/z_1)$.

What will reality make of the above, or in some way improved, theoretical predictions?

References

1. J. Clayden, N. Greeves, S. Warren, P. Wothers, *Organic Chemistry* (Oxford University Press, Oxford, 2001)
2. O.K. Rice, H.C. Ramsperger, J. Am. Chem. Soc. **49**, 1617 (1927); **50**, 617 (1927); L.S.J. Kassel, J. Phys. Chem. **32**, 225 (1928); R.A. Markus, O.K. Rice, J. Phys. Colloid. Chem. **55**, 894 (1951)
3. E.R. Andrew, R.G. Eades, Proc.R. Soc. Lond. A **216**, 398 (1953)

---

[3] It might be necessary, though, to consider different values of $(z^\#_{1,r}/z_1)$ for the two types of transitions.

Chapter 14
Equilibrium (ETG) and Non-equilibrium (NETG) Thermodynamic Theory of Glasses

14.1 Introductory Notes on Quantum Mechanics and Statistical Physics

For amorphous materials at *constant volume*, we may approximately assume that, as for stiff crystalline structures, the terms of the third and higher order in the Taylor expansion of the system's potential energy (in terms of the displacements of the constituent atoms or molecules) may be neglected. In fact, at equilibrium, the first-order terms are of course zero, because the potential energy will be at a minimum at zero displacements, and all third- and other odd-order terms could also be neglected, because at constant volume the potential energy may be expected and assumed to be closely symmetrical with respect to each individual displacement relative to that minimum. The assumption at the start of this paragraph would thus reduce to neglecting the fourth and all higher even-order terms in the potential energy expansion.

In these conditions, it is known that a set of *normal* or *principal coordinates*, q_i, could in principle be defined, such that they would become explicit linear combinations of the ordinary Cartesian coordinates of the system's constitutive particles, and that the system's potential energy, $\mathcal{V}$, would be approximately expressed as a linear combination of the squares of those normal coordinates, without any cross-terms,[1] i.e.

$$\mathcal{V} = \mathcal{V}_0 + \frac{1}{2} \sum_i \lambda_i q_i^2, \tag{14.1}$$

where $\mathcal{V}_0$ is the potential energy when the displacements are zero, and the λ_i stand for generalized force constants, related to the force constants in Cartesian coordinates.

---

[1] Cf. Denbigh [1].

© The Author(s), under exclusive license to Springer Nature Switzerland AG 2024
J. J. Cruz Pinto and J. R. dos Santos André, *Analytical Molecular Dynamics of Amorphous Condensed Matter*, Springer Series in Materials Science 342,
https://doi.org/10.1007/978-3-031-56517-5_14

On the above conditions, the system's overall wave function, ψ, could be considered as a product of wave functions, ψ_i, , for each of the above normal coordinates, and therefore the system's energy would become separable into energy contributions from each of the normal coordinates or modes. Now, providing that the normal coordinates (once identified) would be those used in any ensuing analysis of the system's physical behavior, some of them would be purely vibrational (and would yield corresponding vibrational contributions to the system's total energy), and some would be associated with large scale motions at the molecular scale, relevant to the glass transition and other complex relaxation processes, whose frequencies and energy contributions would of course not be describable by the same well known quantum mechanical formalism valid for linear harmonic oscillators.

The first motions give rise to a vibrational spectrum that contributes to the system's energy and specific heat capacity as predicted by Debye's theory or interpolation formula, while the wide scale cooperative motions of the latter type will have to be modelled and formulated differently. This must of course be one of the specific objectives of any proposed comprehensive theory.

So, the above discussion may justify, at least as a first approximation, the assumption that a set of wide scale cooperative motions (or modes) may be identified such as to be independent, and therefore separable, from the vibrational ones and, of course, also separable from any electronic, nuclear or other contributions, given that their excited levels will not be occupied. The system's overall wave function, ψ, could thus be written approximately as $\psi = \psi_{coop}\psi_{coop}$ (T, *timescale*) where T is the temperature, and ψ_{coop} is the non-cooperative part of ψ during any ordinary (thermal of forced) physical excitation/experiment on the material, whereby large-scale cooperative motions might be involved.

The aim of the intended theory will of course not be the formulation of ψ_{coop} (T, *timescale*), but will concentrate on a realistic approximate formulation of the corresponding cooperative component of the system's partition function, Z_{coop}, for both equilibrium and non-equilibrium[2] conditions, and of the system's dynamics, namely its characteristic cooperative frequencies and frequency spectrum. The above discussion should justify that, as a first approximation—we stress—this cooperative spectrum might be added or superposed to any other (e.g. vibrational) contributions to the overall frequency spectrum. Such cooperative spectrum, however, should vary significantly with temperature and, in the non-equilibrium case, also with the excitation's timescale, as it is indeed known to happen in most physical excitations (thermal, mechanical, dielectric, etc.).

---

[2] Non-equilibrium conditions should in principle be describable by being able to specify which of the system's microscopic motions are readily accessible and which are not, within the timescale of the observation/experiment. An accurate future theoretical description, though, should be able to also account for states that are partly accessible to varying degrees.

It is again stressed that this separation of the vibrational and cooperative components of ψ and Z is an approximation, as their mutual interference cannot strictly be avoided. Chapter 8 and the present one, however, make clear that the various cooperative, cluster-specific, components themselves should indeed be separable from each other, perhaps to a much better approximation than from the vibrational ones.[3]

As explained in the main text of Chap. 8, the assumed relaxor association in clusters of varying but specific sizes must at any given instant in time occur at the expense of, and in competition with, their association in clusters of any *other* sizes, as well as of the independent, uncorrelated, transition/relaxation of their individual constitutive or primitive relaxors. Clustering must therefore be viewed and described as leading to, first of all, (1) complete *loss of independence by the primitive relaxors that participate in any given cluster* (by definition, of course) and, at the same time, (2) also to *independence of behavior relative to any other clusters or single relaxors* at the same instant in time and at any different location within the structure, otherwise any assumed cluster would not in fact exist and actively respond, and would instead need to be redefined as a still larger cluster.

Recalling Debye's specific heat capacity formulation for rigid (amorphous or crystalline) solids, it is well known that their internal energy may be written as

$$U = 3N \int_0^{\nu_M} \left[\frac{1}{2}h\nu + \frac{h\nu}{e^{\frac{h\nu}{(k_B T)}} - 1} \right] f(\nu)d\nu = k_B T^2 \left(\frac{\partial \ln Z}{\partial T} \right)_V, \qquad (14.2)$$

where $3N$ is the crystal's number of (independent) normal modes[4] and $f(\nu)$ is its frequency or state density distribution, ν_M being the maximum frequency. With a less restrictive, closer to experimental $f(\nu)$, this limit might extend to very high values. Thus, though the solid's vibrational partition function may [as obtained from (14.2)] be written as

$$Z = e^{\left[3N \int_0^{\nu_M} f(\nu) \ln z(\nu)d\nu \right]}, \qquad (14.3)$$

---

[3] In the case of crystalline solids, the small vibrations along each of the normal modes are truly independent (allowing the factorization of the partition function in factors for each of those modes), but they nevertheless refer to motions involving the whole crystal or, more precisely, standing waves propagating through the crystal. What happens with the cooperative transitions in amorphous solids is not formally different. Thus, it should again be possible to factorize the cooperative part of the partition function in factors for each type of clusters of primitive relaxors, and each set of clusters of any given size in fact also include and refer to the whole solid, inasmuch as they populate it entirely, though in a spatially discontinuous way and intermittently in time. Their independence is physically even clearer than that of the normal modes of crystals (which may perhaps appear of a purely mathematical nature), in view of their real physical distinguishability and intermittency, as each primitive relaxor cannot spatially coincide with, and simultaneously participate in, any other type of clustered or un-clustered transition.

[4] [Accurately $3N - 6$, where N is the number of "lattice" sites].

where $z(\nu) = e^{-h\nu/(2k_B T)}/\left[1 - e^{-h\nu/(k_B T)}\right]$ is the well known partition function associated to the vibrational mode of frequency ν,[5] should the spectrum be discretized into any small frequency increments (such that $f(\nu)_i \Delta \nu_i = f_i$), the partition function could then be factorized as

$$Z = \prod_i (z_i)^{3Nf_i}. \qquad (14.4)$$

So, a straightforward argument is that the above type of representation would also be valid when formulating the overall partition function for a discontinuous spectrum of not purely vibrational frequencies, ν_i, each associated to the appropriate cluster size-dependent partition function, z_i, where all active clusters and single uncorrelated primitive relaxors must be dealt with as independent "particles". In that case, the theory sought should yield the appropriate, *non-vibrational*, cluster-specific partition functions, z_i, as well as the corresponding relative weights, f_i, in addition to the frequencies, ν_i. This is a very ambitious but necessary objective for any comprehensive future theory, to which CTMD intends to be a first contribution.

In the framework of CTMD, as already suggested in Sect. 8.2.3, it may thus be physically acceptable to factorize the cooperative component of the structure's partition function (namely, but not exclusively, that pertaining to equilibrium conditions, Z^*_{coop}) in as many factors as the number of different possible cluster sizes, raised to powers dependent on their *actual* numbers and physical weights (which will also have to be determined), where each of those different cluster sizes i will be responsible for its own *specific normal mode* and *frequency, ν_i^*.*

Denying that possibility would be equivalent to stating that the system's wave and partition functions would not be separable in any way, in the sense explained above and that, in the limit, they could only be formulated as one single factor and that no clusters of any finite size would influence the behavior, except the one including the entire material/sample. We argue that this would not be physically reasonable nor useful.

This question of the *separability* of the partition functions warrants some further discussion. In Chap. 8, we found that the effects of clustering emerged as *attenuation factors* of the non-cooperative (vibrational, etc.) underlying partition functions, and it is now shown that such attenuation is consistent with, and indeed a consequence of, the above separability. If we formulate the logarithmic average of the total partition function of each cluster (activated and non-activated) of n primitive relaxors at equilibrium (*), with all energies referred to the zero-level of the non-activated states, what we may write is[6]

---

[5] [Already used in Chap. 8, (8.3) to formulate the partition function $z_{1,v,t}$ associated with the transition-relevant degree of freedom of each primitive relaxor].

[6] Remember that activated states, not only undoubtedly exist (cf. [2]), but could even be counted ($N_n^{\#}$), as well as the total number of the corresponding non-activated and activated clusters of each size (N_n). Also remember that all $N_n^{\#}/N_n$ are temperature-dependent, as made clear in Chap. 8, which is not explicitly indicated above to avoid cluttered expressions in (14.5) and (14.6).

$$z^*_{total,n}(T) = \left[z^*_{coop,n}(T)\right]^{\frac{N_n - N_n^\#}{N_n}} \times \left[z_n^{\#''}(T)\right]^{\frac{N_n^\#}{N_n}},\qquad(14.5)$$

where $z^*_{coop,n}(T)$ is the non-cooperative component and $z_n^{\#''}(T)$ is the total partition function of each activated cluster (relative to the same zero-level), which may be also expressed by $z_n^{\#''}(T) = z^*_{coop,n}(T) \times z_n^{\#'}(T)$, with $z^*_{coop,n} = z_1^n$, because (*by definition*) activated ($^\#$) clusters gained independence from all other clusters and primitive relaxors, and the non-activated primitive relaxors of course retain their independence of other primitive relaxors and their clusters. So, we will also have

$$z^*_{total,n}(T) = z^*_{coop,n}(T) \times z^*_{coop,n}(T),\quad \text{with } z^*_{coop,n}(T) = \left[z_n^{\#'}(T)\right]^{\frac{N_n^\#}{N_n}},\quad(14.6)$$

where [as we have seen in the paragraph following the one including (8.10) in Chap. 8] $z_n^{\#'} = v_n^*(T)/v_1^\#$ and also $N_n^\#/N_n = v_n^*(T)/v_1^\#$, thus yielding (8.10), and that is how the $z^*_{coop,n}(T)$ emerge as attenuation factors (at $\leq 1\ T \leq T_c$) of the partition function calculated if *no* primitive relaxors were involved in any clustered degrees of freedom and transitions, $z^*_{coop,n}(T)$.

The separability of the cooperative and non-cooperative partition functions is well expressed by (14.5), where each of the factors on the right-hand-side are greater than 1. What happens is that the number of significantly populated states is less after clustering than before. No surprise: as expected, not only clustering is expected to decelerate the approach to equilibrium, but it may also change the equilibrium itself by means of *extra energy requirements* and likely *lower overall entropy gains*, compared to those associated with the response of uncorrelated primitive relaxors. These lower overall entropy gains should however not be confused with the accelerating effect of the significant *local* activation entropies involved in the formation of the largest clusters at temperatures close to and higher than the crossover, T_c, which thus plays the role of a compensation temperature.

The discussed attenuation effect of clustering on the overall partition function may seem to contradict the usual idea that *additional* degrees of freedom (not less) will always result from clustering, being thus responsible for increased materials' heat capacities. The overall number of degrees of freedom (cooperative *plus* vibrational) will be expected to always increase with temperature (and so the heat capacity) but, while below each of the $T_{z_{min},k\geq 1}$ (defined in Chap. 8) the marginal increases in the number of vibrational degrees of freedom will be reduced by the effect of the slowly growing numbers of the cooperative ones due to each of those cluster sizes, above those temperatures (where thermal energy exceeds the activation free energies of clusters of increasing sizes) those clusters do not limit as strongly the growth of the associated vibrational degrees of freedom. Heat capacities will thus be expected to always increase with temperature, in principle more significantly at increasing temperatures.

However, as the various size-dependent $T_{z_{min},k\geq 1}$ are reached one after the other, at temperatures increasing with k (cf. Fig. 8.6), it may be expected that the heat capacities will grow smoothly without any sharp variation at some temperature

below the crossover, and so the existence of thermodynamic glass transition temperatures remain in question. Whatever equilibrium and non-equilibrium thermodynamic
theory of glasses (ETG and NETG) might be developed and validated in the future,
it must be able to answer this and many other open questions.

No restrictions should be imposed within the intended theory on the sizes and
numbers of any clusters, i.e. the cluster size distribution should, ideally, also be one
of the explicit results of the theory. What such theory must seek is the factorization of
the material's global partition function into its independent factors (and *only these*),
some of which will be associated with and characterize large fractions of the whole
macroscopic structure. The number and relative value of the above factors should be
shown to depend on temperature and timescale, which should then also emerge as
one of the specific and significant results of the theory.

14.2 A Conjecture on a Strategy for Developing an Equilibrium (ETG) and Non-equilibrium (NETG) Thermodynamic Theory of Glasses

One may thus conjecture on a scheme for the possible future development of a
generalized equilibrium (ETG) and non-equilibrium (NETG) theory of glasses by a
conventional stepwise procedure, from the partial cluster partition functions of (8.10)
to the various thermodynamic functions and properties. The non-equilibrium version
might one day be developed by assuming a straightforward first-order time- and
temperature-dependent approach to the equilibrium value, $a^*_{coop,i}(T)$, by the average
cooperative Helmholtz free energy contribution per potential cluster, $a^{\pm}_{coop,i}(T, r)$,
with characteristic time $\tau_i^*(T) = 1/[2\pi \nu_i^*(T)]$,[7] according to

$$\frac{d\left(a^{\pm}_{coop,i}(T, r)\right)}{dT} = \mp \frac{\left(a^{\pm}_{coop,i}(T, r) - a^*_{coop,i}(T)\right)}{|r|\,\tau_i^*(T)}, \tag{14.7}$$

where the $\pm$ superscripts and the $\mp$ signs refer to heating or cooling scans,
respectively, and r is the true material's temperature scanning rate, dT/dt.

Equation (14.7) might then prove useful to predict the behavior on cooling and
re-heating (hysteresis) cycles from an initial high temperature equilibrium state. For
an ideal sudden quench from equilibrium, one could consider the use of

$$\frac{d\left(a^-_{coop,i}(t)\right)}{dt} = -\frac{\left(a^-_{coop,i}(t) - a^*_{coop,i}\right)}{\tau_n^*(T_f)}, \tag{14.7a}$$

---

[7] $\tau_i^*(T)$ being an equilibrium characteristic time, representative of i-relaxor cluster fluctuations, this
approach should show consistency with the fluctuation–dissipation (theorem or) relation—FDT/
FDR, which is referred in Chaps. 7, 9 and 10 of the book.

where T_f is the final temperature, and now $a^*_{coop,i}$ is the final ($t \to \infty$) equilibrium, and not (as above) the instantaneous, value for each cluster of size i. However, this assumes that the sample would reach T_f instantaneously, which is a very questionable idealization.[8] As a matter of fact, (14.7) would always be valid, providing the *true instantaneous sample temperature* can be measured or predicted during the whole thermal treatment. This would require detailed heat balances within the material's sample.

Of course, any real problem would then involve the simultaneous numerical solution of a very large, open but finite, set of the above equations (from $i = 1$ to some i_{max}), weighed by a corresponding temperature-dependent set of statistical weights, like those formulated in Sect. 9.1, for each cluster size—an enormous, but still manageable, computation task in present computing platforms. The underlying (even if overoptimistic) expectation is that an interpretation of all non-equilibrium thermodynamic functions would then, eventually, become possible in terms of the (totally and partly) active or accessible modes, each of which should contribute to the cooperative component of the partition function with a specific factor to a product analogous to that of (14.4), but excluding/scaling down the totally/partly inactive modes.

The scheme below (Fig. 14.1) tries to summarize the proposed future strategy to develop the sought ETG and NETG theories, where the indicated cluster weights, $F_i(T)$ [9] might be those obtained in Sect. 9.1 (or some future better alternatives).

The bottom half of Fig. 14.1 (dealing with hysteresis), however, requires further clarification as to the actual layout of the prospective calculation, as follows:

(1)   Start the simulation of two independent cooling experiments from equilibrium at temperatures (a) $T_0 < T_c$ and (b) $T_0 > T_c$, in the latter case to provide one of the possible tests of CTMD's specific and still very conjectural predictions above the crossover.

(2)   Calculate all the $v_i^*(T_0)$ and $\tau_i^*(T_0) = 1/[2\pi v_i^*(T_0)]$ necessary, up to some i_{max}, to obtain the initial cluster size distribution (cf. Sect. 9.1) such as to make both $F_{i_{max}}/\sum_{i=1}^{i_{max}} F_i$ and $F_{i_{max}} a^*_{coop,i_{max}}/\sum_{i=1}^{i_{max}}\left(F_i a^*_{coop,i}\right)$ less than the set (desirably full machine) precision, where $\sum_{i=1}^{i_{max}} F_i = 1$ to within that same precision.

(3)   Calculate all corresponding $z^*_{coop,i}(T_0)$ and, by (8.13), the resulting overall cooperative partition function, $Z^*_{coop}(T_0)$, and corresponding cooperative Helmholtz free energy per mol of primitive relaxors, $\overline{A}^*_{coop}(T_0) =$

---

[8] Further, it may be argued that such idealization would even be somewhat inconsistent with cluster Helmholtz free energy contributions, $a^-_{coop,i}(t)$, different from the equilibrium ones, $a^*_{coop,i}$.

[9] As in Sect. 8.3.3 (Eq. 8.13), we very strongly suspect and suggest that the correct cluster weights to be used in these thermodynamic calculations should instead be those accounting for the respective *active times*, assumed in Sect. 9.1.2 proportional to the equilibrium temperature-dependent response times. Once (and if) confirmed, this will of course greatly increase the complexity of the calculations of the simplified scheme of Fig. 14.1, but their detailed specification is best left to future comprehensive tests and updates of the theory.

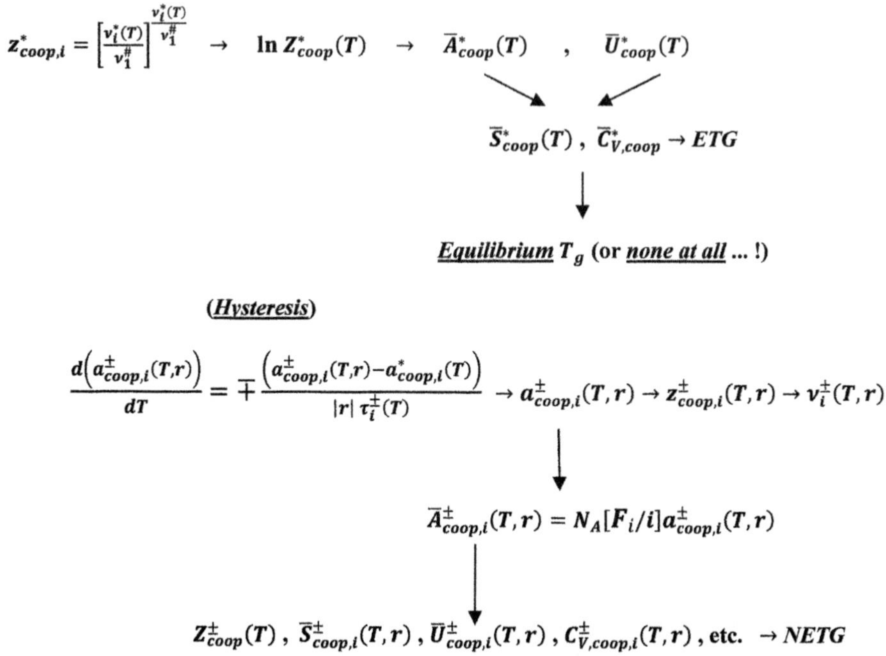

$$z^*_{coop,i} = \left[\frac{v^*_i(T)}{v^{\#}_1}\right]^{\frac{v^*_i(T)}{v^{\#}_1}} \quad \to \quad \ln Z^*_{coop}(T) \quad \to \quad \overline{A}^*_{coop}(T) \quad , \quad \overline{U}^*_{coop}(T)$$

$$\overline{S}^*_{coop}(T) \, , \, \overline{C}^*_{V,coop} \to ETG$$

$$Equilibrium\ T_g\ (or\ none\ at\ all\ …\ !)$$

$$(Hysteresis)$$

$$\frac{d\left(a^{\pm}_{coop,i}(T,r)\right)}{dT} = \mp\, \frac{\left(a^{\pm}_{coop,i}(T,r) - a^*_{coop,i}(T)\right)}{|r|\,\tau^{\pm}_i(T)} \to a^{\pm}_{coop,i}(T,r) \to z^{\pm}_{coop,i}(T,r) \to v^{\pm}_i(T,r)$$

$$\overline{A}^{\pm}_{coop,i}(T,r) = N_A[F_i/i]a^{\pm}_{coop,i}(T,r)$$

$$Z^{\pm}_{coop}(T) \, , \, \overline{S}^{\pm}_{coop,i}(T,r) \, , \, \overline{U}^{\pm}_{coop,i}(T,r) \, , \, C^{\pm}_{V,coop,i}(T,r) \, , \, etc. \ \to NETG$$

Fig. 14.1 Siomplified scheme of a proposed strategy to develop an equilibrium (ETG) and non-equilibrium (NETG) theory of glasses

$-k_B T_0 \ln\left[Z^*_{coop}(T_0)\right]$, as well as all other relevant cooperative components of thermodynamic functions and properties, by known statistical physics relationships. The above will define the entire set of the system's initial conditions.

(4)   Start the simultaneous numerical solution procedure of the whole set of differential equations (14.7), giving at each temperature the average cooperative Helmholtz free energy contributions per potential cluster of each size.

(5)   While advancing in that solution, at every adjusted small (differential) step in time and decreased temperature (such as to comply with the condition 5a below), calculate each of the new individual equilibrium, $z^*_{coop,i}(T)$, and non-equilibrium, $z^-_{coop,i}(T)$, clusters' partition functions, corresponding frequencies, v^*_i and v^-_i,[10] and statistical weights, $F_i(T)$, following the requirements specified below in (5a), the new cluster partial—$a^-_{coop,i}(T)$—and overall—$\overline{A}^-_{coop}(T)$—cooperative Helmholtz free energy contributions to that of 1 mol of primitive relaxors, and then the corresponding overall non-equilibrium

---

[10] [By numerical solution of $z^*_{coop,i} = \left[\dfrac{v^*_i(T)}{v^{\#}_1}\right]^{\frac{v^*_i(T)}{v^{\#}_1}}$ and $z^-_{coop,i} = \left[\dfrac{v^-_i(T)}{v^{\#}_1}\right]^{\frac{v^-_i(T)}{v^{\#}_1}}$ for $v^*_i(T)$ and $v^-_i(T)$, respectively, which may be difficult only when those partition functions are close to their minimum value, $(1/e)^{(1/e)}$ (cf. Sect. 8.2.3)].

cooperative partition function, $Z^-_{coop}(T)$, by

$$\ln Z^-_{coop}(T) = N_A \sum_{i=1}^{i_{max}} \frac{F_i(T)}{i} \ln z^-_{coop,i}(T), \qquad (14.8)$$

assuming one ascertained that all $F_{i>imax} \cong 0$ within machine precision, where all the log-factors in that product, with the (desirable) exception of the last one (pertaining to i_{max}), involve equilibrium values of $z^*_{coop,i}(T)$. Finally, all relevant thermodynamic functions and properties could then in principle be obtained.

(5a)   At each of the above adjustable steps, one should try to find that only one extra i value (the new i_{max}) will be necessary to ensure that the sum of all previous F_i with the new F_{imax} equals 1 within the same set precision, and update the value of the number of equations (previous $+1$) to be solved there from.[11,12,13]

(6)   Proceed with the above solution algorithm—(5) and (5a) until the final lower temperature is reached, and then continue the procedure, now into the increasing temperature stage ($^+$) of the simulation, recalculating all frequencies (whose number may now, if so whished, be made progressively smaller, as well as the number of differential equations, for faster computations), partial and overall cooperative partition functions, and the cooperative components of thermodynamic functions and properties at each temperature, until the original high temperature, T_0, is reached. At each step, a so-called *effective temperature*, T_{eff} (often but in our opinion wrongly named *"fictive temperature"*), may also be calculated as the temperature at which $v^*_{avg}(T_{eff}) = v^{\pm}_{avg}(T)$.

(7)   Plot, analyze, and physically discuss all results.

There is a problem hidden in (14.7). From equilibrium down to lower temperatures, the response times used in the denominator of its right-hand-side will be correct until, for some large i, its $z^-_{coop,i}$ will differ from $z^*_{coop,i}$ (and $v^-_i \neq v^*_i$), meaning that the corresponding cluster mode fell out of equilibrium. From then onwards, as temperature decreases, the response time to be used in the mentioned denominator will have to be changed accordingly, such that $\tau^-_i = 1/(2\pi v^-_i)$, the

---

[11] From the ratios $z^-_{coop,i}(T)/z^*_{coop,i}(T)$ for the various cluster at each calculation step, one might arguably obtain what may be considered their respective effective "activity" or "accessibility", while for the smaller clusters those ratios will be 1 or close to 1. This of course will just try to portray the gradual, cluster mode by cluster mode, breaking-off from equilibrium of the system, because the real frequency spectrum will turn out continuous or smoothed, not discontinuous, as pointed out towards the end of Sect. 9.1.3.

[12] This will stand as the numerical "materialization" of the increase of the average and maximum cooperativities as the temperature is lowered.

[13] The solution of the extra equations may proceed with the corresponding $a^-_{coop,i}(T)$ as initial values, determined by solving the extra ($i > i_{max}$) (14.7) equation(s) strictly necessary from the corresponding $a^*_{coop,i}$.

same gradually happening with other smaller clusters as temperature decreases. In heating, the procedure will be reversed and, while one will start with a significant number (or even all) response times given by $\tau_i^+ = 1/(2\pi\nu_i^+)$ (the first of which will be identical to the last τ_i^- in cooling, assuming no aging period at the lowest temperature[14]), similar tests will have to be made, and a growing number of τ_i^- will then need to be replaced by the corresponding $\tau_i^* = 1/(2\pi\nu_i^*)$ at equilibrium.

Such simulation, if at all feasible, correct, and eventually successful, might be expected to reveal the onset of a *dynamic glass transition* at decreasing temperatures as the scanning rate decreases, both in cooling and in heating, as well as yield two separate curves in heating and colling (i.e. *hysteresis*), for any of the cooperative components of relevant thermodynamic functions and properties, analogous to those of the qualitative example of Fig. 6.7, after being combined with their purely vibrational counterparts.

The eventual reasonableness (or otherwise) of the results of the above very complex calculations may provide the stringiest possible test of CTMD and/or (assuming no disproportionate "wishful thinking" in all the above) suggest pathways towards its upgrade or correction. How soon one will be able to carry out such comprehensive and elaborate test is still beyond any present forecasts.

14.3   A Final Conjecture

What about the *thermodynamic glass transition*? Will it really exist?

We may be living in (and belong to) an un-paradoxical nature. However, valuable but approximate theories of nature's workings (like so many other more mundane human activities), may sometimes generate and sustain puzzling paradoxes. One of them, specific to this field, is *Kauzmann's paradox* [3], related to a purported *negative entropy catastrophe* when the estimated equilibrium properties of a liquid are (inaccurately) extrapolated to temperatures well below T_c, unless a true second-order thermodynamic phase transition comes to the rescue. But the existence of such thermodynamic transition raises growing experimental and theoretical doubts [4–12].

Science has been solving several other paradoxes, and it was argued [13, 14] that an adequate *equilibrium theory of glasses* would be a necessary pre-requisite for also solving this one. Will CTMD allow, and the upper half of Fig. 14.1 illustrate, a viable attempt?

If we accept that the various cluster processes that have been analyzed in this book appear to be the only possible contributors to changes in the system's configurational entropy, and that all their characteristic frequencies smoothly tend to zero as the temperature approaches 0 K, the result will be, as already pointed out, *a unique completely frozen structural configuration at* 0 K. One may thus expect that, given

---

[14] With aging at the lowest temperature, the process could be followed by using (14.7a), with the appropriate τ_i^* replaced by the corresponding τ_i^- obtainable from the ν_i^- of Footnote 10

that all $v_n^*(T)$ and the associated entropies increase with T even at infinitely long timescales, nowhere ($T \geq 0$) ought the total entropy (vibrational *plus* cooperative) turn out negative. Thus, Kauzmann's paradox *may* not be real. At all temperatures below T_{EM} (for a given forced frequency v or temperature scanning rate $r = dT/dt$), down to 0 K, the configurational entropy should gradually become negligible and then exactly zero; and, at v or $r = 0$, T_{EM} and T_{EB} themselves would be both expected to reduce to 0 K, in agreement with (8.8). This would then be an appropriate definition of the (ideal but experimentally unattainable[15]) equilibrium at 0 K. What is relevant and not at all strange, is that this definition, as well as a possible generalized equilibrium (and non-equilibrium) theory of glasses, might one day result from a purely kinetic theory like CTMD, or some better one.

References

1. K. Denbigh, *The Principles of Chemical Equilibrium*, 4th edn (Cambridge University Press, 1997) (Chapter 13)
2. J.C. Polanyi, A.H. Zewail, Acc. Chem. Res. **28**, 119 (1995)
3. W. Kauzmann, Chem. Rev. **43**, 219 (1948)
4. M. Wolfgardt, J. Baschnagel, W. Paul, K. Binder, Phys. Rev. E **54**, 1535 (1996)
5. J. Baschnagel, M. Wolfgardt, W. Paul, K. Binder, J. Res. Nat. Inst. Stand. Technol. **102**, 159 (1997)
6. M. Pyda, B. Wunderlich, Macromolecules **32**, 2044 (1999)
7. G.P. Johari, J. Chem. Phys. **265**, 217 (2000)
8. L. Sante, W. Krauth, Nature (London) **405**, 550 (2000)
9. G.P. Johari, J. Non-Cryst. Solids **288**, 148 (2001)
10. F. H. Stillinger, P. G. Debenedetti, T. M. Truskett, J. Phys. Chem. B **105**, 11809 (2001)
11. M. Pyda, B. Wunderlich, J. Polym. Sci. Phys. Ed. **40**, 1245 (2002)
12. S.L. Simon, G.B. Mckenna, J. Non-Cryst. Solids **355**, 672 (2009)
13. E. A. Di Marzio, Ann. N. Y. Acad, Sci. **371**, 1 (1981)
14. E.A. Di Marzio, Comput. Mat. Sci. **4**, 317 (1995)

---

[15] [Because, as $T \to 0$, all cluster frequencies will approach zero and "everything will gradually freeze" in whatever experiment's timescale]. However, bearing in mind the extremely low frequencies accessible to *calculations* within CTMD (even much lower than those plotted in Figs. 8.3 and 8.4), such *extrapolation* might only become feasible (if and when an ETG may come to light) within the limitations of one's processor's number representation (cf. also Footnote 28 of Chap. 8).

Chapter 15
How CTMD Might Be Applied to Free Volume Relaxation at Constant Temperature and at Controlled Rate of Cooling

15.1 Introduction

In Sects. 6.10, 9.11 and 12.2, respectively, discussions have been offered about (1) the classical approaches to the influence of molecular packing and free volume on the response of amorphous materials, including Cohen and Grest's one, (2) which effects may already be most easily identified but only approximately described by CTMD, and (3) which prospects may be anticipated of applying CTMD to a comprehensive and quantitative account of the mentioned effects of molecular packing. A more detailed discussion of the latter prospects than that of the last paragraphs of Sect. 9.11 is the object of the present chapter.

15.2 Free Volume Relaxation at Equilibrium at Constant Temperature

Let us assume that (somewhat unrealistically) a material is kept at constant temperature and allowed to evolve from a hypothetical initial non-equilibrium configuration of *local free volumes* around its primitive relaxors (as defined within CTMD). Let us assume that an initial fraction, $f_{u,0}$, of the material's primitive relaxors will be "unrelaxed" (u), with an average free volume $v_{f,u}$, while all others have a "relaxed" (r) structure or configuration with average free volume $v_{f,r} < v_{f,u}$. Following the suggestion of Sect. 9.11, let us also consider that, in the framework of CTMD's *activated transitions approach* developed in Chaps. 8 and 9, the average loss of free volume in each $u \rightarrow r$ transition of a primitive relaxor is about twice its activation volume, $2v_1^{\#}$, such as to make $v_{f,r} = v_{f,u} - 2v_1^{\#}$. Being v_f the overall average free volume per primitive relaxor at any time during the ensuing volume relaxation process, we may then write the following relationships between the above variables:

© The Author(s), under exclusive license to Springer Nature Switzerland AG 2024

J. J. Cruz Pinto and J. R. dos Santos André, *Analytical Molecular Dynamics of Amorphous Condensed Matter*, Springer Series in Materials Science 342, https://doi.org/10.1007/978-3-031-56517-5_15

$v_f = f_u v_{f,u} + (1 - f_u) v_{f,r} = v_{f,u} - 2v_1^\# f_r$, where f_u and $f_r = 1 - f_u$ denote the instantaneous fractions of unrelaxed and relaxed primitive relaxors, respectively.

Similarly, the initial and final equilibrium values of v_f will be such that $v_{f,0} - v_{f,\infty} = 2v_1^\# (f_{r,\infty} - f_{r,0})$, meaning of course that the coefficient of proportionality between the fractional increase of relaxed primitive relaxors and their individual loss of free volume will be $a_1' = (2v_1^\#)^{-1}$. Further, let $v_{f,k}'(t)$ be the instantaneous average free volume per primitive relaxor within the active clusters of size k.

We are now ready to formulate the kinetics of the free volume relaxation process. The problem in its fullness is simpler than the corresponding stress relaxation one, as we now have only two kinds of transitions, $u \rightleftharpoons r$, and we may proceed to considering that the primitive relaxors within any of the clusters will be binomially distributed between the u and r states, as described in Appendix B for the lower part of Fig. 9.3 of Chap. 9, such that, under the same simplifying assumption of identical activation energies of the direct and reverse transitions, the effective rate of free volume relaxation per primitive relaxor within the clusters of size k during their undisturbed lifetime[1] will be

$$-\frac{dv_{f,k}'}{dt} = 2v_1^\# \frac{\omega_k}{2} e^{-\frac{kE_{a,1}}{(k_B T)}} \frac{1}{k} \sum_{k^+=0}^{k} (k^+ - k^-) \binom{k}{k^+} f_{u,0}^{k^+} [1 - a_1'(v_{f,k,0}' - v_{f,k}')]^{k^+}$$

$$\times \left\{ 1 - f_{u,0}[1 - a_1'(v_{f,k,0}' - v_{f,k}')] \right\}^{k^-}, \tag{15.1}$$

where, like the suggestion in Appendix B, $p_k^+ = f_{u,k} = f_{u,0}\left[1 - a_1'\left(v_{f,k,0}' - v_{f,k}'\right)\right]$ with $v_{f,k,0}' = v_{f,0}$, and $p_k^- = f_{r,k} = 1 - p_k^+$ are the probabilities of unrelaxed and relaxed primitive relaxors, respectively, within the active clusters of size k, which is equivalent to

$$-\frac{dv_{f,k}'}{dt} = v_1^\# \omega_k e^{-kE_{a,1}/(k_B T)} \left(p_k^+ - p_k^- \right), \tag{15.2}$$

if we consider the first moments (averages) of the binomial distribution. Being $p_k^+ - p_k^- = (2f_{u,0} - 1) - 2f_{u,0}a_1'\left(v_{f,k,0}' - v_{f,k}'\right)$, one concludes from the final equilibrium condition of no further volume relaxation (valid to all cluster sizes) that $a_1' = \dfrac{2f_{u,0}-1}{2f_{u,0}\left(v_{f,k,0}' - v_{f,k,\infty}'\right)}$ with $v_{f,k,\infty}' = v_{f,\infty}$[2] and, being[3] $v_1^\# = \dfrac{v_{f,k,0}' - v_{f,k,\infty}'}{2(f_{r,k,\infty} - f_{r,k,0})} = \dfrac{v_{f,0} - v_{f,\infty}}{2f_{u,0} - 1}$,

---

[1] [Excluding their destruction and the redistribution of their primitive relaxors among new clusters].

[2] [Because (1) all primitive relaxors may be assumed to reach the same average final state, (2) even when jointly contributing to the response, they have the opportunity to intermittently participate in the whole range of cluster sizes, and (3) we are assuming that they start at the same average or very similar initial states, $v_{f,k,0} = v_{f,0}$].

[3] As in stress relaxation (cf. Appendixes B and C), $f_{u,k}$ and $f_{r,k}$ will depend on k at the same value of time (both when, as assumed, these clusters of size k are taken as the only contributors to the response, and when they intermittently compete with other clusters containing the same primitive relaxors), because one expects that smaller and faster clusters will contain higher fractions of

the effective undisturbed rates of free volume relaxation per primitive relaxor within the clusters of size k become

$$-\frac{d\upsilon'_{f,k}}{dt} = \omega_k e^{-kE_{a,1}/(k_BT)}\left(\upsilon'_{f,k} - \upsilon_{f,\infty}\right), \tag{15.3}$$

which has the same general form of any first order reversible process, as highlighted in Chap. 9, Study Question 9.1.

From (15.3), we conclude that each primitive relaxor of the clusters of size k^4 would on average relax their average free volume according to $\upsilon'_{f,k}(t) = \upsilon_{f,\infty} + \left(\upsilon_{f,0} - \upsilon_{f,\infty}\right)e^{-t/\theta_k^*(k,T)}$, with the same relaxation time obtained in Chap. 9 for stress relaxation in the linear viscoelastic domain. Finally, each of the entire set of primitive relaxors, through their many and intermittent participations from time zero to t within the whole range of cluster sizes, should end up with an average free volume of

$$\upsilon_f(t) = \sum_{k\geq 1} F'_k(k,T)\upsilon'_{f,k}(t) = \upsilon_{f,\infty} + \left(\upsilon_{f,0} - \upsilon_{f,\infty}\right)\sum_{k\geq 1} F'^*_k(k,T)e^{-t/\theta_k^*(k,T)}, \tag{15.4}$$

or

$$\upsilon_f(t) = \underbrace{\left(\upsilon_{f,u} - \upsilon_1^{\#}\right)}_{\upsilon_{f,\infty}} + \overbrace{\upsilon_1^{\#}\left[1 - 2\left(1 - f_{u,0}\right)\right]}^{\left(\upsilon_{f,0}-\upsilon_{f,\infty}\right)}\sum_{k\geq 1} F'^*_k(k,T)e^{-t/\theta_k^*(k,T)}, \tag{15.5}$$

where the $F'^*_k(k,T)$ are the equilibrium cluster weights determined in Sects. 9.1 and 9.2 (or other future better alternatives), which already consider the clusters' intermittency that leads to dynamic heterogeneity.

The above values of $\upsilon_{f,\infty}$ and $\upsilon_{f,0} - \upsilon_{f,\infty}$ are easily understood as the values they would respectively have if the primitive relaxors were totally unrelaxed subtracted from the volume lost $(2\upsilon_1^{\#})$ by those that relaxed until equilibrium is reached (with $f_{r,\infty} = 1/2$, given the assumption of equal activation energies of the direct and reverse transitions), and the elementary volume lost per transition $(2\upsilon_1^{\#})$ multiplied by the fractional increase in the relaxed primitive relaxors $(1/2 - f_{r,0})$.

So, it appears possible within CTMD to approximately model free volume relaxation at constant temperature. The above simple analytical solution of the problem only requires taking both the average free volume of the unrelaxed primitive relaxors, $\upsilon_{f,u}$, and the activation volume, $\upsilon_1^{\#}$, as known or adjustable average constants, which is reasonable at constant temperature. It is also seen that the parameters of the formulation are CTMD's υ_c, T_c and $E_{a,1}$, all of them specific dynamical properties (known

---

relaxed primitive relaxors if $f_{u,0} > 1/2$ (assuming equal activation energies of the direct and reverse transitions).

[4] [If, as mentioned in Footnote 1, they relaxed continually in time, without being intermittently dismantled and replaced by other smaller or larger clusters].

or accessible by independent measurements [1, 2]), plus the initial composition of the system, $f_{u,0}$. From the material presented in Chaps. 9 and 11, the reader will be able to recognize that, from stress relaxation or other experiments and CTMD calculations, an estimate of the minimum activation volume, $v_1^{\#}$, might be possible from the best fitted total volume of each primitive relaxor, v_1, and the calculated value of the parameters c (Chap. 9) or a_1 (Appendixes B and C).

From (15.5), one confirms that, with equal activation energies for the direct and reverse transitions, (1) if the initial equilibrium composition is $f_{u,0} = 1/2$, neither the composition nor the average free volume will change, (2) if $f_{u,0} < 1/2$, both v_f and f_u will increase and (3) if $f_{u,0} > 1/2$, both v_f and f_u will decrease.

15.3   Free Volume Relaxation at Controlled Rate of Cooling

The analysis just presented requires significant modification under controlled cooling from some equilibrium at a reasonably high temperature, for which the tentative strategy sketched in Chap. 14, Sect. 14.2 for the thermodynamic properties might be fruitful. While the material will be at equilibrium, the behavior may be described by k_{max} equations (such that $\sum_{k=1}^{k_{max}} F_k'^* = 1$ to within a chosen high precision) of the form $-\frac{dv_{f,k}'}{dT} = \frac{\left(v_{f,k}'-v_{f,k,\infty}'\right)}{\left|\frac{dT}{dt}\right|\theta_k^*(k,T)}$5 but, once one or more clusters fall out of equilibrium, it will become necessary to adopt (or at least try) the strategy suggested in the above chapter and section for the formulation of non-equilibrium thermodynamics (NETG), by integrating along time (and temperature) the changes in the whole range of cooperative cluster free energies, and obtain therefrom the non-equilibrium θ_k values at each temperature, that should replace the equilibrium θ_k^*. At each temperature, the values of $v_f(T)$ could then in principle be obtained by weighing all $v_{f,k}'$ by the $F_k'^*$ at equilibrium, or F_k' out of equilibrium,6 and adding them.7

Closing Note

From this book's analyses and discussions of the effects of temperature and molecular packing on amorphous condensed matter dynamics, we may reasonably conclude that the most promising approach might be to consider that both effects are simultaneous and mutually interacting: temperature fluctuations (and forced physical excitations) drive all *activated transitions*, including volume changes and fluctuations, while these impact on the relevant effective *activation barriers* that must be locally overcome.

---

5 [To be solved numerically and *iteratively*, given the variation of θ_k^*, $v_{f,u}$ and $v_1^{\#}$ with temperature, the latter two subject to a thermal contraction coefficient related to/upscaled from dv_f/dt].

6 Section 15.2 suggests how the gradual fall from equilibrium of each cluster size might be detected, from the largest to the smallest, as cooling progresses.

7 It should be noted that, when trying this strategy to describe non-equilibria at *constant pressure*, the cluster partition functions used to estimate all θ_k^* and θ_k should be modified as suggested in Footnote 31 of Chap. 8, Sect. 8.3.3.

References

1. E.-J. Donth, *The Glass Transition—Relaxation Dynamics in Liquids and Disordered Materials.* Springer Series in Materials Science, vol. 48 (Springer, Berlin, Heidelberg, 2001)
2. M. Beiner, H. Huth, K. Schröter, J. Non-Cryst. Solids **279**, 126 (2001)

Correction to: Analytical Molecular Dynamics of Amorphous Condensed Matter

Correction to:
J. J. Cruz Pinto and J. R. dos Santos André, *Analytical Molecular Dynamics of Amorphous Condensed Matter,* **Springer Series in Materials Science 342, https://doi.org/10.1007/978-3-031-56517-5**

The original version of this book has been updated with the following belated corrections: The author corrections provided at proof stage have been incorporated. The book has been updated with the changes.

The updated version of the book can be found at
https://doi.org/10.1007/978-3-031-56517-5

Afterword

Amorphous condensed matter physics is a most puzzling and tricky subject. Many specialists recognize that they know little, but nevertheless staunchly defend and promote their own *"pet-theories"*. It might appear that this book ends up bringing forth just another one of the same kind but, instead, what is really intended is to convey an assertive list of our *"pet-questions"*, most of which will still warrant better and more complete answers than those given in this book.

We conjectured a lot—who didn't, and who doesn't?—and adopted (and recommend) the following *"pet-method"*:

1. Avoid excessive "anti-phenomenology" bias. [Otherwise, your conjectures may end up wide off the mark.]
2. Be accurate, but realistic. [Think as a scientist and mathematician,[1] but proceed as an engineer. First: Does it make sense? Then: Does it work? How well ... *and ... how badly?*].
3. Bring in ideas, knowledge and methods from related (and even unrelated) disciplines.
4. Test and be critical of your own ideas and "pet-theories", ... and remain open to entirely new approaches.
5. Do not side with hardliners on which effects or variables dominate any physical behavior, lest they all contribute and mutually interact.

---

[1] [Even when doing "philosophy"].

J. J. Cruz Pinto and J. R. dos Santos André, *Analytical Molecular Dynamics of Amorphous Condensed Matter*, Springer Series in Materials Science 342, https://doi.org/10.1007/978-3-031-56517-5

Appendix A

Calculation of the Fraction of Compact Clusters, $f(k)$, in the Total Number of Clusters of Size k Containing a Given Primitive Relaxor

At the beginning of Chap. 9 the above fraction, $f(k)$, was estimated as $(N_A - k)!/(N_A - 1)!$, with N_A representing the total number of primitive relaxors in the system,[2] apparently neglecting a significant number of the sparse clusters containing a given primitive relaxor that could be obtained from any specific compact cluster (the so-called "anchor" cluster, with all primitive relaxors adjacent to each other) by exchanging any number but one (from 1 to $k - 1$) of its primitive relaxors by an equal number of the other $N_A - 1$ relaxors of the system.

Considering all mentioned exchanges of those "anchor"'s primitive relaxors by an equal number of other relaxors (including some of the same "anchor" cluster, but avoiding any intersecting or adjacent compact clusters), f_k will be such that its reciprocal may be calculated as

$$\frac{1}{f_k} = \frac{k - 1}{1!}\left(N_A' - 1\right) + \frac{(k - 1)(k - 2)}{2!}\left(N_A' - 1\right)\left(N_A' - 2\right)$$

$$+ \cdots + \frac{(k - 1)!}{(k - 1)!}\left(N_A' - 1\right)\left(N_A' - 2\right)\cdots\left[N_A' - (k - 1)\right], \tag{A.1}$$

(not just limited to the last term of the above sum, as in the original formulation), where $N_A' = N_A - n_a$, thus now discounting the n_a primitive relaxors adjacent to the original "anchor" cluster. The F_k defined and initially formulated in Chap. 9 will then become

---

[2] [The system's bulk in thermodynamic behavior, the resisting area in stress relaxation, or along a given reference length in creep, in a first simplified formulation, awaiting the possible adoption of the developments at the end of Sect. 9.4].

J. J. Cruz Pinto and J. R. dos Santos André, *Analytical Molecular Dynamics of Amorphous Condensed Matter*, Springer Series in Materials Science 342, https://doi.org/10.1007/978-3-031-56517-5

$$F_k \propto \frac{1}{(k-1)!} \frac{(N_A - 1)!}{(N_A - k)!} f_k$$

$$\propto \frac{1}{(k-1)!} \frac{(N_A - 1)!}{(N_A - k)!} \frac{1}{\frac{k-1}{1!}(N_A' - 1) + \cdots + \frac{(k-1)!}{(k-1)!}(N_A' - 1) \cdots [N_A' - (k-1)]},$$

$$(A.2)$$

where $\frac{(N_A-1)!}{(N_A-k)!}$ is a product of $k-1$ factors proportional to N_A and the denominator of the third fraction in the right-hand-side of (A.2) is a sum of $k-1$ terms proportional to $N_A'^{\,i}$, with $i = 1$ to $k-1$. It may then be recognized that, for very large N_A, and being $n_a \ll N_A$, the last two factors of the right-hand-side of (A.2) will turn out very close to 1, and we thus recover, after normalization, the same final result of Chap. 9,

$$F_k = \frac{1}{e(k-1)!}. \qquad (A.3)$$

So, neglecting in the original formulation of Chap. 9 the sparse clusters that would result from the substitution of only a fraction of the $k-1$ primitive relaxors of the "anchor" cluster turned out as "irrelevant" numerically as the implicit (though wrong) inclusion of significant numbers of compact clusters intersecting, or adjacent to, the "anchor" one.

Complementary Clarification

In Chap. 9, $\frac{(N_A-1)!}{k!(N_A-k)!}$ includes all distinguishable (sparse and/or compact) clusters of size k containing a given primitive relaxor—in very large numbers, in particular when sparse—and the objectives are (1) to determine how many sparse *and* compact clusters exist per compact cluster, and then (2) obtain the average fraction of compact clusters by taking the reciprocal of the value obtained in (1).

Imagining the replacement of up to $k-1$ primitive relaxors other than the reference relaxor of the "anchor" compact cluster by any of $N_A - 1 - n_a$ other primitive relaxors (thereby excluding any of the n_a relaxors adjacent to the "anchor" one), one guarantees that no other compact cluster will result except the "anchor" cluster we took as reference.

Appendix B
Accounting in Cooperative Materials Dynamics for a Binomial Distribution of Unrelaxed and Relaxed Primitive Relaxors Within Each Cluster

B.1   Introduction

This Appendix discusses the same problems analyzed in Chap. 9—the non-linear stress relaxation of amorphous polymers (and also creep, towards the end), in the framework of our recent *cooperative theory of materials dynamics* (CTMD), considering again the non-affine nature of the materials' local stresses and strains relative to their average overall values but, in an upgrade of the initial approach, it is taken into account that each of the individual primitive relaxors participating in any given cluster may in fact contribute differently to the overall process.

That is what is physically expected, as some of those primitive relaxors may lead to a local stress reduction (or increased strain in creep), while others to a local stress increase (or strain decrease in creep), though each having the same or very similar absolute value, depending on their respective previous states—un-relaxed (or undeformed), or relaxed (or deformed)—at any given instant in time. For simplicity and clarity, this presentation will still concentrate on the particular case of a simple *two-state system*, where all relaxors are initially assumed un-relaxed *and* fully relaxable—lower part of the diagram of Fig. 9.3 in Chap. 9 (or the rightmost part of Fig. 9. 5 in the case of creep), with no cross-links or crystallinity, corresponding to assuming $f_{u_1,0} = 1$ for both stress relaxation and creep. This condition means that we will be dealing here with a homogeneous, 100% un-oriented and amorphous material, without significant cross-links or other physical hindrances to local relaxation or deformation. The effect of cross-linking or partial crystallization (where $f_{u_1,0} < 1$, but still assuming a two-state system) may, if needed, be considered in a way similar to the outline presented in Sect. 9.5.

© The Editor(s) (if applicable) and The Author(s), under exclusive license
to Springer Nature Switzerland AG 2024
J. J. Cruz Pinto and J. R. dos Santos André, *Analytical Molecular Dynamics of Amorphous Condensed Matter*, Springer Series in Materials Science 342,
https://doi.org/10.1007/978-3-031-56517-5

As to the much more complex application of the same modeling strategy to a four-state system ($f_{u_1,0} < 1$, but then following the complete diagram of Fig. 9.3 in Chap. 9 for stress relaxation and Fig. 9.5 for creep), it may be dealt with as suggested in Appendix C for stress relaxation.

Some review and clarifying notes may be useful about the general modeling strategy of the response of materials before the specific formulation of the following sections. Like in a parallel (or series) combination of standard linear or non-linear solids in stress relaxation (or in creep), where each of the elements respond independently and their local stresses (or strains) combine to yield the corresponding total average value, the same happens and must be considered with the different clusters of primitive ralaxors within a real material, which respond independently, and intermittently, originating the so-called *dynamic heterogeneity*.

In the specific case of the stress relaxation of a real material, what occurs is a sequence of positive or negative transitions at random times by the primitive relaxors occupying or crossing the resisting area, where some of the transitions occur simultaneously with those of other (mainly adjacent) primitive relaxors in the same or in the reverse direction, as *single* collective (*clustered*) *transitions*, with specific frequencies (calculated in Chap. 8) that differ from those of single uncorrelated primitive relaxors. Therefore, those sporadic participations of the primitive relaxors in different single and clustered transitions with a variable number of other primitive relaxors at a wide range of frequencies may and will in fact result in local stresses that fluctuate in time[3] (between those characterizing relaxed and un-relaxed relaxors) but yield an average measurable stress that slowly decreases in time.

The method of CTMD in modeling *stress relaxation* is to consider and quantify the time-dependent average contributions of every single and cooperative transitions that occur within the time scale of the experiment or calculation and combine them as if one had a *parallel association of independent and intermittent[4] standard non-linear solids*. The only difference is that each "would-be elementary non-linear solid" is made of, or represented by, the set of all possible clusters of primitive relaxors of each specific size, capable of intermittently respond (in single independent transitions, randomly distributed in time according to Poisson's distribution[5]), with an average frequency (in Hz) given by $\nu_k = \frac{1}{(2\pi\theta_k)}$, where θ_k is the corresponding characteristic relaxation time as estimated in Chap. 9 and in this Appendix.

---

[3] [At the microscopic scale of each primitive relaxor].

[4] [And mutually excluding, as each primitive relaxor can only participate in *one* cluster at any instant in time].

[5] [Like in the decay of radioactive elements].

As before (cf. *Highlight* 9.2), once obtained the differential constitutive equations of each set of k primitive relaxors, of the form $-\frac{d\sigma_k}{dt} = \frac{(\sigma_k - \sigma_{k,\infty})}{\theta_k} = \frac{(\sigma_0 - \sigma_\infty)}{\theta_k} e^{-t/\theta_k}$, and given the intermittence of the underlying processes and their true statistical independence,[6] they may, and indeed should, be separately integrated and the results combined, such that $\sigma = \sum_{k \geq 1} F_k' \sigma_k$,[7] to obtain the total stress and the stress relaxation modulus as functions of time (cf. Sect. 9.2.5).

B.2 Detailed Model Upgrade for Uniaxial Tensile Stress Relaxation

We recall that the initial modeling approach of Sect. 9.2 looked at the detailed behavior of each set of individual type of relaxors or clusters (primitive, if $k = 1$, or not, if $k > 1$, k being the number of primitive relaxors in each specific type of cluster) to obtain its characteristic relaxation time, as if no other different structures had any simultaneous participation in the process, and then (2) proceeded to their combination, by weighing and summing-up their estimated individual contributions, assuming that these were in all cases exactly proportional to the relaxor or cluster size, i.e. to k. This of course implied that all primitive relaxors taking part in any of the cooperative cluster transitions, in the relevant resisting area of the bulk material, had been assigned the same (either positive *or* negative) contribution to the overall stress reduction.

Whereas that raises no problem when describing the contribution of the transitions by single primitive relaxors, in the case of any of the cooperative transitions ($k > 1$), of the k primitive relaxors involved (depending on their individual structure and instantaneous state), some number $k^+ \leq k$ will only be allowed a positive transition (with a local stress reduction) and $k^- = k - k^+$ only a negative transition (with a local stress increase). For a simplified two-state system, with $f_{u_1,0} = f_{u_1,k,0} = 1$ at $t = 0$, the distribution at any time $t > 0$ of the primitive relaxors in any cluster of each specific size k among un-relaxed ($^+$) and relaxed ($^-$) will thus be *binomial*, with probabilities $p_k^+ = f_{u_1,k,0}\left[1 - a_1(\sigma_{k,0} - \sigma_k)\right] = 1 - a_1(\sigma_{k,0} - \sigma_k)$ and $p_k^- = 1 - p_k^+ = a_1(\sigma_{k,0} - \sigma_k)^8$ with $\sigma_{k,0} = \sigma_0$, assuming that the un-relaxed and relaxed primitive relaxors are uniformly distributed within the whole material. The

---

[6] Cf. Footnote 49 in Sect. 9.2.5 (relative to *Highlight* 9.2).

[7] Both σ_k and σ are taken as average local stresses (i.e. per primitive relaxor), σ_k being the average contribution to σ by each of the primitive relaxors taking part in the clusters of size $k > 1$. The F_k' are in this book assigned the role of the (intermittent) fractional resisting areas covered or crossed by the primitive relaxors of these specific clusters. How physically reasonable and at least approximate this will be remains an explicitly recognized open question.

[8] We have seen in Chap. 9 that $a_1 = \left(2E_0 v_1^\# / v_1\right)^{-1}$. Note that the cluster size-specific instantaneous p_k^- and p_k^+ will differ from the overall values for any other different cluster sizes or for the system as a whole, $p^- = 1 - p^+ = a_1(\sigma_0 - \sigma)$ at the same time t, because the various clusters provide contributions different from those of the other partial and overall stress relaxations and yield different changes to the ratio of relaxed to un-relaxed primitive relaxors.

physical justification of the above two exclusive states and corresponding transition possibilities is the obvious fact that, of the k primitive relaxors involved in any cooperative transition (except at $t = 0$), $k^- \leq k$ will have resulted from the same number of previous positive transitions (within other clusters of any size), making impossible any immediately subsequent positive transitions. As to the above probabilities p_k^+ and p_k^-, they of course have to, respectively, decrease or increase in exact proportion to the contribution by all k-clusters to the overall stress reduction, $(\sigma_{k,0} - \sigma_k)$, up to any given time t, because each and every primitive relaxor transition will lead, as assumed in the previous formulation, to the same local (negative or positive, respectively) local stress variation equal to $(2E_0 v_1^\# / v_1) = 1/a_1$ as given by (9.10), where the volumes $v_1^\#$ and v_1 refer here to a primitive relaxor.

On this basis, and assuming $f_{u_1,0} = 1$, the exact formulation of the specific contribution, $-\frac{d\sigma_k}{dt}$, to the overall average stress reduction at each primitive relaxor within the resisting area, $-\frac{d\sigma}{dt}$, by the whole set of possible clusters of size k (as mentioned, such as to make $\sigma = \sum_{k \geq 1} F'_k \sigma_k$) will have to be written as

$$-\frac{d\sigma_k}{dt} = \left(2E_0 v_1^\# / v_1\right)[\omega_k(k, T)/2]e^{-kE_{a,1}/(k_B T)} \sum_{k^+ = 0}^{k} \frac{(k^+ - k^-)}{k}\binom{k}{k^+}$$

$$\times \left[1 - a_1(\sigma_{k,0} - \sigma_k)\right]^{k^+}\left[a_1(\sigma_{k,0} - \sigma_k)\right]^{k^-} e^{[k^+\alpha_{11}(T,\varepsilon_0) - k^-\alpha_{21}(T,\varepsilon_0)]}, \quad \text{(B.1)}$$

with $\sigma_{k,0} = \sigma_0 = E_0\varepsilon_0$ (considering uniform initial stresses), where $\binom{k}{k^+}$ is the number of distinguishable combinations of k primitive relaxors k^+ by k^+ (or k^- by k^-), α_{11} and α_{21} are the previous temperature- and strain-dependent variables α_1 and α_2 of (9.6a), but now explicitly referred to a single primitive relaxor, and $\omega_k(k, T)$ is the same cluster-specific angular frequency pre-factor calculated in Chap. 9 by its (9.18a) and (9.18b).[9] As to the product $(2E_0 v_1^\# / v_1) \cdot (k^+ - k^-)/k$, it is the average local stress relaxation on the clusters of size k, per primitive relaxor, due to each set of k^+ direct and k^- reverse transitions by their primitive relaxors, given that $(2E_0 v_1^\# / v_1)$ is the average local absolute stress relaxation after each positive or negative transition of a single primitive relaxor.

The above formulation takes into account that, to the k^+ positive elementary transitions in a cluster of primitive relaxors corresponds an overall activation energy given by $k^+\left[E_{a,1} - \alpha_{11}(k_B T)\right]$ and, to the remaining k^- negative transitions, the activation energy $k^-\left[E_{a,1} + \alpha_{21}(k_B T)\right]$, giving a total of $kE_{a,1} - (k^+\alpha_{11} - k^-\alpha_{21})(k_B T)$ for the correct total activation energy of that individual *cooperative transition* of k primitive relaxors (of mixed nature), which should not be confused with those of a set of k^+ positive and k^- negative *independent* (or *uncorrelated*) transitions because,

---

[9] Note that, in the above and following rate equations, the pre-factor considered is already $\omega_k(k, T)/2$, as justified in Sect. 9.2.4.

in the case considered, they occur simultaneously within the same cluster, and so they make up *one single cooperative transition*, whose frequency differs from that of the uncorrelated primitive relaxors, as we have seen in Chap. 8. In these conditions, being $k^+ + k^- = k$, (B.1) may be condensed in the form

$$-\frac{d\sigma_k}{dt} = \left(\frac{2E_0\upsilon_1^\#}{\upsilon_1}\right)[\omega_k(k,T)/2]e^{-\frac{kE_{a,1}}{(k_BT)}}$$
$$\times \frac{1}{k}\left\{e^{-k\alpha_{21}}\mu_{k,1}^+\left[k^+e^{k^+(\alpha_{11}+\alpha_{21})}\right] - e^{k\alpha_{11}}\mu_{k,1}^-\left[k^-e^{-k^-(\alpha_{11}+\alpha_{21})}\right]\right\}, \quad (B.2)$$

where $\mu_{k,1}^\pm[\cdots]$ are the indicated binomial averages (or first moments) of the quantities shown as arguments.

It may be suspected that the right-hand-side of (B.2) may in general not yield a linear expression in σ_k (the specific average contribution of the whole set of the k-clusters to the overall average stress), to allow the simplest possible analytical solution. But, at least in the linear viscoelastic limit, as well as an approximation in the non-linear viscoelastic domain, we will next see that (B.2) do yield, as before, linear differential equations of *standard non-linear solids*.

Solution in the Limit of Very Low Strains

With $\varepsilon_0 \to 0$, and therefore α_{11} and α_{21} close or equal to 0, one may see that, by substitution of the resulting binomial averages of k^+ and k^- in (B.2), whose difference is $k(p_k^+ - p_k^-) = k(2p_k^+ - 1) = k[1 - 2a_1(\sigma_{k,0} - \sigma_k)]$, that equation would then simplify to[10]

$$-\frac{d\sigma_k}{dt} = \left(\frac{2E_0\upsilon_1^\#}{\upsilon_1}\right)[\omega_k(k,T)/2]e^{-kE_{a,1}/(k_BT)}[1 - 2a_1(\sigma_{k,0} - \sigma_k)], \quad (B.2a)$$

making possible to pursue the analytical development as in the approximate model of Chap. 9. As the same stress stationary condition at $t = \infty$ must be satisfied for each average contribution, σ_k, of that set of clusters to the overall stress—because, at equilibrium, the frequencies of all individual types of positive transitions (of whatever relaxors or clusters) should equal those of the corresponding negative transitions,[11]— one obtains in this limit ($\varepsilon_0 \to 0$) $a_1 = 1/[2(\sigma_{k,0} - \sigma_{k,\infty})]$, with $\sigma_{k,\infty} = \sigma_\infty = E_\infty\varepsilon_0$ (the final average, uniform, equilibrium state) in this linear limit case. We should note that now the inconsistencies pointed out in Footnotes 38 and 39 in

---

[10] Note that, at any finite time, t, during the process, the $f_{r_1,k}$, and therefore $(\sigma_{k,0} - \sigma_k) = (\sigma_0 - \sigma_k)$ and σ_k, will vary with k, because the various cluster sizes will provide different contributions to stress relaxation—smaller clusters respond faster and a higher proportion of their primitive relaxors would have relaxed fully after any given time. What one is considering in the above formulation (as in Chap. 9) is the separate response of the set of clusters of size k if no other smaller or larger clusters were allowed to form and contribute to the response.

[11] [Which does not exclude, as explained in the introductory section, that local stresses may show fast fluctuations in time].

Chap. 9 do not show up. This will also imply that, with $f_{u_1,k,0} = 1$ and at $\varepsilon_0 \to 0$, at the final equilibrium at any temperature all $f_{r_1,k,\infty} = f_{r_1,\infty} = 1/2$, a logical result of the assumed equality of the unstrained activation energies, $E_{ur} = E_{ru}$, corresponding to u and r states of identical energies. With $E_{ur} \neq E_{ru}$, the final composition would have to be different (even for $\epsilon_0 = 0$), favoring the state of lower energy.

By substituting a_1, (B.2a) then become

$$
\begin{aligned}
-\frac{d\sigma_k}{dt} &= \left(2E_0 v_1^{\#}/v_1\right)[\omega_k(k,T)/2]e^{-kE_{a,1}/(k_BT)}\frac{\sigma_k - \sigma_{k,\infty}}{\sigma_{k,0} - \sigma_{k,\infty}} \\
&= \frac{1}{a_1}[\omega_k(k,T)/2]e^{-kE_{a,1}/(k_BT)}\frac{\sigma_k - \sigma_{k,\infty}}{\sigma_{k,0} - \sigma_{k,\infty}},
\end{aligned}
\tag{B.3}
$$

because $\frac{2E_0 v_1^{\#}}{v_1} = \frac{2E_0 v_k^{\#}}{v_k} = \sigma_{k,0} - \sigma'_{k,\infty} = \frac{\sigma_{k,0}-\sigma_{k,\infty}}{f_{r_1,k,\infty}} = 2\left(\sigma_{k,0} - \sigma_{k,\infty}\right) = \frac{1}{a_1}$, as may be seen from a partial stress balance over the entire set of relaxors of size k (analogous to that made for primitive relaxors in (9.9), but for the two-state model), where $\sigma'_{k,\infty} < \sigma_{k,\infty}$ would be the contribution to the final local relaxed stress *if* all k primitive relaxors in the clusters ended up fully relaxed.

It is finally seen that (B.3) yield the constitutive equations of *standard non-linear solids* with *relaxation times*

$$
\theta_k = \frac{1}{\omega_k(k,T)}e^{kE_{a,1}/(k_BT)},
\tag{B.4}
$$

identical to the ones in Chap. 9, also in the limit $\varepsilon_0 \to 0$. By integrating (B.3) between $(0, \sigma_{k,0})$ and (t, σ_k), with $\sigma_{k,0} = E_{k,0}\varepsilon_0$, $\sigma_{k,\infty} = E_{k,\infty}\varepsilon_0$ and $\sigma_k(t) = E_{r,k}(t) \cdot \varepsilon_0$, one obtains $E_{r,k}(\varepsilon_0, T, t) = E_{k,\infty} + \left(E_{k,0} - E_{k,\infty}\right)e^{-t/\theta_k(\varepsilon_0,T)}$ and, for the overall stress relaxation modulus,

$$
E_r(\varepsilon_0, T, t) = E_\infty + (E_0 - E_\infty)\sum_{k\geq1} F'_k e^{-t/\theta_k(\varepsilon_0,T)},
\tag{B.5}
$$

with all $E = \sum_{k\geq1} F'_k E_k$ (because $\sigma = \sum_{k\geq1} F'_k \sigma_k$), $E_{k,0} = E_0$ and $E_{k,\infty} = E_\infty$. As before, the number of terms in the above summations must guarantee that $F'_{k_{max}}\sigma_{k_{max}}/\sigma$ is below machine precision, like $F'_{k,max}/\sum_{k\geq1} F'_k$, as calculated in Chap. 9.

It may be observed that, in cases where α_{21} might turn out equal to $-\alpha_{11}$,[12] the above (otherwise approximate) results will be the exact solutions of (B.2).

First Approximation in the Non-linear Viscoelastic Domain with $\alpha_{21} \neq -\alpha_{11}$

Expanding the exponentials in the arguments of the averages $\mu_k^{\pm}[\cdots]$ shown in (B.2), they yield for its right-hand-side a series of terms proportional to the various positive and negative moments of the binomial distribution of the primitive relaxors. Thus,

---

[12] In Chap. 9, Sect. 9.2.3, it has been shown and justified that α_{21} may in fact turn out negative.

depending on the actual value of $(\alpha_{11} + \alpha_{21})$—proportional to ε_0^2—that series, whose terms are thus proportional to growing powers of ε_0^2 (ε_0^2 for the second term, ε_0^4 for the third, ε_0^6 for the fourth, etc.), may converge within the required precision after just a few terms, which however will not yield linear first-order differential equations in the σ_k.

From the moment-generating function of the binomial distribution referred to any cluster of k primitive relaxors, $M_j^k(t) = \sum_{j=0}^{k} e^{jt} \binom{k}{j} p^j q^{k-j} = (q + pe^t)^k$, the various moments of the distribution may be obtained by $\mu_l = \left[d^l M_j^k(t)/dt^l \right]_{t=0}$, where $p = p_k^+$ when calculating the moments μ_k^+, and $p = p_k^- = 1 - p_k^+$ in the case of the moments μ_k^-, for the un-relaxed $(^+)$ and relaxed $(^-)$ primitive relaxors, respectively. For example, the first two moments will be given by $\mu_{k,1}^+ = kp_k^+ = k\left[1 - a_k(\sigma_{k,0} - \sigma_k)\right]$ and $\mu_{k,1}^- = kp_k^- = ka_k(\sigma_{k,00} - \sigma_k)$, already used to obtain (B.2a), $\mu_{k,2}^+ = kp_k^+(kp_k^+ + p_k^-) = k^2 - k(2k - 1)a_k(\sigma_{k,0} - \sigma_k) + k(k - 1)a_k^2(\sigma_{k,0} - \sigma_k)^2$. and $\mu_{k,2}^- = ka_k(\sigma_{k,0} - \sigma_k) + k(k - 1)a_k^2(\sigma_{k,0} - \sigma_k)^2$, and it may therefore be recognized that just the need to include the second term of the series will result in non-linear differential constitutive equations for all clusters (except for single primitive relaxors). It is of course not surprising at all that the accurate formulation of cooperative *non-linear stress relaxation* should lead to a set of cluster-specific *non-linear* differential constitutive equations.

Nevertheless, in the limit of very low strains, the approximate solution represented by (B.3) and (B.4) may in fact still be slightly improved, because – if $f_{r_1,k,\infty}$ falls below $\sim 3/4$ [cf. the second of (9.15)], as it does, being (as we have seen) 1/2 for $f_{u_1,0} = 1$ at the limit of very low initial strain, α_{21} becomes negative,[13] and we may then approximate to 1 the exponentials in the arguments of the binomial averages in (B.2) proportional to $(\alpha_{11} + \alpha_{21})$, but not necessarily $e^{-k\alpha_{21}}$ and $e^{k\alpha_{11}}$, and therefore more accurate expressions for a_k result as $a_1 = \dfrac{e^{-k\alpha_{21}}}{(\sigma_{k,0} - \sigma_{k,\infty})(e^{k\alpha_{11}} + e^{-k\alpha_{21}})}$, which should remain k-independent and equal to $a_1 = \left(2E_0 v_1^{\#}/v_1\right)^{-1}$, because it is the reciprocal of the local absolute stress change after each transition by one single primitive relaxor.[14] Thus, instead of (B.3), one has

$$-\frac{d\sigma_k}{dt} = \frac{1}{a_1} e^{-k\alpha_{21}} [\omega_k(k, T)/2] e^{-kE_{a,1}/(k_B T)} \frac{(\sigma_k - \sigma_{k,\infty})}{\sigma_{k,0} - \sigma_{k,\infty}}$$

$$= [\omega_k(k, T)/2] e^{-kE_{a,1}/(k_B T)} \left(e^{k\alpha_{11}} + e^{-k\alpha_{21}}\right)(\sigma_k - \sigma_{k,\infty}), \tag{B.6}$$

---

[13] This ($\alpha_{21} < 0$) means that, as pointed out in Sect. 9.2.3, the stress σ_∞' may be locally negative, decreasing (rather than increasing) the effective activation energy of the local reverse transitions (as when releasing a compressed spring).

[14] However, in this case, though $\sigma_{k,\infty}$ may still be k-independent, the $f_{r_1,k,\infty}$, will not be exactly 1/2 (cf. (9.11) for similarity), because the direct and reverse *effective activation energies* will not be the same as at non-zero applied strain, even when $E_{ur} = E_{ru}$.

the constitutive differential equations of *standard non-linear solids*, with relaxation times

$$\theta_k = \frac{2}{\omega_k(k,T)} e^{kE_{a,1}/(k_B T)} \left(e^{k\alpha_{11}} + e^{-k\alpha_{21}} \right)^{-1}$$

$$= \frac{1}{\omega_k(k,T)} e^{kE_{a,1}/(k_B T)} e^{-k(\alpha_{11}-\alpha_{21})/2} \mathrm{sech}[k(\alpha_{11}+\alpha_{21})/2]. \qquad (B.7)$$

These results are as those of (9.13) and (9.14) after replacing there ω_1 (and all ω_k) by $\omega_k/2$, but they are only valid for sufficiently low strains.

A still more accurate solution to (B.2) for not too low strains, would have to include in the expansion of the right-hand-sides of those equations the higher moments of the binomial distribution (at least the second moments $\mu_{k,2}^+$ and $\mu_{k,2}^-$) and the equation would then require numerical integration. This would be much more complex,[15] but the procedure could benefit from starting the calculation with the optimized set of the CTMD parameters (all of them relevant physical properties) determined by adjusting the approximate analytical solution of (B.6) and (B.7) to experimental stress relaxation data.[16] However, we may have to face the still open problem of defining more representative values of θ_k and F_k' consistent with the corresponding non-linear constitutive equations, to replace those of (B.7) above and (9.2) and (9.2a) in Chap. 9, needed to combine their σ_k solutions.

A possible (but as yet untested) way out might prove to be assuming that the relations $-\frac{d\sigma_k}{dt} = \frac{(\sigma_k-\sigma_{k,\infty})}{\theta_k} = \frac{(\sigma_0-\sigma_\infty)}{\theta_k} e^{-t/\theta_k}$ maintain their validity (cf. Highlight 9.2), at least within limited time intervals (because the process likely keeps its "first-order reversible" nature[17]), and numerically solving each of them for the corresponding θ_k after the accurate non-linear, numerically obtained, $d\sigma_k/dt$ values are determined from more accurate expansions of the right-hand-side of (B.2). It is then not excluded that shifting θ_k and F_k' values (and therefore shifting relaxation spectra) could result, corresponding to a gradually shifting structure of the clusters involved.[18]

B.3 Summary of the Formulation for Uniaxial Tensile Creep

For the same situation of a binomial distribution of deformed and undeformed primitive relaxors within any of the clusters, the differential constitutive equations for the contribution to the creep strain by the whole set of clusters of size k along a given

---

[15] It would involve the simultaneous numerical solution of a large (and ideally adjustable) number of first-order non-linear differential equations, one for each cluster size (somewhat analogous to the thermodynamic (14.5) in Chap. 14), satisfying the conditions $\sigma = \sum_{k \geq 1} F_k' \sigma_k$ and $F_{k_{max}}' \sigma_{k_{max}}/\sigma \leq$ machine precision. However, as implied in the above main text, we would meet an uncertainty in the definition of the θ_k and F_k' values.

[16] [Like the calculations of Chap. 11 using the approximate formulations of Chap. 4].

[17] The reader may here also recall Study Question 9.1 at the end of Sect. 9.3.2.

[18] Not at all surprising, if one considers that (B.2) cannot be accurately made equivalent to the above or to Highlight 9.2's and Study Question 9.1's first-order differential equations with specifiable and constant θ_k values.

length [considering the model approximately represented by the equivalent of (9.21)] may be written as

$$\frac{d\varepsilon_k}{dt} = \left(\frac{2v_1^{\#}}{v_1}\right)\frac{\omega_k}{2}e^{-\frac{kE_{ud,1}}{k_BT}}\sum_{k^+=0}^{k}\frac{(k^+ - k^-)}{k}\binom{k}{k^+}$$

$$\times \left[1 - b_k(\varepsilon_k - \varepsilon_{k,0})\right]^{k^+}\left[b_k(\varepsilon_k - \varepsilon_{k,0})\right]^{k^-}e^{(k^+ - k^-)\alpha_1} \tag{B.8}$$

with $\alpha_1 = \sigma_0 v_1^{\#}/(k_BT)$, $E_{ud,1} = E_{a,1}$, the minimum activation energy, and the fraction of undeformed primitive relaxors in the clusters of size k, again assuming $f_{u_1,0} = 1$, given by $f_{u_1,k} = f_{u_1,0}\left[1 - b_k(\varepsilon_k - \varepsilon_{k,0})\right]$, or

$$\frac{d\varepsilon_k}{dt} = \left(\frac{2v_1^{\#}}{v_1}\right)\frac{\omega_k}{2}e^{-\frac{kE_{ud,1}}{k_BT}}\frac{1}{k}\left[e^{-k\alpha_1}\mu_{k,1}^{+}\left(k^+e^{2k^+\alpha_1}\right) - e^{k\alpha_1}\mu_{k,1}^{-}\left(k^-e^{-2k^+\alpha_1}\right)\right], \tag{B.8a}$$

involving again the binomial averages of the indicated expressions.

At the *linear viscoelastic limit* of quasi-equilibrium conditions (almost zero applied stress), we only need to consider the averages of the fractions of undeformed and deformed primitive relaxors, and the above equations reduce to

$$\frac{d\varepsilon_k}{dt} = \left(\frac{2v_1^{\#}}{v_1}\right)\frac{\omega_k}{2}e^{-\frac{kE_{ud,1}}{k_BT}}\left[1 - 2b_k(\varepsilon_k - \varepsilon_{k,0})\right], \tag{B.9}$$

where, from the equilibrium condition at very long times, one obtains $b_k = \left[2(\varepsilon_{k,\infty} - \varepsilon_{k,0})\right]^{-1} = \left[2(\varepsilon_\infty - \varepsilon_0)\right]^{-1} = b_1$ and, considering that, from a strain balance over the whole set of relaxors of size k,

$$\frac{2v_1^{\#}}{v_1} = \varepsilon_{k,\infty}' - \varepsilon_{k,0} = \frac{\varepsilon_{k,\infty} - \varepsilon_{k,0}}{f_{d_1,k,\infty}} = \frac{1}{b_k} = \frac{1}{b_1}, \tag{B.10}$$

where $f_{d_1,k,\infty} = f_{d_1,\infty} = 1/2$ is the final (equilibrium) fraction of deformed primitive relaxors among the clusters of any size,[19] one obtains the differential constitutive equations of *standard non-linear solids* as

$$\frac{d\varepsilon_k}{dt} = \omega_k e^{-\frac{kE_{ud,1}}{k_BT}}\left(\varepsilon_{k,\infty} - \varepsilon_{k,0}\right), \tag{B.11}$$

with retardation times

---

[19] At any finite time, however, the $f_{d_1,k}$, like the $f_{r_1,k}$ in stress relaxation, will not be the same and will be higher for each set of smaller clusters, because they respond faster and a higher proportion of their primitive relaxors would in principle have deformed fully.

$$\tau_k = \frac{1}{\omega_k(k,\,T)}e^{kE_{a,1}/(k_BT)}, \tag{B.12}$$

analogous to the stress relaxation results in the same linear viscoelastic limit,[20] whose solutions are

$$\varepsilon_k = \varepsilon_{k,0} + \left(\varepsilon_{k,\infty} - \varepsilon_{k,0}\right)\left(1 - e^{-t/\tau_k}\right). \tag{B.13}$$

At not negligible stresses, however, the solutions may of course only be obtained numerically.

Like the discussion of creep in Chap. 9, this simplified formulation may arguably be improved by elimination (or reduction) of the exponent $k^+\alpha_1$ in (B.8), while keeping the exponents $k^-\alpha_1$, thus increasing the activation energies of the direct transitions. The reader is invited to work out that this modification yields for the modified b_k the expressions $\left[\left(1 + e^{k\alpha_1}\right)\left(\varepsilon_{k,\infty} - \varepsilon_{k,0}\right)\right]^{-1}$ and, for the retardation time,

$$\tau'_k = \frac{2}{\omega_k(k,\,T)}e^{kE_{a,1}/(k_BT)}\left(1 + e^{-k\alpha_1}\right)^{-1}, \tag{B.14}$$

analogous to those of (9.23). However, in the non-linear viscoelastic domain, the solutions may only be obtained by numerical integration of the whole (temperature-dependent) set of (B.8a), with the same difficulty found in the stress relaxation case to assign values to the retardation time and statistical weight of each cluster size.

As pointed out at the end of Sect. 9.1, the most critical issue in the development of CTMD and its many possible applications (as described in Chaps. 8, 9, 11, 12, 14 and 15, and Appendixes B and C) is the validity of the estimates of the statistical weights of the various cluster sizes (or characteristic time spectra) $F'_k(\theta_k)$ or $F'_k(\tau_k)$ in equilibrium and non-equilibrium conditions, which may well require revising the formulations of Sect. 9.1 and Chap. 12.

It is explicitly recognized that this book is just one of many likely necessary attempts at solving the complex and challenging problems of the equilibrium and non-equilibrium behavior of amorphous condensed matter.

---

[20] This *apparent* equality of the retardation and relaxation times and their possible correction has been discussed in some detail in Chap. 9, not considering the additional possibility of different effective values of $E_{a,1}$ in creep and stress relaxation.

Appendix C
A Four-State Model of Stress Relaxation

C.1 Introduction

In Chap. 9 and Appendix B, a simple two-state model was assumed to formulate the stress relaxation (and creep) dynamics, considering in the first case the highly improbable situation where all primitive relaxors of the system are initially unrelaxed (or undeformed, in the case of creep). Even when in Appendix B we accounted for a binomial distribution of the primitive relaxors in any of the clusters among the unrelaxed and the relaxed states, or when we formulated the behavior of partly crosslinked or crystalline materials (in Chap. 9), we have assumed that all amorphous primitive relaxors were initially unrelaxed ($t = 0 : f_{u,0} = 1$, or 1 minus the fraction of crosslinked and/or crystallized primitive relaxors and $f_{r,0} = 0$).

The real initial structural configuration of any amorphous material in stress relaxation is most often the one represented by Fig. 9.3's scheme, or here by Fig. C.1, where only minor changes are made in the symbols used for the various states.

As a result, the primitive relaxors of any family of clusters of a given size k are distributed among the four available states (I_u, I_r, II_r and II_u, as shown) according to a *quadrinomial* distribution, while assuming that they are uniformly distributed within the material as a whole, with probabilities $p_{I_u} = p_I^+ = f_{I,u}$, $p_{I_r} = p_I^- = f_{I,r}$, $p_{II_r} = p_{II}^- = f_{II,r}$ and $p_{II_u} = p_{II}^+ = f_{II,u}$, where the + or − superscripts, and u and r subscripts, denote (as in Appendix B) that the indicated primitive relaxors can only be involved in ensuing direct (+) *or* reverse (−) transitions, respectively. At $t = 0$, $p_{I_u,0} = f_{u,0} < 1$, $p_{II_r,0} = f_{r,0} = 1 - f_{u,0}$, $p_{I_r,0} = p_{II_u,0} = 0$ corresponding to their initial fractions in the whole set of primitive relaxors, where the latter relationships, together with the inital strain, ϵ_0, and stress, $\sigma_0 = E_0\epsilon_0$, define the initial conditions of the stress relaxation problem.

J. J. Cruz Pinto and J. R. dos Santos André, *Analytical Molecular Dynamics of Amorphous Condensed Matter*, Springer Series in Materials Science 342, https://doi.org/10.1007/978-3-031-56517-5

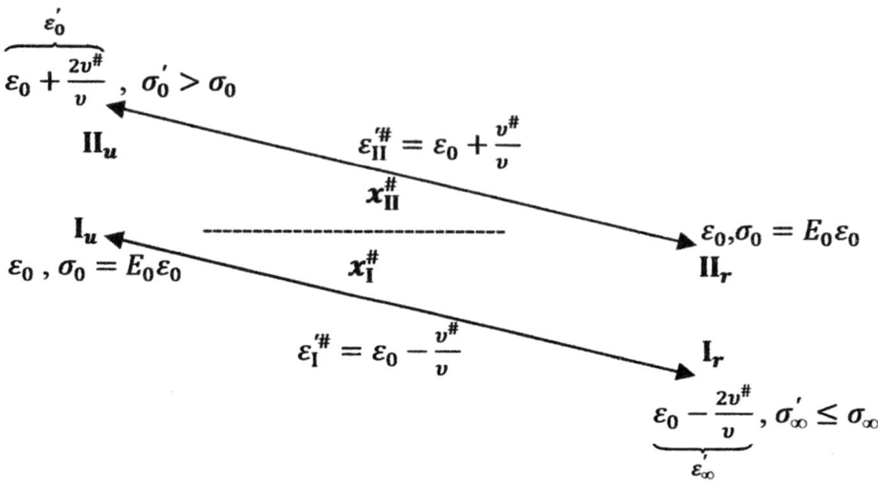

Fig. C.1 Re-edition of Fig. 9.3 with the two subsystems of lower and higher skeleton strain and stresses represented by the symbols I and II, respectively, and all variables referred to the primitive relaxors with $v = v_1$ (cf. Fig. 9.3's caption and Sect. 9.2.1's text for the meaning of all other symbols)

The above system is thus made up of two subsystems, I and II, within which two reversible transitions may occur, $\mathbf{I}_u \rightleftharpoons \mathbf{I}_r$ and $\mathbf{II}_r \rightleftharpoons \mathbf{II}_u$, respectively, no reversible or irreversible transitions being possible between subsystems I and II, which are strictly independent. Subsystem I is the main contributor to stress relaxation, while subsystem II counteracts the process, because the stress increases from $\mathbf{II}_r$ (the only populated state in II at $t = 0$) to $\mathbf{II}_u$. The transitions between the two states of subsystem II cannot be neglected, except perhaps in extreme situations with much lower activation energies of the direct transitions, E_{ur}, relative to the reverse ones, E_{ru}. As in all formulations in this book, we only consider the special, much simplified, case of $E_{ur} = E_{ru}$. The general case would require the use of an additional parameter specifying the difference of the above two activation energies, identical to the difference between the u and r states themselves without any applied strain.

C.2 Basic Relationships

Like in the formulation of Appendix B, and considering the obvious linear relationships of the above fractional populations with the associated local stress relaxations (in I) or amplifications (in II), we may write the following "defining" equations for the specified fractional populations or probabilities:

$$p_I^+ = f_{I,u} = f_{u,0}[1 - a_1(\sigma_0 - \sigma_I)], \qquad (C.1)$$

$$p_{\mathrm{I}}^{-} = f_{\mathrm{I},r} = f_{u,0}a_1(\sigma_0 - \sigma_{\mathrm{I}}), \tag{C.2}$$

$$p_{\mathrm{I}}^{-} - p_{\mathrm{II}}^{+} = f_{\mathrm{I},r} - f_{\mathrm{II},u} = A(\sigma_0 - \sigma), \tag{C.3}$$

in the latter case because this difference must be proportional to the drop in the overall average stress per primitive relaxor, and

$$p_{\mathrm{II}}^{+} = f_{\mathrm{II},u} = A'(\sigma_{\mathrm{II}} - \sigma_0), \tag{C.4}$$

because $f_{\mathrm{II},u}$ must be proportional to the local average stress increase over the initial value, σ_0, within the primitive relaxors belonging to subsystem. In the above relationships, σ is the overall average stress per primitive relaxor, irrespective of its state, expressing the overall average result of local stress relaxations or amplifications up to time t due to the joint behavior of both subsystems I and II, σ_{I} is the specific average local stress per primitive relaxor within subsystem I (σ_{II} being the corresponding value within subsystem II), and the proportionality constants A and A' must ensure consistency of all the above relationships with $\sigma(t) = f_{u,0}\sigma_{\mathrm{I}}(t) + (1 - f_{u,0})\sigma_{\mathrm{II}}(t)$.

The latter formulation of the overall *average* local stress per primitive relaxor, $\sigma(t)$, is itself consistent with its previous formulation (cf. Chap. 9) in terms of the contributions by the various clusters when they are allowed to form, $\sigma = \sum_{k\geq 1} F_k'\sigma_k$). The reader should bear in mind that, subsystems I and II being independent, the average local stresses in the primitive relaxors belonging to those subsystems are those denoted here by $\sigma_{\mathrm{I}}(t)$ and $\sigma_{\mathrm{II}}(t)$, respectively.

Substituting (C.4) into (C.3) and taking the above definition of $\sigma(t)$, one sees that $A = a_1$ and $A' = a_1(1 - f_{u,0})$. Therefore,

$$p_{\mathrm{II}}^{+} = f_{\mathrm{II},u} = a_1(1 - f_{u,0})(\sigma_{\mathrm{II}} - \sigma_0), \tag{C.4a}$$

and

$$p_{\mathrm{II}}^{-} = f_{\mathrm{II},r} = (1 - f_{u,0})[1 - a_1(\sigma_{\mathrm{II}} - \sigma_0)], \tag{C.5}$$

and it may be checked that

$$p_{\mathrm{I}}^{+} + p_{\mathrm{I}}^{-} + p_{\mathrm{II}}^{+} + p_{\mathrm{II}}^{-} = 1. \tag{C.6}$$

We now need to relate the constant a_1 with system properties, which may be done (and confirmed) in two ways:

(1) $(1)p_{\mathrm{I}}^{-} = f_{\mathrm{I},r} = f_{u,0}a_1(\sigma_0 - \sigma_{\mathrm{I}})$ by (C.2), and $f_{\mathrm{I},r}$ multiplied by $\left(\frac{2E_0v_{\mathrm{I}}^{\#}}{v_{\mathrm{I}}}\right)$, the elementary local stress change per primitive relaxor transition $\mathrm{I}_u \rightleftharpoons \mathrm{I}_r$ (as formulated in Chap. 9), must yield $f_{u,0}(\sigma_0 - \sigma_{\mathrm{I}})$, the overall average stress drop

up to time t per primitive relaxor due to subsystem I, and that gives $a_1 = \left(\frac{2E_0 v_1^{\#}}{v_1}\right)^{-1}$.

(2) $p_{\text{II}}^{+} = f_{\text{II},u} = a_1(1 - f_{u,0})(\sigma_{\text{II}} - \sigma_0)$, by (C.4), and $f_{\text{II},u}$ multiplied by the same $\left(\frac{2E_0 v_1^{\#}}{v_1}\right)$, the assumed elementary local stress change per primitive relaxor transition $\text{II}_r \rightleftharpoons \text{II}_u$, must give $(1 - f_{u,0})(\sigma_{\text{II}} - \sigma_0)$, the overall average stress increase up to time t per primitive relaxor due to subsystem II, and that confirms that $a_1 = \left(\frac{2E_0 v_1^{\#}}{v_1}\right)^{-1}$, as in the previous formulations of Chap. 9 and Appendix B.

To formulate the specific contributions of every set of clusters of a given size k (if we exclude the formation of any other clusters of different size), we must consider that, consistent with the uniform distribution of all primitive relaxors among the above four states at any given time, as expressed by (C.1), (C.2), (C.4) and (C.5), their instantaneous uniform distribution among those same states *within the clusters of each specific size* may nevertheless differ from the above overall one, because smaller clusters respond faster and are thus expected to have a lower average fraction of unrelaxed primitive relaxors at any time t. This possibility had already been left implicit in Footnote 39 of Chap. 9 and is also explicitly mentioned in Footnote 8 of Appendix B. As a result, by analogy to (C.1), (C.2), (C.4) and (C.5), we may be justified to formulate the mentioned distribution among the four states of the primitive relaxors within the whole set of clusters of each specific size k as[21]

$$p_{\text{I},k}^{+} = f_{\text{I},u,k} = f_{u,0}\left[1 - a_1(\sigma_0 - \sigma_{\text{I},k})\right], \tag{C.7}$$

$$p_{\text{I},k}^{-} = f_{\text{I},r,k} = f_{u,0}a_1(\sigma_0 - \sigma_{\text{I},k}), \tag{C.8}$$

$$p_{\text{II},k}^{+} = f_{\text{II},u,k} = a_1(1 - f_{u,0})(\sigma_{\text{II},k} - \sigma_0) \tag{C.9}$$

and

$$p_{\text{II},k}^{-} = f_{\text{II},r} = 1 - f_{u,0} - p_{\text{II},k}^{+} = (1 - f_{u,0})\left[1 - a_1(\sigma_{\text{II},k} - \sigma_0)\right], \tag{C.10}$$

with the same a_1 and $f_{u,0}$, because they all relate to individual primitive relaxors in an assumed initially uniform material.

---

[21] This induction may appear problematic, because the $\sigma_{\text{I},k}$ and $\sigma_{\text{II},k}$ within any of the clusters of each given size are stochastic in time and space, and thus "ill-defined", in as much as they are the result of multiple previous transitions of variable nature (within clusters of variable sizes). However, other than by a full Monte-Carlo approach and computation, there is no alternative to the proposed (physically reasonable) formulation.

C.3 Formulation of the Dynamics of Uniaxial Tensile Stress Relaxation of Uncorrelated Primitive Relaxors

As done in Chap. 9, Sect. 9.2, we first develop the complete formulation of stress relaxation assuming that uncorrelated primitive relaxors are the only contributors to the process (thus neglecting all clustering), before considering in the next section (though only for the limit of very low strains) the procedure to generalize it to the joint contribution of the whole range of cluster sizes.

Bearing in mind (9.10) and the foregoing expressions for the fractional populations of primitive relaxors in each state, the replacement of (9.12) for this four-state model (when $f_{u,0} < 1$) is

$$-\frac{d\sigma}{dt} = \left[-f_{u,0}\frac{d\sigma_{\mathrm{I}}}{dt} - (1-f_{u,0})\frac{d\sigma_{\mathrm{II}}}{dt} \right]$$

$$= \frac{\omega_1}{2} \underbrace{\left(\frac{\sigma_0 - \sigma_\infty}{f_{\mathrm{I},r,\infty} - f_{\mathrm{II},u,\infty}} \right)}_{\frac{2E_0 v_1^\#}{v_1} = \frac{1}{a_1}}$$

$$\times \left\{ \overbrace{\begin{array}{c} \left[f_{u,0}e^{-E'_{\mathrm{I},ur}/(k_BT)}[1 - a_1(\sigma_0 - \sigma_{\mathrm{I}})] - f_{u,0}e^{-E'_{\mathrm{I},ru}/(k_BT)}a_1(\sigma_0 - \sigma_{\mathrm{I}})\right] \\[4pt] \underbrace{\left[-e^{-E'_{\mathrm{II},ru}/(k_BT)}(1 - f_{u,0})[1 - a_1(\sigma_{\mathrm{II}} - \sigma_0)] + (1 - f_{u,0})e^{-E'_{\mathrm{II},ur}/(k_BT)}a_1(\sigma_{\mathrm{II}} - \sigma_0)\right]}_{\mathrm{II}} \end{array}}^{\mathrm{I}} \right\}$$

$$\tag{C.11}$$

a too clumsy expression which should soon be separated into its two independent parts, where the two top terms within {} refer to subsystem Iand the other two to subsystem II, $E'_{\mathrm{II},ru}$ and $E'_{\mathrm{II},ur}$ are the modified, strain-dependent, activation energies of the transitions in subsystem II, additional to those of subsystem I, $E'_{\mathrm{I},ur}$ and $E'_{\mathrm{I},ru}$. The latter were previously obtained from (9.4) and (9.5),[22] and the reader may likewise verify (with reference to Fig. C.1 and the changes in the stored elastic energies between each of the states and the relevant activated ones) that $E'_{\mathrm{II},ur}$ and $E'_{\mathrm{II},ru}$ obey the relationships

$$E'_{\mathrm{II},ur} - E_{ur} = -\frac{v_1}{2}\left(\sigma'_0\epsilon'_0 - \sigma'^\#_{\mathrm{II}}\epsilon'^\#_{\mathrm{II}} \right) = -E_0 v_1^\#\left(\epsilon_0 + \frac{3}{2}v_1^\#/v_1 \right) = -k_BT\beta_1,$$

$$\tag{C.12}$$

$$E'_{\mathrm{II},ru} - E_{ru} = \frac{v_1}{2}\left(\sigma'^\#_{\mathrm{II}}\epsilon'^\#_{\mathrm{II}} - \sigma_0\epsilon_0 \right) = E_0 v_1^\#\left(\epsilon_0 + \frac{1}{2}v_1^\#/v_1 \right) = k_BT\beta_2, \tag{C.13}$$

where $E_{ur} = E_{ru} = E_{a,1}$ is the activation energy of each primitive relaxor in the absence of any applied strain, and $v_1^\#/v_1$ may be seen to be [from (9.10)]

---

[22] [Enabling the definition of the variables α_1 and α_2, such that $E'_{\mathrm{I},ur} - E_{ur} = -k_BT\alpha_1$ and $E'_{\mathrm{I},ru} - E_{ur} = k_BT\alpha_2$].

$$\frac{v_1^{\#}}{v_1} = \frac{1}{2}\frac{(E_0 - E_\infty)}{E_0\big(f_{\mathrm{I},r,\infty} - f_{\mathrm{II},u,\infty}\big)}\epsilon_0. \tag{C.14}$$

Because now $f_{\mathrm{II},u,\infty}$ will not be zero, (9.15) will change to

$$\alpha_1 = \frac{v_1}{8k_BT}\epsilon_0^2\frac{E_0 - E_\infty}{E_0\big(f_{\mathrm{I},r,\infty} - f_{\mathrm{II},u,\infty}\big)^2}\Big\{\big[4(f_{\mathrm{I},r,\infty} - f_{\mathrm{II},u,\infty}) - 1\big]E_0 + E_\infty\Big\},$$

$$\tag{C.15}$$

$$\alpha_2 = \frac{v_1}{8k_BT}\epsilon_0^2\frac{E_0 - E_\infty}{E_0\big(f_{\mathrm{I},r,\infty} - f_{\mathrm{II},u,\infty}\big)^2}\Big\{\big[4(f_{\mathrm{I},r,\infty} - f_{\mathrm{II},u,\infty}) - 3\big]E_0 + 3E_\infty\Big\},$$

$$\tag{C.16}$$

and the reader may also obtain, by expansion of (C.12) and (C.13) using (C.14),

$$\beta_1 = \frac{v_1}{8k_BT}\epsilon_0^2\frac{E_0 - E_\infty}{E_0\big(f_{\mathrm{I},r,\infty} - f_{\mathrm{II},u,\infty}\big)^2}\Big\{\big[4(f_{\mathrm{I},r,\infty} - f_{\mathrm{II},u,\infty}) + 3\big]E_0 - 3E_\infty\Big\},$$

$$\tag{C.17}$$

$$\beta_2 = \frac{v_1}{8k_BT}\epsilon_0^2\frac{E_0 - E_\infty}{E_0\big(f_{\mathrm{I},r,\infty} - f_{\mathrm{II},u,\infty}\big)^2}\Big\{\big[4(f_{\mathrm{I},r,\infty} - f_{\mathrm{II},u,\infty}) + 1\big]E_0 - E_\infty\Big\},$$

$$\tag{C.18}$$

and observe that $\beta_1 - \beta_2 = \alpha_1 - \alpha_2 = \frac{E_0 v_1^{\#2}}{v_1 k_B T} = \frac{v_1\epsilon_0^2}{4k_BT}\frac{(E_0-E_\infty)^2}{E_0\big(f_{\mathrm{I},r,\infty}-f_{\mathrm{II},u,\infty}\big)^2}$ [cf. (9.4), (9. 5), (C.12), (C.13) and (C.15)–(C.18)]. These are the four dimensionless variables (with only three of them independent) that will determine the non-linearity, or strain dependence, of the stress relaxation modulus.

As the I and II average component stresses have been defined relative to each primitive relaxor crossing the resisting area, such as to make $\sigma(t) = f_{u,0}\sigma_\mathrm{I}(t) + \big(1 - f_{u,0}\big)\sigma_\mathrm{II}(t)$, and subsystems I and II are independent, it is practical to separately expand [substituting the activation energies by those of Chap. 9, also given here in Footnote 22, and of (C.12) and (C.13)], integrate each of the rates $-d\sigma_\mathrm{I}/dt$ and $d\sigma_\mathrm{II}/dt$,[23] and finally combine the results.

At the final equilibrium ($t \rightarrow \infty$), the direct and reverse transitions equilibrate in *both* subsystems, and so we obtain for subsystem I

$$a_1 = \frac{e^{\alpha_1}}{(e^{\alpha_1} + e^{-\alpha_2})(\sigma_0 - \sigma_{\mathrm{I},\infty})}, \tag{C.19}$$

leading to

---

[23] Note that, while $d\sigma_\mathrm{I}/dt$ is negative, $d\sigma_\mathrm{II}/dt$ is positive.

$$-\frac{d\sigma_{\mathrm{I}}}{dt} = \frac{\omega_1}{2} e^{-\frac{E_{a,1}}{k_B T}} \frac{1}{a_1} \left\{ e^{\alpha_1}[1 - a_1(\sigma_0 - \sigma_{\mathrm{I}})] - e^{-\alpha_2} a_1(\sigma_0 - \sigma_{\mathrm{I}}) \right\}$$

$$= \frac{\omega_1}{2} e^{-\frac{E_{a,1}}{k_B T}} \left(e^{\alpha_1} + e^{-\alpha_2} \right) \left(\sigma_{\mathrm{I}} - \sigma_{\mathrm{I},\infty} \right), \tag{C.20}$$

which integrates to

$$\sigma_{\mathrm{I}}(t) = \sigma_{\mathrm{I},\infty} + \left(\sigma_0 - \sigma_{\mathrm{I},\infty} \right) e^{-t/\theta_{\mathrm{I},1}}, \tag{C.21}$$

with

$$\theta_{\mathrm{I},1} = \frac{2}{\omega_1} e^{\frac{E_{a,1}}{k_B T}} \left(e^{\alpha_1} + e^{-\alpha_2} \right)^{-1}. \tag{C.21a}$$

Likewise, we obtain for subsystem II

$$\frac{d\sigma_{\mathrm{II}}}{dt} = \frac{\omega_1}{2} e^{-\frac{E_{a,1}}{k_B T}} \frac{1}{a_1} \left[e^{-\beta_2} - a_1\left(e^{\beta_1} + e^{-\beta_2} \right)(\sigma_{\mathrm{II}} - \sigma_0) \right] \tag{C.22}$$

and, at equilibrium, we see that σ_{II} must satisfy

$$a_1 = \frac{e^{-\beta_2}}{\left(e^{\beta_1} + e^{-\beta_2} \right)\left(\sigma_{\mathrm{II},\infty} - \sigma_0 \right)}, \tag{C.23}$$

leading to

$$\frac{d\sigma_{\mathrm{II}}}{dt} = \frac{\omega_1}{2} e^{-\frac{E_{a,1}}{k_B T}} \left(e^{\beta_1} + e^{-\beta_2} \right)\left(\sigma_{\mathrm{II},\infty} - \sigma_{\mathrm{II}} \right), \tag{C.24}$$

which integrates to

$$\sigma_{\mathrm{II}}(t) = \sigma_{\mathrm{II},\infty} - \left(\sigma_{\mathrm{II},\infty} - \sigma_0 \right) e^{-t/\theta_{\mathrm{II},1}}, \tag{C.25}$$

with

$$\theta_{\mathrm{II},1} = \frac{2}{\omega_1} e^{\frac{E_{a,1}}{k_B T}} \left(e^{\beta_1} + e^{-\beta_2} \right)^{-1}. \tag{C.25a}$$

The total average stress per primitive relaxor that results is then

$$\sigma(t) = f_{u,0}\sigma_{\mathrm{I}}(t) + \left(1 - f_{u,0}\right)\sigma_{\mathrm{II}}(t) = \sigma_\infty + \sigma_0\left[f_{u,0} e^{-t/\theta_{\mathrm{I},1}} + \left(1 - f_{u,0}\right) e^{-t/\theta_{\mathrm{II},1}} \right]$$

$$- f_{u,0}\sigma_{\mathrm{I},\infty} e^{-\frac{t}{\theta_{\mathrm{I},1}}} - \left(1 - f_{u,0}\right)\sigma_{\mathrm{II},\infty} e^{-\frac{t}{\theta_{\mathrm{II},1}}}, \tag{C.26}$$

showing that two different (though expectedly similar[24]) characteristic relaxation times control the process, one per subsystem.[25] However, at the linear viscoelastic limit of negligible applied strain, these two relaxation times are identical and reduce to $\theta_{1,0} = \frac{1}{\omega_1} e^{\frac{E_{a,1}}{k_B T}}$, the average stress will be $\sigma(t) = \sigma_\infty + (\sigma_0 - \sigma_\infty)e^{-\frac{t}{\theta_{1,0}}}$, with $\sigma_{\mathrm{I},\infty} = \sigma_0 - 1/(2a_1)$, $\sigma_{\mathrm{II},\infty} = \sigma_0 + 1/(2a_1)$ and $\sigma_\infty = \sigma_0 - \frac{2f_{u,0}-1}{2a_1}$. As expected, with $E_{ur} = E_{ru}$, no relaxation will be possible when $f_{u,0} = 1/2$, because the system will already be at equilibrium at $t = 0$; stress will only relax when $f_{u,0} > 1/2$. As shown in Chap. 15, free volume relaxation is expected to show the same limitations for any given system.

Illustrations of the numerical results obtained from the foregoing formulation (only valid for uncorrelated primitive relaxors, in the absence of any clustering, we stress) and the calculation procedure outlined below just after (C.28) (including those mentioned in Footnote 24) are provided in Figs. C.2 and C.3, for PMMA of two initial compositions ($f_{u,0} = 0.9$ and 0.8) and three applied strains ($\epsilon_0 = 0, 0.01$ and 0.02), at the low temperature of 200 K. The opposite effects of subsystems I and II are made clear (with the development of small compressive stresses in I and some amplification, rather than relaxation, of tensile stresses in II), yielding the physically expected, non-linear, behavior of the compound system. However, similar results could not be obtained for lower initial fractions of unrelaxed primitive relaxors, $f_{u,0}$, or higher applied strains, ϵ_0. We could not establish if this was due to amplification of numerical errors[26] or to a physical incompatibility of the assumed material properties and conditions (e.g. $E_{ur} = E_{ru}$, E_0, E_∞, etc.[27]) with actual stress relaxation. The fact is that, in those unsuccessful calculations, the a_1 and $f_{\mathrm{I},r,\infty} - f_{\mathrm{II},u,\infty}$ values [cf. (C.27) and (C.28)] gradually reduced to zero, meaning no changes in the populations of the four states of the system, and no stress relaxation.

---

[24] For example, for the primitive relaxors of the PMMA studied in Chap. 11, at the assumed temperature of 200 K and tensile strain of 1% we obtained $\theta_{\mathrm{I},1} = 0.2275\,\mathrm{s}$ and $\theta_{\mathrm{II},1} = 0.2220\,\mathrm{s}$ for the example of $f_{u,0} = 0.9$. [Remember that such short response times will however be strongly weighed down by negligible values of F_1' (cf. Chap. 9) when clustering is operating and dominating at low temperatures.] From the relationships between α_1, α_2, β_1 and β_2, it may be anticipated that $\theta_{\mathrm{II},1} < \theta_{\mathrm{I},1}$ always, but subsystem II *counteracts* stress relaxation, because $\sigma_{\mathrm{II}}(t)$ will be growing above σ_0. On the other hand, $\sigma_{\mathrm{I}}(t)$ is always decreasing, and $\sigma_{\mathrm{I},\infty}$ will drop below the final average σ_∞ and may be negative. So, depending on the material's initial structure (and even in the absence of any initial residual stresses), as pointed out in Chap. 9, local compressive stresses may be generated during tensile stress relaxation! [In the above PMMA calculations, the final average E_∞ agreed with the $1.000\ldots$ MPa assumed and previously obtained value (cf. Chap. 11) to within (not 16 but) 12 significant figures.]

[25] Remember that subsystems I and II are completely independent.

[26] [As may be suggested by the already mentioned convergence of E_∞ to within 12, but not 16, significant figures].

[27] In addition to the arbitrary simplifying assumption $E_{ur} = E_{ru}$, the assumed physical properties of PMMA were those numerically adjusted to limited stress relaxation data, using the approximate model of (9.7) and (9.8) with truncated log-normal relaxation spectra.

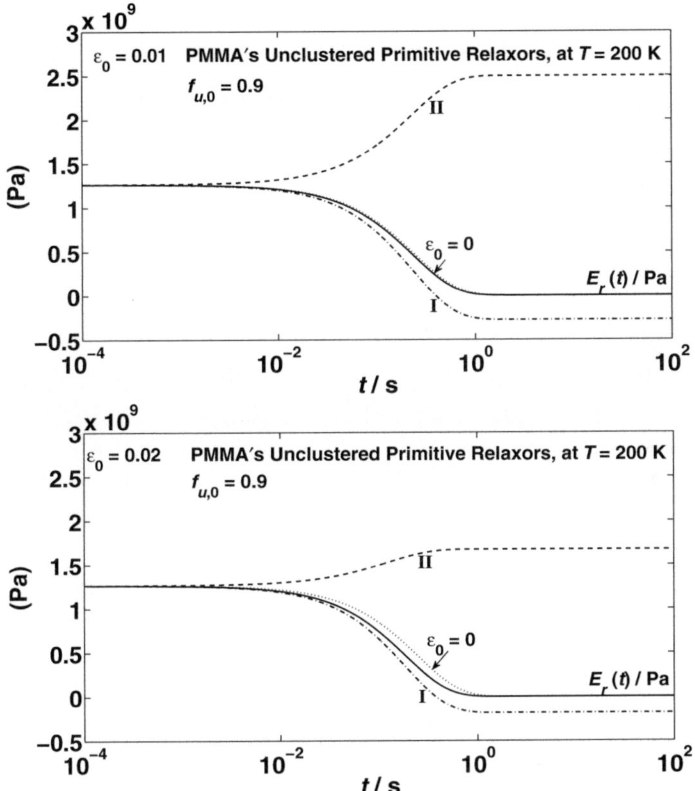

Fig. C.2 Subsystems I and II specific and compound responses to stress relaxation of a PMMA with $f_{u,0} = 0.9$ at $T = 200$ K, $\epsilon_0 = 0, 0.01$ (top) and $\epsilon_0 = 0.02$ (bottom) according to the four-state model, assuming *no clustering*

In the non-linear case, the resulting equilibrium average stresses per primitive relaxor become $\sigma_{\mathrm{I},\infty} = \sigma_0 - \frac{1}{a_1} \frac{e^{\alpha_1}}{e^{\alpha_1}+e^{-\alpha_2}}$ and $\sigma_{\mathrm{II},\infty} = \sigma_0 + \frac{1}{a_1} \frac{e^{-\beta_2}}{e^{\beta_1}+e^{-\beta_2}}$ and thus

$$
a_1 = \frac{\overbrace{f_{u,0}\,\dfrac{e^{\alpha_1}}{e^{\alpha_1}+e^{-\alpha_2}}}^{f_{\mathrm{I},r,\infty}} - \overbrace{\left(1-f_{u,0}\right)\dfrac{e^{-\beta_2}}{e^{\beta_1}+e^{-\beta_2}}}^{f_{\mathrm{II},u,\infty}}}{(E_0 - E_\infty)\epsilon_0}.
\tag{C.27}
$$

Equations (C.15)–(C.18), giving α_1, α_2, β_1 and β_2, (C.27) and the one giving

$$
f_{\mathrm{I},r,\infty} - f_{\mathrm{II},u,\infty} = a_1(E_0 - E_\infty)\epsilon_0
\tag{C.28}
$$

are six equations in six unknowns—α_1, α_2, β_1, β_2, $\left(f_{\mathrm{I},r,\infty} - f_{\mathrm{II},u,\infty}\right)$ and a_1—the known (or experimentally adjustable) parameters being E_0, E_∞, $f_{u,0}$, υ_1 and ϵ_0, in

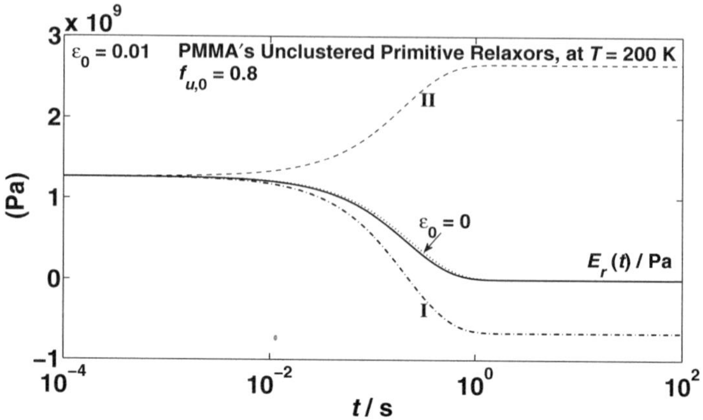

Fig. C.3 Subsystems I and II specific and compound responses to stress relaxation of a PMMA with $f_{u,0} = 0.8$ at $T = 200$ K, $\epsilon_0 = 0$ and 0.01 according to the four-state model, assuming *no clustering*

addition to the CTMD-specific physical properties ν_c, $E_{a,1}$ and T_c. This single system of six equations in six unknowns should be (and *has been*) iteratively solved to within machine precision, to enable the calculation of the stress relaxation modulus, $E_r(t)$, at any given time, t, for each set of parameters or physical properties, such as those determined in Chap. 11 (though there only for the $f_{u,0} = 1$ case) from experimental stress relaxation data. The initial iterating value may be taken as $\left(f_{I,r,\infty} - f_{II,u,\infty}\right) = f_{u,0}$, followed by β_1, β_2 (from $\beta_1 - \beta_2$), α_2, α_1 (from $\beta_1 - \beta_2 = \alpha_1 - \alpha_2$), and finally by a corrected $\left(f_{I,r,\infty} - f_{II,u,\infty}\right)$, repeating the procedure until convergence is reached.

The above treatment would only be the accurate one if no clusters of primitive relaxors were allowed to form, and one sees that, with $f_{u,0} = 1$, the result is identical to that of (9.13) and (9.14). Unfortunately, a similar accurate and analytical treatment for any clusters of primitive relaxors ($k > 1$) will not be possible, as shown below.

C.4 Formulation of the Dynamics of Uniaxial Tensile Stress Relaxation of Clusters of Size $k > 1$

Of course, we could generalize the treatment of the preceding section to any cluster size, assuming that we could take all primitive relaxors within any of the individual clusters as being at the same unrelaxed or unrelaxed state at any given time t. However, we have seen that this is a very crude approximation, as the primitive relaxors within any of the clusters distribute themselves among the four states specified in Fig. C.1 and in the foregoing discussion. We will then now more accurately formulate the stress relaxation of clusters of any specific size, k, but just in the limit of very low applied strains, because we have already seen in Appendix B (for the simplified two-state model) that the binomial distribution of the primitive relaxors, and so surely the present quadrinomial one, determine that the exact model can only be solved numerically. We must thus recognize that we have greatly different capabilities of modelling the behavior of (1) uncorrelated primitive relaxors (hardly representative of any real behavor) and (2) the whole range of their clusters.

Considering the quadrinomial distribution of the primitive relaxors among the different states, the rate of stress relaxation per primitive relaxor due to clusters of size k (and no others) in the whole system (I and II) is now not possible to separate into two independent contributions by each of the subsystems, because in any of the clusters and at any time both subsystems may be simultaneously present. So, the specific size-dependent contributions to the local stress relaxation rate per primitive relaxor will have to be written as

$$-\frac{d\sigma_k}{dt} = \frac{1}{a_1}\frac{\omega_k}{2}e^{-\frac{kE_{a,1}}{(k_BT)}}\frac{1}{k}\left\{\sum_{k_{I,II}^{\pm}=0}^{k}\left(k_I^+ + k_{II}^+ - k_I^- - k_{II}^-\right)\right.$$

$$\left. \times \frac{k!}{k_I^+!k_I^-!k_{II}^+!k_{II}^-!}p_{I,k}^{+k_I^+}\,p_{I,k}^{-k_I^-}\,p_{II,k}^{+k_{II}^+}\,p_{II,k}^{-k_{II}^-}e^{\left(k_I^+\alpha_{11}-k_I^-\alpha_{21}+k_{II}^+\beta_{11}-k_{II}^-\beta_{21}\right)}\right\},\quad \text{(C.29)}$$

with $k_I^+ + k_{II}^+ + k_I^- + k_{II}^- = k$, where $\frac{1}{a_1} = \frac{2E_0\upsilon_1^{\#}}{\upsilon_1}$ is the mentioned elementary local stress change in any of the transitions of a primitive relaxor, the various probabilities are those given by (C.7)–(C.10), and the activation energies of each specific cooperative (cluster) transition are $kE_{a,1} - k_BT\left(k_I^+\alpha_{11} - k_I^-\alpha_{21} + k_{II}^+\beta_{11} - k_{II}^-\beta_{21}\right)$, depending on the numbers of primitive relaxors in each of the *four states inside each cluster*. The parameters α_{11}, α_{21}, β_{11} and β_{21} are the same temperature- and strain-dependent variables specified in the preceding section of this Appendix,[28] where the first two were already used in Chap. 9 with $f_{u,0} = 1$.

By contrast with the more complete presentation of the non-linear stress relaxation process of Appendix B for the two-state model, we will not explore the full non-linear behavior for this four-state model and, as already stated, will limit the present discussion to the linear viscoelastic domain, in the limit of extremely low strains, to avoid (at the present, still exploratory, stage) the additional complexity brought about by the strain-dependent exponential factor in (C.29), which determines that a more accurate solution can only be obtained numerically.[29] In these conditions, the exponential inside the expression within { } in (C.29) becomes unity (because its argument is proportional to the strain squared), and so that expression will reduce to $k\left(p_{I,k}^+ + p_{II,k}^+ - p_{I,k}^- - p_{II,k}^-\right) = k\left[(2f_{u,0} - 1) - 2a_1(\sigma_0 - \sigma_k)\right]$, by definition of the

---

[28] As in Appendix B, we have only added a second subscript "1" to specify that they pertain to single primitive relaxors.

[29] Cf. Appendix B, Sect. B.2, including its Footnote 15, to recall the real complexity of the problem, even for the two-state model. Further, within the present four-state model, remember that transitions involving subsystem II's uncorrelated primitive relaxors will have different response times from those that involve subsystem I's uncorrelated primitive relaxors.

various averages of the quadrinomial distribution and substitution of (C.7)–(C.10), and we see that, at equilibrium ($t \to \infty$), the contribution of these clusters to the fully relaxed stress will be such that $\sigma_0 - \sigma_{k,\infty} = (2f_{u,0} - 1)/(2a_1)$, corresponding to zero rate of stress relaxation. This leads to the final stress relaxation rate due to the clusters of size k as

$$-\frac{d\sigma_k}{dt} \cong \omega_k e^{-kE_{a,1}/(k_B T)} \left(\sigma_k - \sigma_{k,\infty}\right), \tag{C.30}$$

whose solution is that of a standard linear[30] solid,

$$\sigma_k(t) = \sigma_{k,\infty} + \left(\sigma_0 - \sigma_{k,\infty}\right)e^{-t/\theta_k}, \tag{C.31}$$

with a relaxation time equal to that of (9.16) with $\varepsilon_0 = 0$.

C.5 The Overall Uniaxial Tensile Stress Relaxation Modulus

So, in the same limiting condition of very low applied strain, the contribution of each cluster size to the stress relaxation modulus will be $E_{r,k}(t) = E_{r,\infty} + \left(E_0 - E_{r,\infty}\right)e^{-t/\theta_k}$ and the final overall stress relaxation modulus[31]

$$E_r(t) = E_{r,\infty} + \left(E_0 - E_{r,\infty}\right)\sum_{k\geq 1} F'_k e^{-t/\theta_k}. \tag{C.32}$$

As throughout the development of CTMD's consequences in Chap. 9, all cluster sizes are considered to relax from the same initial strain and stress state (ϵ_0, σ_0) down to the same ultimate *average* stress, $\sigma_{k,\infty} = \sigma_\infty$, both when they are imagined responding independently of any other clusters (as in the foregoing formulation) and when they intermittently compete with other clusters and primitive relaxors, because all their primitive relaxors have multiple and identical possibilities to participate in the whole range of single and clustered transitions.

---

[30] [Because we are explicitly taking $\varepsilon_0 \cong 0$].

[31] Remember that the F'_k, as formulated at the beginning of Chap. 9 (or by some future better alternative), are (and should be) meant to account for the intermittent activity of all clusters and single primitive relaxors. However, in a more accurate (numerical) solution of the four-state model for non-negligible strain, following the developments in Sect. C.2, one would meet the need to allow for a range of relaxation times (not just a single one) characterizing each set of clusters of any given size, because (1) there will be a distribution of their primitive relaxors among the two subsystems, and (2) the primitive relaxors belonging to subsystem II will counteract the response of their home clusters. Such development will be extremely complex and cannot yet be addessed at the present stage of CTMD.

Given the various mentioned limitations of the proposed treatments and the complexity of the behavior, there can be no doubt that we still remain very far from a complete and accurate theoretical and numerical solution of a stress relaxation problem in amorphous condensed matter.

Index